D1263942

The
Communications
Facility Design
HANDBOOK

Jerry C. Whitaker

ELECTRONICS HANDBOOK SERIES

Series Editor:
Jerry C. Whitaker
Technical Press
Morgan Hill, California

PUBLISHED TITLES

AC POWER SYSTEMS HANDBOOK, SECOND EDITION
Jerry C. Whitaker

THE ELECTRONIC PACKAGING HANDBOOK
Glenn R. Blackwell

POWER VACUUM TUBES HANDBOOK, SECOND EDITION
Jerry C. Whitaker

FORTHCOMING TITLES

ELECTRONIC SYSTEMS MAINTENANCE HANDBOOK
Jerry C. Whitaker

FORMULAS FOR THERMAL DESIGN OF ELECTRONIC EQUIPMENT
Ralph Remsberg

THE RESOURCE HANDBOOK OF ELECTRONICS
Jerry C. Whitaker

The Communications Facility Design HANDBOOK

Jerry C. Whitaker

National Association of
NAB BROADCASTERS

CRC

CRC Press

Boca Raton   London   New York   Washington, D.C.

Library of Congress Cataloging-in-Publication Data

Catalog information is available from the Library of Congress.

No claim to original U.S. Government works
International Standard Book Number 0-8493-0908-5
Printed in the United States of America 1 2 3 4 5 6 7 8 9 0
Printed on acid-free paper

Preface

From the earliest days of electronics, the concepts of system design have evolved and engineering practices have been developed. Signal parameters, connector and cable specifications, and equipment-mounting dimensions have all been standardized. Most of the equipment and hardware used to assemble systems today are available from a number of manufacturers; end-users do not have to custom-build their components. These advances have helped to significantly reduce the engineering design time required for a given project. Many systems of advanced design with superior performance and improved operating efficiency have resulted. Veteran engineers and technical managers are familiar with these practices and standards. However, this is not necessarily the case for less experienced engineers or new engineers who are just entering the electronics industry.

This handbook has been written to establish a foundation for designing, installing, operating, and maintaining audio, video, computer, and radio frequency systems and facilities. It describes the important steps required to take a project from basic design to installation and completion.

This handbook examines the tasks and functions for which the *system engineer* will generally be responsible. It discusses steps required to complete complex projects. For smaller projects, these steps can be implemented easily and—in some cases—certain steps and documentation can be simplified or eliminated without compromising the success of the project.

Although small projects can be completed by a single engineer, larger projects require the system engineer to work with many other people. The reader will realize that the structure of different organizations within companies varies greatly, as do the responsibilities of the individuals who make up the organization.

Within any company, the function of the engineer will vary. A thorough understanding of electronics fundamentals and the workings of a project organization can help engineers understand their responsibilities and deal with the many issues involved in facility design. Many organizations have engineering departments that have established standards for building systems for internal use or, in the case of system integrators who build turnkey systems for their clients, for installation at the client's facility. Either way, this handbook will serve as a valuable reference.

The system engineer is responsible for specifying all of the details of how a facility will be built, and it is that person's responsibility to communicate those details to the contractors, craftsmen, and technicians who will actually build and install the hardware and software. The system engineer is further responsible for installation quality and ultimate performance.

Successful execution of these responsibilities requires an understanding of the underlying technologies and the applicable quality standards and methods for achieving them. *The Communications Facility Design Handbook* is dedicated to that effort.

For updated information on this and other engineering books, visit the author's
Internet site
www.technicalpress.com

About the Author

Jerry Whitaker is a technical writer based in Morgan Hill, California, where he operates the consulting firm *Technical Press*. Mr. Whitaker has been involved in various aspects of the communications industry for more than 25 years. He is a Fellow of the Society of Broadcast Engineers and an SBE-certified Professional Broadcast Engineer. He is also a member and Fellow of the Society of Motion Picture and Television Engineers, and a member of the Institute of Electrical and Electronics Engineers. Mr. Whitaker has written and lectured extensively on the topic of electronic systems installation and maintenance.

Mr. Whitaker is the former editorial director and associate publisher of *Broadcast Engineering* and *Video Systems* magazines. He is also a former radio station chief engineer and TV news producer.

Mr. Whitaker is the author of a number of books, including:

- *Power Vacuum Tubes Handbook*, 2nd edition, CRC Press 1999.

- *AC Power Systems*, 2nd edition, CRC Press, 1998.

- *DTV: The Revolution in Electronic Imaging*, 2nd edition, McGraw-Hill, 1999.

- Editor-in-Chief, *NAB Engineering Handbook*, 9th edition, National Association of Broadcasters, 1999.

- Editor-in-Chief, *The Electronics Handbook*, CRC Press, 1996.

- Coauthor, *Communications Receivers: Principles and Design*, 2nd edition, McGraw-Hill, 1996.

- *Electronic Displays: Technology, Design, and Applications*, McGraw-Hill, 1994.

- Coeditor, *Standard Handbook of Video and Television Engineering*, 3rd edition, McGraw-Hill, 2000.

- Coeditor, *Information Age Dictionary*, Intertec/Bellcore, 1992.

- *Maintaining Electronic Systems*, CRC Press, 1991.

- *Radio Frequency Transmission Systems: Design and Operation*, McGraw-Hill, 1990.

- Coauthor, *Television and Audio Handbook for Technicians and Engineers*, McGraw-Hill, 1990.

Mr. Whitaker has twice received a Jesse H. Neal Award *Certificate of Merit* from the Association of Business Publishers for editorial excellence. He also has been recognized as *Educator of the Year* by the Society of Broadcast Engineers.

Acknowledgment

The author wishes to express appreciation to the following contributors for their assistance in the preparation of this book.

K. Blair Benson
E. Stanley Busby
Michael W. Dahlgren
Gene DeSantis
C. Robert Paulson
Richard Rudman

Contents

In memory of

James Kovach

A special person by any measure

Electronics Fundamentals

1.1   Introduction

The atomic theory of matter specifies that each of the many chemical elements is composed of unique and identifiable particles called atoms. In ancient times only 10 were known in their pure, uncombined form; these were carbon, sulfur, copper, antimony, iron, tin, gold, silver, mercury, and lead. Of the several hundred now identified, less than 50 are found in an uncombined, or chemically free, form on earth.

Each atom consists of a compact nucleus of positively and negatively charged particles (protons and electrons, respectively). Additional electrons travel in well-defined orbits around the nucleus. The electron orbits are grouped in regions called *shells*, and the number of electrons in each orbit increases with the increase in orbit diameter in accordance with quantum-theory laws of physics. The diameter of the outer orbiting path of electrons in an atom is in the order of one-millionth (10^{-6}) millimeter, and the nucleus, one-millionth of that. These typical figures emphasize the minute size of the atom.

1.2   Electrical Fundamentals

The nucleus and the free electrons for an iron atom are shown in the schematic diagram in Figure 1.1. Note that the electrons are spinning in different directions. This rotation creates a magnetic field surrounding each electron. If the number of electrons with positive spins is equal to the number with negative spins, then the net field is zero and the atom exhibits no magnetic field.

In the diagram, although the electrons in the first, second, and fourth shells balance each other, in the third shell five electrons have clockwise positive spins, and one a counterclockwise negative spin, which gives the iron atom in this particular electron configuration a cumulative *magnetic effect*.

The parallel alignment of electron spins over regions, known as *domains*, containing a large number of atoms. When a magnetic material is in a demagnetized state, the direction of magnetization in the domain is in a random order. Magnetization by an external field takes place by a change or displacement in the isolation of the domains, with

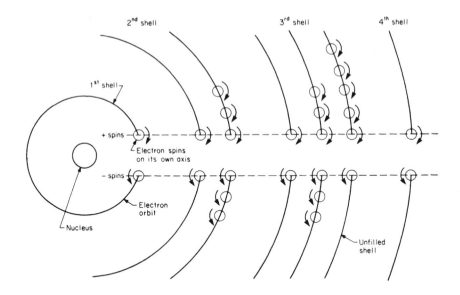

Figure 1.1 Schematic of the iron (Fe) atom.

the result that a large number of the atoms are aligned with their charged electrons in parallel.

1.2.1 Conductors and Insulators

In some elements, such as copper, the electrons in the outer shells of the atom are so weakly bound to the nucleus that they can be released by a small electrical force, or voltage. A voltage applied between two points on a length of a metallic conductor produces the flow of an electric current, and an electric field is established around the conductor. The conductivity is a constant for each metal that is unaffected by the current through or the intensity of any external electric field.

In some nonmetallic materials, the free electrons are so tightly bound by forces in the atom that, upon the application of an external voltage, they will not separate from their atom except by an electrical force strong enough to destroy the insulating properties of the material. However, the charges will realign within the structure of their atom. This condition occurs in the insulating material (dielectric) of a capacitor when a voltage is applied to the two conductors encasing the dielectric.

Semiconductors are electronic conducting materials wherein the conductivity is dependent primarily upon impurities in the material. In addition to negative mobile charges of electrons, positive mobile charges are present. These positive charges are called *holes* because each exists as an absence of electrons. Holes (+) and electrons (−), because they are oppositely charged, move in opposite directions in an electric field. The conductivity of semiconductors is highly sensitive to, and increases with, temperature.

1.2.2 Direct Current (dc)

Direct current is defined as a unidirectional current in which there are no significant changes in the current flow. In practice, the term frequently is used to identify a voltage source, in which case variations in the load can result in fluctuations in the current but not in the direction.

Direct current was used in the first systems to distribute electricity for household and industrial power. For safety reasons, and the voltage requirements of lamps and motors, distribution was at the low nominal voltage of 110. The losses in distribution circuits at this voltage seriously restricted the length of transmission lines and the size of the areas that could be covered. Consequently, only a relatively small area could be served by a single generating plant. It was not until the development of alternating-current systems and the voltage transformer that it was feasible to transport high levels of power at relatively low current over long distances for subsequent low-voltage distribution to consumers.

1.2.3 Alternating Current (ac)

Alternating current is defined as a current that reverses direction at a periodic rate. The average value of alternating current over a period of one cycle is equal to zero. The effective value of an alternating current in the supply of energy is measured in terms of the root mean square (rms) value. The rms is the square root of the square of all the values, positive and negative, during a complete cycle, usually a sine wave. Because rms values cannot be added directly, it is necessary to perform an rms addition as shown in the equation:

$$V_{rms\ total} = \sqrt{V_{rms\ 1}^2 + V_{rms\ 2}^2 + \mathsf{L}\ V_{rms\ n}^2} \qquad (1.1)$$

As in the definition of direct current, in practice the term frequently is used to identify a voltage source.

The level of a sine-wave alternating current or voltage can be specified by two other methods of measurement in addition to rms. These are *average* and *peak*. A sine-wave signal and the rms and average levels are shown in Figure 1.2. The levels of complex, symmetrical ac signals are specified as the peak level from the axis, as shown in the figure.

1.3   Electronic Circuits

Electronic circuits are composed of elements such as resistors, capacitors, inductors, and voltage and current sources, all of which may be interconnected to permit the

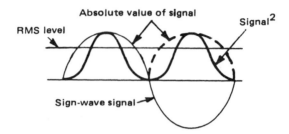

Figure 1.2 Root mean square (rms) measurements. The relationship of rms and average values is shown.

flow of electric currents. An *element* is the smallest component into which circuits can be subdivided. The points on a circuit element where they are connected in a circuit are called *terminals*.

Elements can have two or more terminals, as shown in Figure 1.3. The resistor, capacitor, inductor, and diode shown in the Figure 1.3*a* are two-terminal elements; the transistor in Figure 1.3*b* is a three-terminal element; and the transformer in Figure 1.3*c* is a four-terminal element.

Circuit elements and components also are classified as to their function in a circuit. An element is considered *passive* if it absorbs energy and *active* if it increases the level of energy in a signal. An element that receives energy from either a passive or active element is called a *load*. In addition, either passive or active elements, or components, can serve as loads.

The basic relationship of current and voltage in a two-terminal circuit where the voltage is constant and there is only one source of voltage is given in Ohm's law. This states that the voltage V between the terminals of a conductor varies in accordance with the current I. The ratio of voltage, current, and resistance R is expressed in Ohm's law as follows:

$$E = I \times R \tag{1.2}$$

Using Ohm's law, the calculation for power in watts can be developed from $P = E \times I$ as follows:

$$P = \frac{E^2}{R} \quad and \quad P = I^2 \times R \tag{1.3}$$

A circuit, consisting of a number of elements or components, usually amplifies or otherwise modifies a signal before delivering it to a load. The terminal to which a signal is applied is an *input port*, or *driving port*. The pair or group of terminals that delivers a signal to a load is the *output port*. An element or portion of a circuit between two termi-

(a)

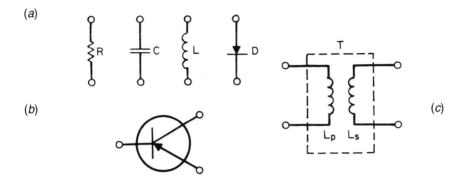

(b) 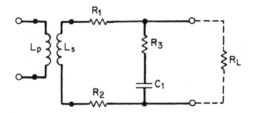 (c)

Figure 1.3 Schematic examples of circuit elements: (a) two-terminal element, (b) three-terminal element, (c) four-terminal element.

Figure 1.4 Circuit configuration composed of several elements and branches, and a closed loop (R_1, R, C_1, R_2, and L_s).

nals is a *branch*. The circuit shown in Figure 1.4 is made up of several elements and branches. R_1 is a branch, and R_1 and C_1 make up a two-element branch. The secondary of transformer T, a voltage source, and R_2 also constitute a branch. The point at which three or more branches join together is a *node*. A series connection of elements or branches, called a *path*, in which the end is connected back to the start is a *closed loop*.

1.3.1 Circuit Analysis

Relatively complex configurations of *linear circuit elements*, that is, where the signal gain or loss is constant over the signal amplitude range, can be analyzed by simplification into the equivalent circuits. After the restructuring of a circuit into an equivalent form, the current and voltage characteristics at various nodes can be calculated using network-analysis theorems, including Kirchoff's current and voltage laws, Thevenin's theorem, and Norton's theorem.

- **Kirchoff's current law (KCL).** The algebraic sum of the instantaneous currents entering a node (a common terminal of three or more branches) is zero. In other

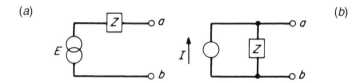

Figure 1.5 Equivalent circuits: (*a*) Thevenin's equivalent voltage source, (*b*) Norton's equivalent current source. (*After* [1].)

words, the currents from two branches entering a node add algebraically to the current leaving the node in a third branch.

- **Kirchoff's voltage law (KVL).** The algebraic sum of instantaneous voltages around a closed loop is zero.

- **Thevenin's theorem.** The behavior of a circuit at its terminals can be simulated by replacement with a voltage E from a dc source in series with an impedance Z (see Figure 1.5*a*).

- **Norton's theorem.** The behavior of a circuit at its terminals can be simulated by replacement with a dc source I in parallel with an impedance Z (see Figure 1.5*b*).

AC Circuits

Vectors are used commonly in ac circuit analysis to represent voltage or current values. Rather than using waveforms to show phase relationships, it is accepted practice to use vector representations (sometimes called *phasor diagrams*). To begin a vector diagram, a horizontal line is drawn, its left end being the *reference point*. Rotation in a counterclockwise direction from the reference point is considered to be positive. Vectors may be used to compare voltage drops across the components of a circuit containing resistance, inductance, and/or capacitance. Figure 1.6 shows the vector relationship in a series RLC circuit, and Figure 1.7 shows a parallel RLC circuit

Power Relationship in AC Circuits

In a dc circuit, power is equal to the product of voltage and current. This formula also is true for purely resistive ac circuits. However, when a reactance—either inductive or capacitive—is present in an ac circuit, the dc power formula does not apply. The product of voltage and current is, instead, expressed in volt-amperes (VA) or kilovoltamperes (kVA). This product is known as the *apparent power*. When meters are used to measure power in an ac circuit, the apparent power is the voltage reading multiplied by the current reading. The *actual power* that is converted to another form of energy by the circuit is measured with a wattmeter, and is referred to as the *true power*. In ac power-system design and operation, it is desirable to know the ratio of

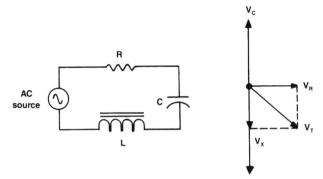

Figure 1.6 Voltage vectors in a series RLC circuit.

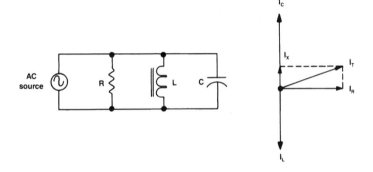

Figure 1.7 Current vectors in a parallel RLC circuit.

true power converted in a given circuit to the apparent power of the circuit. This ratio is referred to as the *power factor.*

Complex Numbers

A complex number is represented by a *real part* and an *imaginary part*. For example, in $A = a + jb$, A is the complex number; a is real part, sometimes written as $\mathrm{Re}(A)$; and b is the imaginary part of A, often written as $\mathrm{Im}(A)$. It is a convention to precede the imaginary component by the letter j (or i). This form of writing the real and imaginary components is called the *Cartesian form* and symbolizes the complex (or s) plane, wherein both the real and imaginary components can be indicated graphically [2]. To illustrate this, consider the same complex number A when represented graphically as shown in Figure 1.8. A second complex number B is also shown to illustrate the fact that the real and imaginary components can take on both positive and negative values. Figure 1.8 also shows an alternate form of representing complex numbers. When a

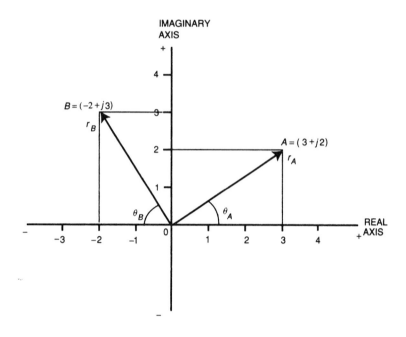

Figure 1.8 The *s* plane representing two complex numbers. (*From* [2]. *Used with permission.*)

complex number is represented by its magnitude and angle, for example, $A = r_A \angle \theta_A$, it is called the *polar representation*.

To see the relationship between the Cartesian and the polar forms, the following equations can be used:

$$r_A = \sqrt{a^2 + b^2} \tag{1.4}$$

$$\theta_A = \tan^{-1} \frac{b}{a} \tag{1.5}$$

Conceptually, a better perspective can be obtained by investigating the triangle shown in Figure 1.9, and considering the trigonometric relationships. From this figure, it can be seen that

$$a = \mathrm{Re}(A) = r_A \cos(\theta_A) \tag{1.6}$$

$$b = \mathrm{Im}(A) = r_A \sin(\theta_A) \tag{1.7}$$

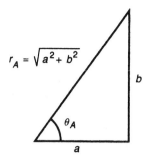

Figure 1.9 The relationship between Cartesian and polar forms. (*From* [2]. *Used with permission.*)

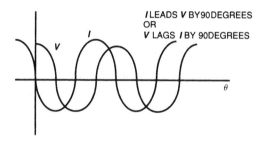

Figure 1.10 Waveforms representing leading and lagging phasors. (*From* [2]. *Used with permission.*)

The well-known *Euler's identity* is a convenient conversion of the polar and Cartesian forms into an *exponential form*, given by

$$\exp(j\theta) = \cos\theta + j\sin\theta \tag{1.8}$$

Phasors

The ac voltages and currents appearing in distribution systems can be represented by *phasors*, a concept useful in obtaining analytical solutions to one-phase and three-phase system design. A phasor is generally defined as a transform of sinusoidal functions from the time domain into the complex-number domain and given by the expression

$$\mathbf{V} = V\exp(j\theta) = P\{V\cos(\omega t + \theta)\} = V\angle\theta \tag{1.9}$$

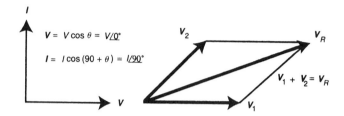

Figure 1.11 Phasor diagram showing phasor representation and phasor operation. (*From* [2]. *Used with permission.*)

where V is the phasor, V is the magnitude of the phasor, and θ is the angle of the phasor. The convention used here is to use boldface symbols to symbolize phasor quantities. Graphically, in the time domain, the phasor V would be a simple sinusoidal wave shape as shown in Figure 1.10. The concept of a phasor leading or lagging another phasor becomes very apparent from the figure.

Phasor diagrams are also an effective medium for understanding the relationships between phasors. Figure 1.11 shows a phasor diagram for the phasors represented in Figure 1.10. In this diagram, the convention of positive angles being read counterclockwise is used. The other alternative is certainly possible as well. It is quite apparent that a purely capacitive load could result in the phasors shown in Figures 1.10 and 1.11.

Per Unit System

In the *per unit system*, basic quantities such as voltage and current, are represented as certain percentages of base quantities. When so expressed, these per unit quantities do not need units, thereby making numerical analysis in power systems somewhat easier to handle. Four quantities encompass all variables required to solve a power system problem. These quantities are:

- Voltage
- Current
- Power
- Impedance

Out of these, only two base quantities, corresponding to voltage (V_b) and power (S_b), are required to be defined. The other base quantities can be derived from these two. Consider the following. Let

V_b = voltage base, kV
S_b = power base, MVA
I_b = current base, A
Z_b = impedance base, Q

Then,

$$Z_b = \frac{V_b^2}{S_b} \ \Omega \tag{1.10}$$

$$I_b = \frac{V_b 10^3}{Z_b} \ A \tag{1.11}$$

1.4 Static Electricity

The phenomenon of static electricity and related potential differences concerns configurations of conductors and insulators where no current flows and all electrical forces are unchanging; hence the term *static*. Nevertheless, static forces are present because of the number of excess electrons or protons in an object. A static charge can be induced by the application of a voltage to an object. A flow of current to or from the object can result from either a breakdown of the surrounding nonconducting material or by the connection of a conductor to the object.

Two basic laws regarding electrons and protons are:

- Like charges exert a repelling force on each other; electrons repel other electrons and protons repel other protons

- Opposite charges attract each other; electrons and protons are attracted to each other

Therefore, if two objects each contain exactly as many electrons as protons in each atom, there is no electrostatic force between the two. On the other hand, if one object is charged with an excess of protons (deficiency of electrons) and the other an excess of electrons, there will be a relatively weak attraction that diminishes rapidly with distance. An attraction also will occur between a neutral and a charged object.

Another fundamental law, developed by Faraday, governing static electricity is that all of the charge of any conductor not carrying a current lies in the surface of the conductor. Thus, any electric fields external to a completely enclosed metal box will not penetrate beyond the surface. Conversely, fields within the box will not exert any force on objects outside the box. The box need not be a solid surface; a conduction cage or grid will suffice. This type of isolation frequently is referred to as a *Faraday shield*.

1.5 Magnetism

The elemental magnetic particle is the spinning electron. In magnetic materials, such as iron, cobalt, and nickel, the electrons in the third shell of the atom (see Figure 1.1) are the source of magnetic properties. If the spins are arranged to be parallel, the atom and its associated domains or clusters of the material will exhibit a magnetic field. The magnetic field of a magnetized bar has lines of magnetic force that extend be-

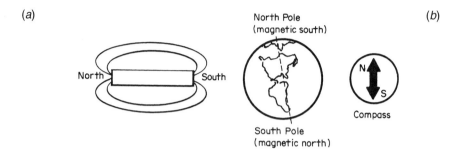

Figure 1.12 The properties of magnetism: (*a*) lines of force surrounding a bar magnet, (*b*) relation of compass poles to the earth's magnetic field.

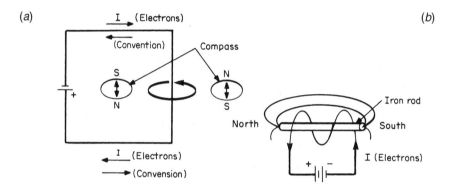

Figure 1.13 Magnetic field surrounding a current-carrying conductor: (*a*) Compass at right indicates the polarity and direction of a magnetic field circling a conductor carrying direct current. *I* indicates the direction of electron flow. Note: The convention for flow of electricity is from + to −, the reverse of the actual flow. (*b*) Direction of magnetic field for a coil or solenoid.

tween the ends, one called the north pole and the other the south pole, as shown in Figure 1.12*a*. The lines of force of a magnetic field are called *magnetic flux lines*.

1.5.1 Electromagnetism

A current flowing in a conductor produces a magnetic field surrounding the wire as shown in Figure 1.13*a*. In a coil or solenoid, the direction of the magnetic field relative to the electron flow (− to +) is shown in Figure 1.13*b*. The attraction and repulsion between two iron-core electromagnetic solenoids driven by direct currents is similar to that of two permanent magnets described previously.

The process of magnetizing and demagnetizing an iron-core solenoid using a current being applied to a surrounding coil can be shown graphically as a plot of the magnetizing field strength and the resultant magnetization of the material, called a *hyster-*

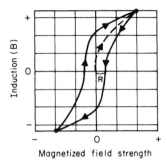

Magnetized field strength

Figure 1.14 Graph of the magnetic hysteresis loop resulting from magnetization and demagnetization of iron. The dashed line is a plot of the induction from the initial magnetization. The solid line shows a reversal of the field and a return to the initial magnetization value. *R* is the remaining magnetization (*remnance*) when the field is reduced to zero.

esis loop (Figure 1.14). It will be found that the point where the field is reduced to zero, a small amount of magnetization, called r*emnance*, remains.

1.5.2 Magnetic Shielding

In effect, the shielding of components and circuits from magnetic fields is accomplished by the introduction of a magnetic short circuit in the path between the field source and the area to be protected. The flux from a field can be redirected to flow in a partition or shield of magnetic material, rather than in the normal distribution pattern between north and south poles. The effectiveness of shielding depends primarily upon the thickness of the shield, the material, and the strength of the interfering field.

Some alloys are more effective than iron. However, many are less effective at high flux levels. Two or more layers of shielding, insulated to prevent circulating currents from magnetization of the shielding, are used in low-level audio, video, and data applications.

1.5.3 Electromagnetic-Radiation Spectrum

The usable spectrum of electromagnetic-radiation frequencies extends over a range from below 100 Hz for power distribution to 1020 for the shortest X-rays. The lower frequencies are used primarily for terrestrial broadcasting and communications. The higher frequencies include visible and near-visible infrared and ultraviolet light, and X-rays.

The standard frequency band designations are listed in Tables 1.1 and 1.2. Alternate and more detailed subdivision of the VHF, UHF, SHF, and EHF bands are given in Tables 1.3 and 1.4.

Table 1.1 Standardized Frequency Bands (*From* [3]. *Used with permission.*)

Extremely low-frequency (ELF) band:	30 Hz up to 300 Hz	(10 Mm down to 1 Mm)
Voice-frequency (VF) band:	300 Hz up to 3 kHz	(1 Mm down to 100 km)
Very low-frequency (VLF) band:	3 kHz up to 30 kHz	(100 km down to 10 km)
Low-frequency (LF) band:	30 kHz up to 300 kHz	(10 km down to 1 km)
Medium-frequency (MF) band:	300 kHz up to 3 MHz	(1 km down to 100 m)
High-frequency (HF) band:	3 MHz up to 30 MHz	(100 m down to 10 m)
Very high-frequency (VHF) band:	30 MHz up to 300 MHz	(10 m down to 1 m)
Ultra high-frequency (UHF) band:	300 MHz up to 3 GHz	(1 m down to 10 cm)
Super high-frequency (SHF) band:	3 GHz up to 30 GHz	(1 cm down to 1 cm)
Extremely high-frequency (EHF) band:	30 GHz up to 300 GHz	(1 cm down to 1 mm)

Table 1.2 Standardized Frequency Bands at 1 GHz and Above (*From* [3]. *Used with permission.*)

L band:	1 GHz up to 2 GHz	(30 cm down to 15 cm)
S band:	2 GHz up to 4 GHz	(15 cm down to 7.5 cm)
C band:	4 GHz up to 8 GHz	(7.5 cm down to 3.75 cm)
X band:	8 GHz up to 12 GHz	(3.75 cm down to 2.5 cm)
Ku band:	12 GHz up to 18 GHz	(2.5 cm down to 1.67 cm)
K band:	18 GHz up to 26.5 GHz	(1.67 cm down to 1.13 cm)
Ka band:	26.5 GHz up to 40 GHz	(1.13 cm down to 7.5 mm)
Q band:	32 GHz up to 50 GHz	(9.38 mm down to 6 mm)
U band:	40 GHz up to 60 GHz	(7.5 mm down to 5mm)
V band:	50 GHz up to 75 GHz	(6 mm down to 4 mm)
W band:	75 GHz up to 100 GHz	(4 mm down to 3.33 mm)

Low-End Spectrum Frequencies (1 to 1000 Hz)

Electric power is transmitted by wire but not by radiation at 50 and 60 Hz, and in some limited areas, at 25 Hz. Aircraft use 400-Hz power in order to reduce the weight of iron in generators and transformers. The restricted bandwidth that would be available for communication channels is generally inadequate for voice or data transmission, although some use has been made of communication over power distribution circuits using modulated carrier frequencies.

Low-End Radio Frequencies (1000 to 100 kHz)

These low frequencies are used for very long distance radio-telegraphic communication where extreme reliability is required and where high-power and long antennas can be erected. The primary bands of interest for radio communications are given in Table 1.5.

Table 1.3 Detailed Subdivision of the UHF, SHF, and EHF Bands (*From* [3]. *Used with permission.*)

L band:	1.12 GHz up to 1.7 GHz	(26.8 cm down to 17.6 cm)
LS band:	1.7 GHz up to 2.6 GHz	(17.6 cm down to 11.5 cm)
S band:	2.6 GHz up to 3.95 GHz	(11.5 cm down to 7.59 cm)
C(G) band:	3.95 GHz up to 5.85 GHz	(7.59 cm down to 5.13 cm)
XN(J, XC) band:	5.85 GHz up to 8.2 GHz	(5.13 cm down to 3.66 cm)
XB(H, BL) band:	7.05 GHz up to 10 GHz	(4.26 cm down to 3 cm)
X band:	8.2 GHz up to 12.4 GHz	(3.66 cm down to 2.42 cm)
Ku(P) band:	12.4 GHz up to 18 GHz	(2.42 cm down to 1.67 cm)
K band:	18 GHz up to 26.5 GHz	(1.67 cm down to 1.13 cm)
V(R, Ka) band:	26.5 GHz up to 40 GHz	(1.13 cm down to 7.5 mm)
Q(V) band:	33 GHz up to 50 GHz	(9.09 mm down to 6 mm)
M(W) band:	50 GHz up to 75 GHz	(6 mm down to 4 mm)
E(Y) band:	60 GHz up to 90 GHz	(5 mm down to 3.33 mm)
F(N) band:	90 GHz up to 140 GHz	(3.33 mm down to 2.14 mm)
G(A) band:	140 GHz up to 220 GHz	(2.14 mm down to 1.36 mm)
R band:	220 GHz up to 325 GHz	(1.36 mm down to 0.923 mm)

Table 1.4 Subdivision of the VHF, UHF, SHF Lower Part of the EHF Band (*From* [3]. *Used with permission.*)

A band:	100 MHz up to 250 MHz	(3 m down to 1.2 m)
B band:	250 MHz up to 500 MHz	(1.2 m down to 60 cm)
C band:	500 MHz up to 1 GHz	(60 cm down to 30 cm)
D band:	1 GHz up to 2 GHz	(30 cm down to 15 cm)
E band:	2 GHz up to 3 GHz	(15 cm down to 10 cm)
F band:	3 GHz up to 4 GHz	(10 cm down to 7.5 cm)
G band:	4 GHz up to 6 GHz	(7.5 cm down to 5 cm)
H band:	6 GHz up to 8 GHz	(5 cm down to 3.75 cm)
I band:	8 GHz up to 10 GHz	(3.75 cm down to 3 cm)
J band:	10 GHz up to 20 GHz	(3 cm down to 1.5 cm)
K band:	20 GHz up to 40 GHz	(1.5 cm down to 7.5 mm)
L band:	40 GHz up to 60 GHz	(7.5 mm down to 5 mm)
M band:	60 GHz up to 100 GHz	(5 mm down to 3 mm)

Medium-Frequency Radio (20 kHz to 2 MHz)

The low-frequency portion of the band is used for around-the-clock communication services over moderately long distances and where adequate power is available to overcome the high level of atmospheric noise. The upper portion is used for AM radio, although the strong and quite variable *sky wave* occurring during the night results

Table 1.5 Radio Frequency Bands (*From* [3]. *Used with permission.*)

Longwave broadcasting band:	150–290 kHz
AM broadcasting band:	550–1640 kHz (1.640 MHz) (107 Channels, 10-kHz separation)
International broadcasting band:	3–30 MHz
Shortwave broadcasting band:	5.95–26.1 MHz (8 bands)
VHF television (channels 2–4):	54–72 MHz
VHF television (channels 5–6):	76–88 MHz
FM broadcasting band:	88–108 MHz
VHF television (channels 7–13):	174–216 MHz
UHF television (channels 14–83):	470–890 MHz

in substandard quality and severe fading at times. The greatest use is for AM broadcasting, in addition to fixed and mobile service, LORAN ship and aircraft navigation, and amateur radio communication.

High-Frequency Radio (2 to 30 MHz)

This band provides reliable medium-range coverage during daylight and, when the transmission path is in total darkness, worldwide long-distance service, although the reliability and signal quality of the latter is dependent to a large degree upon ionospheric conditions and related long-term variations in sun-spot activity affecting sky-wave propagation. The primary applications include broadcasting, fixed and mobile services, telemetering, and amateur transmissions.

Very High and Ultrahigh Frequencies (30 MHz to 3 GHz)

VHF and UHF bands, because of the greater channel bandwidth possible, can provide transmission of a large amount of information, either as television detail or data communication. Furthermore, the shorter wavelengths permit the use of highly directional parabolic or multielement antennas. Reliable long-distance communication is provided using high-power *tropospheric scatter* techniques. The multitude of uses include, in addition to television, fixed and mobile communication services, amateur radio, radio astronomy, satellite communication, telemetering, and radar.

Microwaves (3 to 300 GHz)

At these frequencies, many transmission characteristics are similar to those used for shorter optical waves, which limit the distances covered to line of sight. Typical uses include television relay, satellite, radar, and wide-band information services. (See Tables 1.7 and 1.8.)

Infrared, Visible, and Ultraviolet Light

The portion of the spectrum visible to the eye covers the gamut of transmitted colors ranging from red, through yellow, green, cyan, and blue. It is bracketed by infrared on the low-frequency side and ultraviolet (UV) on the high side. Infrared signals are used in a variety of consumer and industrial equipments for remote controls and sensor circuits in security systems. The most common use of UV waves is for excitation of phosphors to produce visible illumination.

X-Rays

Medical and biological examination techniques and industrial and security inspection systems are the best-known applications of X-rays. X-rays in the higher-frequency range are classified as *hard X-rays* or *gamma rays*. Exposure to X-rays for long periods can result in serious irreversible damage to living cells or organisms.

1.6   Passive Circuit Components

Components used in electrical circuitry can be categorized into two broad classifications as *passive* or *active*. A voltage applied to a passive component results in the flow of current and the dissipation or storage of energy. Typical passive components are resistors, coils or inductors, and capacitors. For an example, the flow of current in a resistor results in radiation of heat; from a light bulb, the radiation of light as well as heat.

On the other hand, an active component either (1) increases the level of electric energy or (2) provides available electric energy as a voltage. As an example of (1), an amplifier produces an increase in energy as a higher voltage or power level, while for (2), batteries and generators serve as energy sources.

1.6.1   Resistors

Resistors are components that have a nearly 0° phase shift between voltage and current over a wide range of frequencies with the average value of resistance independent of the instantaneous value of voltage or current. Preferred values of ratings are given ANSI standards or corresponding ISO or MIL standards. Resistors are typically identified by their construction and by the resistance materials used. Fixed resistors have two or more terminals and are not adjustable. Variable resistors permit adjustment of resistance or voltage division by a control handle or with a tool.

Low-wattage fixed resistors are usually identified by color-coding on the body of the device, as illustrated in Figure 1.15.

Wire-Wound Resistor

The resistance element of most wire-wound resistors is resistance wire or ribbon wound as a single-layer helix over a ceramic or fiberglass core, which causes these re-

Table 1.6 Applications in the Microwave Bands (*From* [3]. *Used with permission.*)

Aeronavigation:	0.96–1.215 GHz
Global positioning system (GPS) downlink:	1.2276 GHz
Military communications (COM)/radar:	1.35–1.40 GHz
Miscellaneous COM/radar:	1.40–1.71 GHz
L-band telemetry:	1.435–1.535 GHz
GPS downlink:	1.57 GHz
Military COM (troposcatter/telemetry):	1.71–1.85 GHz
Commercial COM and private line of sight (LOS):	1.85–2.20 GHz
Microwave ovens:	2.45 GHz
Commercial COM/radar:	2.45–2.69 GHz
Instructional television:	2.50–2.69 GHz
Military radar (airport surveillance):	2.70–2.90 GHz
Maritime navigation radar:	2.90–3.10 GHz
Miscellaneous radars:	2.90–3.70 GHz
Commercial C-band satellite (SAT) COM downlink:	3.70–4.20 GHz
Radar altimeter:	4.20–4.40 GHz
Military COM (troposcatter):	4.40–4.99 GHz
Commercial microwave landing system:	5.00–5.25 GHz
Miscellaneous radars:	5.25–5.925 GHz
C-band weather radar:	5.35–5.47 GHz
Commercial C-band SAT COM uplink:	5.925–6.425 GHz
Commercial COM:	6.425–7.125 GHz
Mobile television links:	6.875–7.125 GHz
Military LOS COM:	7.125–7.25 GHz
Military SAT COM downlink:	7.25–7.75 GHz
Military LOS COM:	7.75–7.9 GHz
Military SAT COM uplink:	7.90–8.40 GHz
Miscellaneous radars:	8.50–10.55 GHz
Precision approach radar:	9.00–9.20 GHz
X-band weather radar (and maritime navigation radar):	9.30–9.50 GHz
Police radar:	10.525 GHz
Commercial mobile COM [LOS and electronic news gathering (ENG)]:	10.55–10.68 GHz
Common carrier LOS COM:	10.70–11.70 GHz
Commercial COM:	10.70–13.25 GHz
Commercial Ku-band SAT COM downlink:	11.70–12.20 GHz
Direct broadcast satellite (DBS) downlink and private LOS COM:	12.20–12.70 GHz
ENG and LOS COM:	12.75–13.25 GHz
Miscellaneous radars and SAT COM:	13.25–14.00 GHz
Commercial Ku-band SAT COM uplink:	14.00–14.50 GHz
Military COM (LOS, mobile, and Tactical):	14.50–15.35 GHz
Aeronavigation:	15.40–15.70 GHz
Miscellaneous radars:	15.70–17.70 GHz
DBS uplink:	17.30–17.80 GHz

Table 1.6 Applications in the Microwave Bands (continued)

Common carrier LOS COM:	17.70–19.70 GHz
Commercial COM (SAT COM and LOS):	17.70–20.20 GHz
Private LOS COM:	18.36–19.04 GHz
Military SAT COM:	20.20–21.20 GHz
Miscellaneous COM:	21.20–24.00 GHz
Police radar:	24.15 GHz
Navigation radar:	24.25–25.25 GHz
Military COM:	25.25–27.50 GHz
Commercial COM:	27.50–30.00 GHz
Military SAT COM:	30.00–31.00 GHz
Commercial COM:	31.00–31.20 GHz
Navigation radar:	31.80–33.40 GHz
Miscellaneous radars:	33.40–36.00 GHz
Military COM:	36.00–38.60 GHz
Commercial COM:	38.60–40.00 GHz

sistors to have a residual series inductance that affects phase shift at high frequencies, particularly in large-size devices. Wire-wound resistors have low noise and are stable with temperature, with temperature coefficients normally between ±5 and 200 ppm/°C. Resistance values between 0.1 and 100,000 W with accuracies between 0.001 and 20 percent are available with power dissipation ratings between 1 and 250 W at 70°C. The resistance element is usually covered with a vitreous enamel, which can be molded in plastic. Special construction includes such items as enclosure in an aluminum casing for heatsink mounting or a special winding to reduce inductance. Resistor connections are made by self-leads or to terminals for other wires or printed circuit boards.

Metal Film Resistor

Metal film, or *cermet*, resistors have characteristics similar to wire-wound resistors except a much lower inductance. They are available as axial lead components in 1/8, 1/4, or ½ W ratings, in chip resistor form for high-density assemblies, or as resistor networks containing multiple resistors in one package suitable for printed circuit insertion, as well as in tubular form similar to high-power wire-wound resistors. Metal film resistors are essentially printed circuits using a thin layer of resistance alloy on a flat or tubular ceramic or other suitable insulating substrate. The shape and thickness of the conductor pattern determine the resistance value for each metal alloy used. Resistance is trimmed by cutting into part of the conductor pattern with an abrasive or a laser. Tin oxide is also used as a resistance material.

Table 1.7 Satellite Frequency Allocations (*From* [3]. *Used with permission.*)

Band	Uplink	Downlink	Satellite Service
VHF		0.137–0.138	Mobile
VHF	0.3120–0.315	0.387–0.390	Mobile
L-Band		1.492–1.525	Mobile
	1.610–1.6138		Mobile, Radio Astronomy
	1.613.8–1.6265	1.6138–1.6265	Mobile LEO
	1.6265–1.6605	1.525–1.545	Mobile
		1.575	Global Positioning System
		1.227	GPS
S-Band	1.980–2.010	2.170–2.200	MSS. Available Jan. 1, 2000
	(1.980–1.990)		(Available in U.S. in 2005)
	2.110–2.120	2.290–2.300	Deep-space research
		2.4835–2.500	Mobile
C-Band	5.85–7.075	3.4–4.2	Fixed (FSS)
	7.250–7.300	4.5–4.8	FSS
X-Band	7.9–8.4	7.25–7.75	FSS
Ku-Band	12.75–13.25	10.7–12.2	FSS
	14.0–14.8	12.2–12.7	Direct Broadcast (BSS) (U.S.)
Ka-Band		17.3–17.7	FSS (BSS in U.S.)
			22.55–23.55 Intersatellite
			24.45–24.75 Intersatellite
			25.25–27.5 Intersatellite
	27–31	17–21	FSS
Q	42.5–43.5, 47.2–50.2	37.5–40.5	FSS, MSS
	50.4–51.4		Fixed
		40.5–42.5	Broadcast Satellite
V	54.24–58.2-		Intersatellite
	59–64		Intersatellite

Sources: Final Acts of the World Administrative Radio Conference (WARC-92), Malaga–Torremolinos, 1992; 1995 World Radiocommunication Conference (WRC-95). Also, see Gagliardi, R.M. 1991. *Satellite Communications,* van Nostrand Reinhold, New York. Note that allocations are not always global and may differ from region to region in all or subsets of the allocated bands.

Carbon Film Resistor

Carbon film resistors are similar in construction and characteristics to axial lead metal film resistors. Because the carbon film is a granular material, random noise may be developed because of variations in the voltage drop between granules. This noise can be of sufficient level to affect the performance of circuits providing high grain when operating at low signal levels.

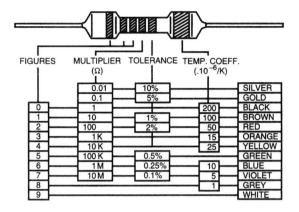

FIGURES	MULTIPLIER (Ω)	TOLERANCE	TEMP. COEFF. ($.10^{-6}$/K)	
	0.01	10%		SILVER
	0.1	5%		GOLD
0	1		200	BLACK
1	10	1%	100	BROWN
2	100	2%	50	RED
3	1 K		15	ORANGE
4	10 K		25	YELLOW
5	100 K	0.5%		GREEN
6	1 M	0.25%	10	BLUE
7	10 M	0.1%	5	VIOLET
8			1	GREY
9				WHITE

Figure 1.15 Color code for fixed resistors in accordance with IEC publication 62. (*From* [3]. *Used with permission.*)

Carbon Composition Resistor

Carbon composition resistors contain a cylinder of carbon-based resistive material molded into a cylinder of high-temperature plastic, which also anchors the external leads. These resistors can have noise problems similar to carbon film resistors, but their use in electronic equipment for the last 50 years has demonstrated their outstanding reliability, unmatched by other components. These resistors are commonly available at values from 2.7 W with tolerances of 5, 10, and 20 percent in 1/8-, 1/4-, 1/2-, 1-, and 2-W sizes.

Control and Limiting Resistors

Resistors with a large negative temperature coefficient, *thermistors*, are often used to measure temperature, limit inrush current into motors or power supplies, or to compensate bias circuits. Resistors with a large positive temperature coefficient are used in circuits that have to match the coefficient of copper wire. Special resistors also include those that have a low resistance when cold and become a nearly open circuit when a critical temperature or current is exceeded to protect transformers or other devices.

Resistor Networks

A number of metal film or similar resistors are often packaged in a single module suitable for printed circuit mounting. These devices see applications in digital circuits, as well as in fixed attenuators or padding networks.

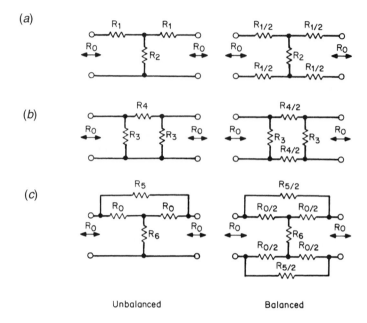

Figure 1.16 Unbalanced and balanced fixed attenuator networks for equal source and load resistance: (a) T configuration, (b) π configuration, (c) bridged-T configuration.

Adjustable Resistors

Cylindrical wire-wound power resistors can be made adjustable with a metal clamp in contact with one or more turns not covered with enamel along an axial stripe. Potentiometers are resistors with a movable arm that makes contact with a resistance element, which is connected to at least two other terminals at its ends. The resistance element can be circular or linear in shape, and often two or more sections are mechanically coupled or ganged for simultaneous control of two separate circuits. Resistance materials include all those described previously.

Trimmer potentiometers are similar in nature to conventional potentiometers except that adjustment requires a tool.

Most potentiometers have a *linear taper*, which means that resistance changes linearly with control motion when measured between the movable arm and the "low," or counterclockwise, terminal. Gain controls however, often have a *logarithmic taper* so that attenuation changes linearly in decibels (a logarithmic ratio). The resistance element of a potentiometer may also contain taps that permit the connection of other components as required in a specialized circuit.

Attenuators

Variable attenuators are adjustable resistor networks that show a calibrated increase in attenuation for each switched step. For measurement of audio, video, and RF equipment, these steps may be decades of 0.1, 1, and 10 dB. Circuits for unbalanced and balanced fixed attenuators are shown in Figure 1.16. Fixed attenuator networks can be cascaded and switched to provide step adjustment of attenuation inserted in a constant-impedance network.

Audio attenuators generally are designed for a circuit impedance of 150 Ω, although other impedances can be used for specific applications. Video attenuators are generally designed to operate with unbalanced 75-Ω grounded-shield coaxial cable. RF attenuators are designed for use with 75- or 50-Ω coaxial cable.

1.6.2 Capacitors

Capacitors are passive components in which current leads voltage by nearly 90° over a wide range of frequencies. Capacitors are rated by capacitance, voltage, materials, and construction.

A capacitor may have two voltage ratings. *Working voltage* is the normal operating voltage that should not be exceeded during operation. Use of the test or *forming voltage* stresses the capacitor and should occur only rarely in equipment operation. Good engineering practice is to use components only at a fraction of their maximum ratings.

The primary characteristics of common capacitors are given in Table 1.8.

Polarized Capacitors

Polarized capacitors can be used in only those applications where a positive sum of all dc and peak-ac voltages is applied to the positive capacitor terminal with respect to its negative terminal. These capacitors include all tantalum and most aluminum electrolytic capacitors. These devices are commonly used in power supplies or other electronic equipment where these restrictions can be met.

Losses in capacitors occur because an actual capacitor has various resistances. These losses are usually measured as the dissipation factor at a frequency of 120 Hz. Leakage resistance in parallel with the capacitor defines the time constant of discharge of a capacitor. This time constant can vary between a small fraction of a second to many hours depending on capacitor construction, materials, and other electrical leakage paths, including surface contamination.

The e*quivalent series resistance* of a capacitor is largely the resistance of the conductors of the capacitor plates and the resistance of the physical and chemical system of the capacitor. When an alternating current is applied to the capacitor, the losses in the equivalent series resistance are the major causes of heat developed in the device. The same resistance also determines the maximum attenuation of a filter or bypass capacitor and the loss in a coupling capacitor connected to a load.

The *dielectric absorption* of a capacitor is the residual fraction of charge remaining in a capacitor after discharge. The residual voltage appearing at the capacitor terminals after discharge is of little concern in most applications but can seriously affect the per-

Table 1.8 Parameters and Characteristics of Discrete Capacitors (*From* [3]. *Used with permission.*)

Capacitor Type	Range	Rated Voltage, V_R	TC ppm/°C	Tolerance, ±%	Insulation Resistance, MΩμF	Dissipation Factor, %	Dielectric Absorption, %	Temperature Range, °C	Comments, Applications	Cost
Polycarbonate	100 pF–30 μF	50–800	±50	10	5×10^5	0.2	0.1	−55/+125	High quality, small, low TC	High
Polyester/Mylar	1000 pF–50 μF	50–600	+400	10	10^5	0.75	0.3	−55/+125	Good, popular	Medium
Polypropylene	100 pF–50 μF	100–800	−200	10	10^5	0.2	0.1	−55/+105	High quality low absorption	High
Polystyrene	10 pF–2.7 μF	100–600	−100	10	10^6	0.05	0.04	−55/+85	High quality, large, low TC, signal filters	Medium
Polysulfone	1000 pF–1 μF	100–600	+80	5	10^5	0.3	0.2	−55/+150	High temperature	High
Parylene	5000 pF–1 μF		±100	10	10^5	0.1	0.1	−55/+125		High
Kapton	1000 pF–1 μF		+100	10	10^5	0.3	0.3	−55/+220	High temperature	High
Teflon	1000 pF–2 μF	50–200	−200	10	5×10^6	0.04	0.04	−70/+250	High temperature lowest absorption	High
Mica	5 pF–0.01 μF	100–600	−50	5	2.5×10^4	0.001	0.75	−55/+125	Good at RF; low TC	High
Glass	5 pF–1000 pF	100–600	+140	5	10^6	0.001		−55/+125	Excellent long-term stability	High
Porcelain	100 pF–0.1 μF	50–400	+120	5	5×10^5	0.10	4.2	−55/+125	Good long-term stability	High
Ceramic (NPO)	100 pF–1 μF	50–400	±30	10	5×10^3	0.02	0.75	−55/+125	Active filters, low TC	Medium
Ceramic	10 pF–1 μF	50–30,000						−55/+125	Small, very popular selectable TC	Low
Paper	0.01 μF–10 μF	200–1600	±800	10	5×10^3	1.0	2.5	−55/+125	Motor capacitors	Low
Aluminum	0.1 μF–1.6 F	3–600	+2500	−10/+100	100	10	8.0	−40/+85	Power supply filters short life	High
Tantalum (Foil)	0.1 μF–1000 μF	6–100	+800	−10/+100	20	4.0	8.5	−55/+85	High capacitance small size, low inductance	High
Thin-film	10 pF–200 pF	6–30	+100	10	10^6	0.01		−55/+125		High
Oil	0.1 μF–20 μF	200–10,000				0.5			High voltage filters, large, long life	
Vacuum	1 pF–1000 pF	2,000–3,600							Transmitters	

formance of *analog-to-digital* (A/D) converters that must perform precision measurements of voltage stored in a sampling capacitor.

The *self-inductance* of a capacitor determines the high-frequency impedance of the device and its ability to bypass high-frequency currents. The self-inductance is determined largely by capacitor construction and tends to be highest in common metal foil devices.

Nonpolarized Capacitors

Nonpolarized capacitors are used in circuits where there is no direct voltage bias across the capacitor. They are also the capacitor of choice for most applications requiring capacity tolerances of 10 percent or less.

Film Capacitors

Plastic is a preferred dielectrical material for capacitors because it can be manufactured with minimal imperfections in thin films. A metal-foil capacitor is constructed by winding layers of metal, plastic, metal, and plastic into a cylinder and then making a connection to the two layers of metal. A *metallized foil capacitor* uses two layers, each of which has a very thin layer of metal evaporated on one surface, thereby obtaining a higher capacity per volume in exchange for a higher equivalent series resistance. Metallized foil capacitors are self-repairing in the sense that the energy stored in the capacitor is often sufficient to burn away the metal layer surrounding the void in the plastic film.

Depending on the dielectric material and construction, capacitance tolerances between 1 and 20 percent are common, as are voltage ratings from 50 to 400 V. Construction types include axial leaded capacitors with a plastic outer wrap, metal-encased units, and capacitors in a plastic box suitable for printed circuit board insertion.

Polystyrene has the lowest dielectric absorption of 0.02 percent, a temperature coefficient of -20 to -100 ppm/°C, a temperature range to 85°C, and extremely low leakage. Capacitors between 0.001 and 2 μF can be obtained with tolerances from 0.1 to 10 percent.

Polycarbonate has an upper temperature limit of 100°C, with capacitance changes of about 2 percent up to this temperature. Polypropylene has an upper temperature limit of 85°C. These capacitors are particularly well suited for applications where high inrush currents occur, such as switching power supplies. Polyester is the lowest-cost material with an upper temperature limit of 125°C. Teflon and other high-temperature materials are used in aerospace and other critical applications.

Foil Capacitors

Mica capacitors are made of multiple layers of silvered mica packaged in epoxy or other plastic. Available in tolerances of 1 to 20 percent in values from 10 to 10,000 pF, mica capacitors exhibit temperature coefficients as low as 100 ppm. Voltage ratings between 100 and 600 V are common. Mica capacitors are used mostly in high-frequency filter circuits where low loss and high stability are required.

Electrolytic Capacitors

Aluminum foil electrolytic capacitors can be made nonpolar through use of two cathode foils instead of anode and cathode foils in construction. With care in manufacturing, these capacitors can be produced with tolerance as tight as 10 percent at voltage ratings of 25 to 100 V peak. Typical values range from 1 to 1000 μF.

Ceramic Capacitors

Barium titanate and other ceramics have a high dielectric constant and a high breakdown voltage. The exact formulation determines capacitor size, temperature range, and variation of capacitance over that range (and consequently capacitor application). An alphanumeric code defines these factors, a few of which are given here.

- Ratings of Y5V capacitors range from 1000 pF to 6.8 μF at 25 to 100 V and vary + 22 to –82 percent in capacitance from –30 to + 85°C.

- Ratings of Z5U capacitors range to 1.5 μF and vary +22 to –56 percent in capacitance from +10 to +85°C. These capacitors quite small in size and are used typically as bypass capacitors.

- X7R capacitors range from 470 pF to 1 μF and vary 15 percent in capacitance from –55 to + 125°C.

Nonpolarized (NPO) rated capacitors range from 10 to 47,000 pF with a temperature coefficient of 0 to +30 ppm over a temperature range of –55 to +125°C.

Ceramic capacitors come in various shapes, the most common being the radial-lead disk. Multilayer monolithic construction results in small size, which exists both in radial-lead styles and as chip capacitors for direct surface mounting on a printed circuit board.

Polarized-Capacitor Construction

Polarized capacitors have a negative terminal—the cathode—and a positive terminal—the anode—and a liquid or gel between the two layers of conductors. The actual dielectric is a thin oxide film on the cathode, which has been chemically roughened for maximum surface area. The oxide is formed with a forming voltage, higher than the normal operating voltage, applied to the capacitor during manufacture. The direct current flowing through the capacitor forms the oxide and also heats the capacitor.

Whenever an electrolytic capacitor is not used for a long period of time, some of the oxide film is degraded. It is reformed when voltage is applied again with a leakage current that decreases with time. Applying an excessive voltage to the capacitor causes a severe increase in leakage current, which can cause the electrolyte to boil. The resulting steam may escape by way of the rubber seal or may otherwise damage the capacitor.

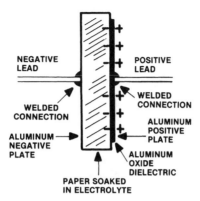

Figure 1.17 The basic construction of an aluminum electrolytic capacitor.

Application of a reverse voltage in excess of about 1.5 V will cause forming to begin on the unetched anode electrode. This can happen when pulse voltages superimposed on a dc voltage cause a momentary voltage reversal.

Aluminum Electrolytic Capacitors

Aluminum electrolytic capacitors use very pure aluminum foil as electrodes, which are wound into a cylinder with an interlayer paper or other porous material that contains the electrolyte. (See Figure 1.17.) Aluminum ribbon staked to the foil at the minimum inductance location is brought through the insulator to the anode terminal, while the cathode foil is similarly connected to the aluminum case and cathode terminal.

Electrolytic capacitors typically have voltage ratings from 6.3 to 450 V and rated capacitances from 0.47 μF to several hundreds of microfarads at the maximum voltage to several farads at 6.3 V. Capacitance tolerance may range from ±20 to +80/–20 percent. The operating temperature range is often rated from –25 to +85°C or wider. Leakage current of an electrolytic capacitor may be rated as low as 0.002 times the capacity times the voltage rating to more than 10 times as much.

Tantalum Electrolytic Capacitors

Tantalum electrolytic capacitors are the capacitors of choice for applications requiring small size, 0.33- to 100-μF range at 10 to 20 percent tolerance, low equivalent series resistance, and low leakage current. These devices are well suited where the less costly aluminum electrolytic capacitors have performance issues. Tantalum capacitors are packaged in hermetically sealed metal tubes or with axial leads in epoxy plastic, as illustrated in Figure 1.18.

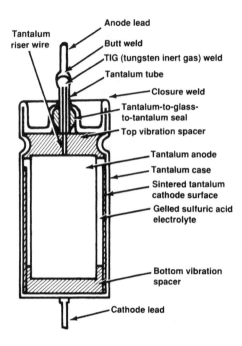

Figure 1.18 Basic construction of a tantalum capacitor.

1.6.3 Inductors and Transformers

Inductors are passive components in which voltage leads current by nearly 90° over a wide range of frequencies. Inductors are usually coils of wire wound in the form of a cylinder. The current through each turn of wire creates a magnetic field that passes through every turn of wire in the coil. When the current changes, a voltage is induced in the wire and every other wire in the changing magnetic field. The voltage induced in the same wire that carries the changing current is determined by the inductance of the coil, and the voltage induced in the other wire is determined by the *mutual induc-tance* between the two coils. A transformer has at least two coils of wire closely coupled by the common magnetic core, which contains most of the magnetic field within the transformer.

Inductors and transformers vary widely in size, weighing less than 1 g or more than 1 ton, and have specifications ranging nearly as wide.

Losses in Inductors and Transformers

Inductors have resistive losses because of the resistance of the copper wire used to wind the coil. An additional loss occurs because the changing magnetic field causes eddy currents to flow in every conductive material in the magnetic field. Using thin magnetic laminations or powdered magnetic material reduces these currents.

Losses in inductors are measured by the Q, or quality, factor of the coil at a test frequency. Losses in transformers are sometimes given as a specific insertion loss in decibels. Losses in power transformers are given as core loss in watts when there is no load connected and as a regulation in percent, measured as the relative voltage drop for each secondary winding when a rated load is connected.

Transformer loss heats the transformer and raises its temperature. For this reason, transformers are rated in watts or volt-amperes and with a temperature code designating the maximum hotspot temperature allowable for continued safe long-term operation. For example, class A denotes 105°C safe operating temperature. The volt-ampere rating of a power transformer must be always larger than the dc power output from the rectifier circuit connected because volt-amperes, the product of the rms currents and rms voltages in the transformer, are larger by a factor of about 1.6 than the product of the dc voltages and currents.

Inductors also have capacitance between the wires of the coil, which causes the coil to have a self-resonance between the winding capacitance and the self-inductance of the coil. Circuits are normally designed so that this resonance is outside of the frequency range of interest. Transformers are similarly limited. They also have capacitance to the other winding(s), which causes *stray coupling*. An electrostatic shield between windings reduces this problem.

Air-Core Inductors

Air-core inductors are used primarily in radio frequency applications because of the need for values of inductance in the microhenry or lower range. The usual construction is a multilayer coil made self-supporting with adhesive-covered wire. An inner diameter of 2 times coil length and an outer diameter 2 times as large yields maximum Q, which is also proportional to coil weight.

Ferromagnetic Cores

Ferromagnetic materials have a permeability much higher than air or vacuum and cause a proportionally higher inductance of a coil that has all its magnetic flux in this material. Ferromagnetic materials in audio and power transformers or inductors usually are made of silicon steel laminations stamped in the forms of letters E or I (Figure 1.19). At higher frequencies, powdered ferric oxide is used. The continued magnetization and remagnetization of silicon steel and similar materials in opposite directions does not follow the same path in both directions but encloses an area in the magnetization curve and causes a hysteresis loss at each pass, or twice per ac cycle.

All ferromagnetic materials show the same behavior; only the numbers for permeability, core loss, saturation flux density, and other characteristics are different. The properties of some common magnetic materials and alloys are given in Table 1.9.

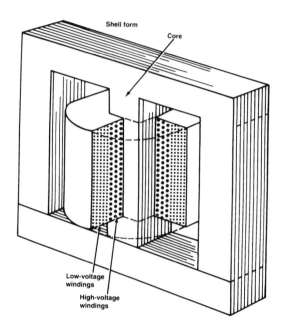

Figure 1.19 Physical construction of an E-shaped power transformer. The low- and high-voltage windings are stacked as shown.

Shielding

Transformers and coils radiate magnetic fields that can induce voltages in other nearby circuits. Similarly, coils and transformers can develop voltages in their windings when subjected to magnetic fields from another transformer, motor, or power circuit. Steel mounting frames or chassis conduct these fields, offering less reluctance than air.

The simplest way to reduce the stray magnetic field from a power transformer is to wrap a copper strip as wide as the coil of wire around the transformer enclosing all three legs of the core. Shielding occurs by having a short circuit turn in the stray magnetic field outside of the core.

1.6.4 Diodes and Rectifiers

A *diode* is a passive electronic device that has a positive anode terminal and a negative cathode terminal and a nonlinear voltage-current characteristic. A *rectifier* is assembled from one or more diodes for the purpose of obtaining a direct current from an alternating current; this term also refers to large diodes used for this purpose. Many types of diodes exist.

Table 1.9 Properties of Magnetic Materials and Magnetic Alloys (*From* [3]. *Used with permission.*)

Material (Composition)	Initial Relative Permeability, μ_i/μ_0	Maximum Relative Permeability, μ_{max}/μ_0	Coercive Force H_c, A/m (Oe)	Residual Field B_r, Wb/m²(G)	Saturation Field B_s, Wb/m²(G)	Electrical Resistivity ρ ×10⁻⁸ Ω·m	Uses
			Soft				
Commercial iron (0.2 imp.)	250	9,000	≈80 (1)	0.77 (7,700)	2.15 (21,500)	10	Relays
Purified iron (0.05 imp.)	10,000	200,000	4 (0.05)	—	2.15 (21,500)	10	
Silicon-iron (4 Si)	1,500	7,000	20 (0.25)	0.5 (5,000)	1.95 (19,500)	60	Transformers
Silicon-iron (3 Si)	7,500	55,000	8 (0.1)	0.95 (9,500)	2 (20,000)	50	Transformers
Silicon-iron (3 Si)	—	116,000	4.8 (0.06)	1.22 (12,200)	2 (20,100)	50	Transformers
Mu metal (5 Cu, 2 Cr, 77 Ni)	20,000	100,000	4 (0.05)	0.23 (2,300)	0.65 (6,500)	62	Transformers
78 Permalloy (78.5 Ni)	8,000	100,000	4(0.05)	0.6 (6,000)	1.08 (10,800)	16	Sensitive relays
Supermalloy (79 Ni, 5 Mo)	100,000	1,000,000	0.16 (0.002)	0.5 (5,000)	0.79 (7,900)	60	Transformers
Permendur (50 Cs)	800	5,000	160 (2)	1.4 (14,000)	2.45 (24,500)	7	Electromagnets
Mn-Zn ferrite	1,500	2,500	16 (0.2)	—	0.34 (3,400)	20 × 10⁶	Core material
Ni-Zn ferrite	2,500	5,000	8 (0.1)	—	0.32 (3,200)	10¹¹	for coils

Source: After Plonus, M.A. 1978. *Applied Electromagnetics.* McGraw–Hill, New York.

Over the years, a great number of constructions and materials have been used as diodes and rectifiers. Rectification in electrolytes with dissimilar electrodes resulted in the *electrolytic rectifier*. The voltage-current characteristic of conduction from a heated cathode in vacuum or low-pressure noble gases or mercury vapor is the basis of vacuum tube diodes and rectifiers. Semiconductor materials such as germanium, silicon, selenium, copper-oxide, or gallium arsenide can be processed to form a pn junction that has a nonlinear diode characteristic. Although all these systems of rectification have seen use, the most widely used rectifier in electronic equipment is the silicon diode. The remainder of this section deals only with these and other silicon two-terminal devices.

The pn Junction

When biased in a reverse direction at a voltage well below breakdown, the diode reverse current is composed of two currents. One current is caused by leakage due to contamination and is proportional to voltage. The intrinsic diode reverse current is independent of voltage but doubles for every 10°C in temperature (approximately). The forward current of a silicon diode is approximately equal to the leakage current multiplied by e (= 2.718) raised to the power given by the ratio of forward voltage divided by 26 mV with the junction at room temperature. In practical rectifier calculations, the reverse current is considered to be important in only those cases where a capacitor must hold a charge for a time, and the forward voltage drop is assumed to be constant at 0.7 V, unless a wide range of currents must be considered.

All diode junctions have a *junction capacitance* that is approximately inversely proportional to the square of the applied reverse voltage. This capacitance rises further with applied forward voltage. When a rectifier carries current in a forward direction, the junction capacitance builds up a charge. When the voltage reverses across the junction, this charge must flow out of the junction, which now has a lower capacitance, giving rise to a current spike in the opposite direction of the forward current. After the *reverse-recovery time*, this spike ends, but interference may be radiated into low-level circuits. For this reason, rectifier diodes are sometimes bypassed with capacitors of about 0.1 mF located close to the diodes. Rectifiers used in high-voltage assemblies use bypass capacitors and high value resistors to reduce noise and equalize the voltage distribution across the individual diodes (Figure 1.20).

Tuning diodes have a controlled reverse capacitance that varies with applied direct tuning voltage. This capacitance may vary over a 2-to-1 to as high as a 10-to-1 range and is used to change the resonant frequency of tuned RF circuits. These diodes find application in radio and television receiver circuits.

Zener Diodes and Reverse Breakdown

When the reverse voltage on a diode is increased to a certain *critical voltage*, the reverse leakage current will increase rapidly or *avalanche*. This breakdown or *zener voltage* sets the upper voltage limit a rectifier can experience in normal operation because the peak reverse currents may become as high as the forward currents. Rectifier and other diodes have a rated *peak reverse voltage*, and some rectifier circuits may de-

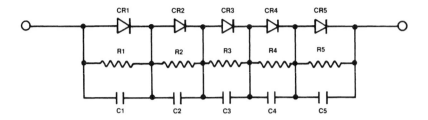

Figure 1.20 A high-voltage rectifier stack.

pend on this reverse breakdown to limit high-voltage spikes that may enter the equipment from the power line. It should also be noted that diode dissipation is very high during these periods.

The reverse breakdown voltage can be controlled in manufacture to a few percent and used to advantage in a class of devices known as *zener diodes*, used extensively in voltage-regulator circuits. It should be noted that the voltage-current curve of a pn junction may go through a region where a *negative resistance* occurs and voltage decreases a small amount while current increases. This condition can give rise to noise and oscillation, which can be minimized by connecting a ceramic capacitor of about 0.02 μF and an electrolytic capacitor of perhaps 100 μF in parallel with the zener diode. Voltage-regulator diodes are available in more than 100 types, covering voltages from 2.4 to 200 V with rated dissipation between 1/4 and 10 W (typical). The forward characteristics of a zener diode usually are not specified but are similar to those of a conventional diode.

Precision voltage or *bandgap reference diodes* make use of the difference in voltage between two diodes carrying a precise ratio of forward currents. Packaged as a two-terminal device including an operational amplifier, these devices produce stable reference voltages of 1.2, 2.5, 5, and 10 V, depending on type.

Current Regulators

The *current regulator diode* is a special class of device used in many small signal applications where constant current is needed. These diodes are *junction field-effect transistors* (FETs) with the gate connected to the source and effectively operated at zero-volt bias. Only two leads are brought out. Current-regulator diodes require a minimum voltage of a few volts for good regulation. Ratings from 0.22 to 4.7 mA are commonly available.

Varistor

Varistors are symmetrical nonlinear voltage-dependent resistors, behaving not unlike two zener diodes connected back to back. The current in a varistor is proportional to applied voltage raised to a power N. These devices are normally made of zinc oxide,

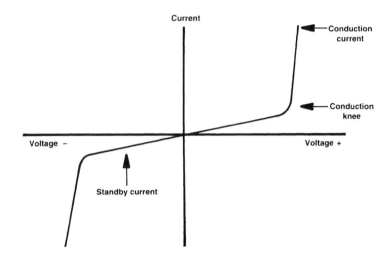

Figure 1.21 The current-vs.-voltage transfer curve for a varistor. Note the *conduction knee.*

which can be produced to have an N factor of 12 to 40. In circuits at normal operating voltages, varistors are nearly open circuits shunted by a capacitor of a few hundred to a few thousand picofarads. Upon application of a high voltage pulse, such as a lightning discharge, they conduct a large current, thereby absorbing the pulse energy in the bulk of the material with only a relatively small increase in voltage, thus protecting the circuit. (See Figure 1.21.) Varistors are available for operating voltages from 10 to 1000 V rms and can handle pulse energies from 0.1 to more than 100 J and maximum peak currents from 20 to 2000 A. Typical applications include protection of power supplies and power-switching circuits, and the protection of telephone and data-communication lines.

1.6.5 Indicators

Indicators are generally passive components that send a message to the operator of the equipment. This message is most commonly a silent visual indication that the equipment is operating in some particular mode, is ready to operate, or is not ready. Indicator lights of different colors illuminating a legend or having an adjacent legend are most commonly used. Alphanumeric codes and complete messages are often displayed on cathode ray tubes or on liquid crystal displays. These more complex displays are computer- or microprocessor-controlled.

Miniature light bulbs are incandescent devices operating at low voltage between 1 and 48 V, with currents from 0.01 to 4 A and total power requirements from 0.04 to more than 20 W, resulting in light output from 0.001 to more than 20 cd. The rated life of normally 10 to 50,000 h will typically decrease to one-tenth of the rating if the lamp voltage is increased to 20 percent above rated value. The resistance of the filament increases with temperature, varying by as much as a factor of 16 from cold to hot.

Solid-state lamps or *light-emitting diodes* (LED) are pn-junction lasers that generate light when diode current exceeds a critical threshold value. Visible red light is emitted from gallium arsenide phosphide junctions. Green or amber light is emitted from doped gallium phosphide junctions. The junctions have a forward voltage drop of 1.7 to 2.2 V at a normal operating current of 10 to 50 mA. Other visible colors are commercially available.

The LED is encased singly in round or rectangular plastic cases or assembled as multiples. A linear array of LEDs is often used in an arrangement similar to a thermometer to indicate volume or transmission level in audio or video circuits. An array, typically seven segments, can form the shapes of numerals and letters by selectively applying power to some or all segments. An array of 35 lamps in a 5×7 matrix can be connected to power to show the shape of letters, numerals, and punctuation marks. Semiconductor integrated circuits are available to achieve such functions with groups of these digit or indicator assemblies.

Light-emitting diodes have a typical operating life of about 50,000 h but have the disadvantage of relatively high current consumption, limited colors and shapes, and reduced visibility in bright light.

Electrons emitted from a heated cathode or a cold cathode can cause molecules of low-pressure gas, such as neon, to ionize and to emit light. Neon lamps require a current-limited supply of at least 90 V to emit orange light. The most frequent use of neon lamps is to indicate the presence of power line voltage. By means of a series resistor, the current is limited to a permissible value.

Emitted electrons can also strike a target connected to an anode terminal and coated with fluorescent phosphors. By directing the electron flow to "flood" different segments, alphanumeric displays can be produced similar to LED configurations with supply voltages found in battery-operated circuits.

When certain solutions of organic chemicals are exposed to an electric field, these crystal-like ions align themselves with the field and cause light of only one polarization to pass through the liquid. A second polarizer of light then causes the assembly to be a voltage-controlled attenuator of light. *Liquid crystal displays* come in many shapes, require low operating power, and can be backlit or used with external light only. These displays are found in many different types of systems, including test equipment, watches, computer terminals, and television sets.

1.7 Active Circuit Components

Active components can generate more alternating signal power into an output load resistance than the power absorbed at the input at the same frequency. Active components are the major building blocks in system assemblies such as amplifiers and oscillators.

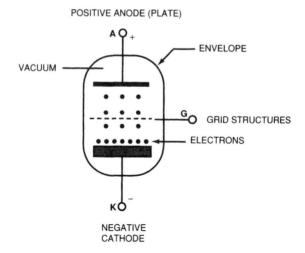

Figure 1.22 Basic operation of the vacuum tube. (*From* [3]. *Used with permission.*)

1.7.1 Vacuum Tubes

Vacuum tubes are the active components that enabled the amplification and control of audio, video, radio frequency, and other signals and helped bring about the growth of the electronics industry from a laboratory curiosity early in the twentieth century to a high state of maturity in the 1960s. Since this time, the transistor and the integrated solid-state circuit have largely replaced vacuum tubes in most low-power applications. The major uses of vacuum tubes today are as displays for television sets and computers, as generators of radio frequency power in selected applications, and the generation of X-rays for medical and industrial use.

A heated cathode coated with rare-earth oxides in a vacuum causes a cloud of electrons to exist near the cathode (Figure 1.22). A positive anode voltage with respect to the cathode causes some of these electrons to flow as a current to the anode. A grid of wires at a location between anode and cathode and biased at a control voltage with respect to the cathode causes a greater or lesser amount of anode current to flow. Other intervening grids also control the anode current and, if biased with a positive voltage, draw grid current from the total cathode current

The three basic types of amplifying vacuum tube devices are the triode, tetrode, and pentode, as illustrated in Figure 1.23.

1.7.2 Bipolar Transistors

A bipolar transistor has two pn junctions that behave in a manner similar to that of the diode pn junctions described previously. These junctions are the base-emitter junction and the base-collector junction. In typical use, the first junction would normally have

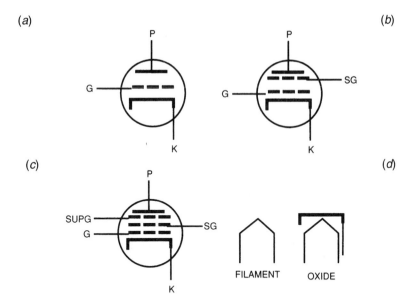

Figure 1.23 Types of vacuum tubes: (*a*) triode, (*b*) tetrode, (*c*) pentode, (*d*) filament designs.

a forward bias, causing conduction, and the second junction would have a reverse bias. If the material of the base were very thick, the flow of electrons into the p-material base junction (of an NPN transistor) would go entirely into the base junction and no current would flow in the reverse-biased collector-base junction.

If, however, the base junction were quite thin, electrons would diffuse in the semiconductor crystal lattice into the base-collector junction, having been injected into the base material of the base-emitter junction. This diffusion occurs because an excess electron moving into one location will bump out an electron in the adjacent semiconductor molecule, which will bump its neighbor. Thus, a collector current will flow that is nearly as large as the injected emitter current.

The ratio of collector to emitter current is *alpha* or the common-base current gain of the transistor, normally a value a little less than 1.000. The portion of the emitter current not flowing into the collector will flow as a base current in the same direction as the collector current. The ratio of collector current to base current, or *beta*, is the conventional current gain of the transistor and may be as low as 5 in power transistors operating at maximum current levels to as high as 5000 in super-beta transistors operated in the region of maximum current gain.

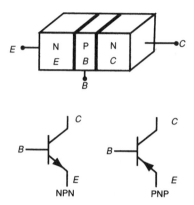

Figure 1.24 Bipolar junction transistor—basic construction and symbols. (*From* [3]. *Used with permission.*)

NPN and PNP Transistors

Bipolar transistors are identified by the sequence of semiconductor material going from emitter to collector. NPN transistors operate normally with a positive voltage on the collector with respect to the emitter, with PNP transistors requiring a negative voltage at the collector and the flow of current being internally mostly a flow of *holes* or absent excess electrons in the crystal lattice at locations of flow. (See Figure 1.24.)

Because the diffusion velocity of holes is slower than that pn electrons, PNP transistors have more junction capacitance and slower speed than NPN transistors of the same size. Holes and electrons in pn junctions are *minority carriers* of electric current as opposed to electrons, which are *majority carriers* and which can move freely in resistors or in the conductive channel of field-effect transistors. Consequently, bipolar transistors are known as *minority carrier devices*.

The most common transistor material is silicon, which permits transistor junction temperatures as high as 200°C. The normal base-emitter voltage is about 0.7 V, and collector-emitter voltage ratings of up to hundreds of volts are available. At room temperature these transistors may dissipate from tens of milliwatts to hundreds of watts with proper heat removal.

Transistors made of gallium arsenide and similar materials are also available for use in microwave and high speed circuits, taking advantage of the high diffusion speeds and low capacitances characteristic of such materials.

Transistor Impedance and Gain

Transistor impedances and gain are normally referred to the *common-emitter* connection, which also results in the highest gain. It is useful to treat transistor parameters first as if the transistor were an ideal device and then to examine degradations resulting from nonideal behavior.

If we assume that the transistor has a fixed current gain, then the collector current is equal to the base current multiplied by the current gain of the transistor, and the emitter current is the sum of both of these currents. Because the collector-base junction is reverse-biased, the output impedance of the ideal transistor is very high.

Actual bipolar transistors suffer degradations from this ideal model. Each transistor terminal may be thought of having a resistor connected in series, although these resistors are actually distributed rather than lumped components. These resistors cause the transistor to have lower gain than predicted and to have a saturation voltage in both input and output circuits. In addition, actual transistors have resistances connected between terminals that cause further reductions in available gain, particularly at low currents and with high load resistances.

In addition to resistances, actual transistors also exhibit stray capacitance between terminals, causing further deviation from the ideal case. These capacitances are—in part—the result of the physical construction of the devices and also the finite diffusion velocities in silicon. The following effects result:

- Transistor current gain decreases with increasing frequency, with the transistor reaching unity current gain at a specific *transition frequency*.

- A feedback current exists from collector to base through the base-collector capacitance.

- Storage of energy in the output capacitance similar to energy storage in a rectifier diode. This stored energy limits the turn-off speed of transistors, a critical factor in certain applications.

Transistor Configurations

Table 1.10 summarizes the most common transistor operating modes. For stages using a single device, the common-emitter arrangement is by far the most common. Power output stages of push-pull amplifiers make use of the common-collector or *emitter-follower* connection. Here, the collector is directly connected to the supply voltage, and the load is connected to the emitter terminal with signal and bias voltage applied to the base terminal. The voltage gain of such a circuit is a little less than 1.000, and the load impedance at the emitter is reflected to the base circuit as if it were increased by the current gain of the transistor.

At high frequencies, the base of a transistor is often grounded for high-frequency signals, which are fed to the emitter of the transistor. With this arrangement, the input impedance of the transistor is low, which is easily matched to radio frequency transmission lines, assisted in part by the minimal capacitive feedback within the transistor.

So far in this discussion, transistor analysis has dealt primarily with the small signal behavior. For operation under large signal conditions, other limitations must be observed. When handling low-frequency signals, a transistor can be viewed as a variable-controlled resistor between the supply voltage and the load impedance. The *quiescent operating point* in the absence of ac signals is usually chosen so that the maximum

Table 1.10 Basic Amplifier Configurations of Bipolar Transistors (*From* [4]. *Used with permission.*)

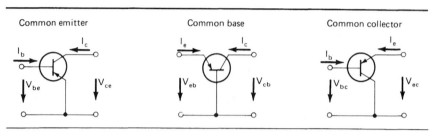

| | Characteristics of basic configurations | | |
	Common emitter	Common base	Common collector
Input impedance Z_1	Medium	Low	High
	Z_{1e}	$Z_{1b} \approx \dfrac{Z_{1e}}{h_{fe}}$	$Z_{1c} \approx h_{fe} R_L$
Output impedance Z_2	High	Very high	Low
	Z_{2e}	$Z_{2b} \approx Z_{2e} h_{fe}$	$Z_{2c} \approx \dfrac{Z_{1e} + R_g}{h_{fe}}$
Small-signal current gain	High	< 1	High
	h_{fe}	$h_{fb} \approx \dfrac{h_{fe}}{h_{fe} + 1}$	$\gamma \approx h_{fe} + 1$
Voltage gain	High	High	< 1
Power gain	Very high	High	Medium
Cutoff frequency	Low	High	Low
	f_{hfe}	$f_{hfb} \approx h_{fe} f_{hfc}$	$f_{hfc} \approx f_{hfe}$

signal excursions in both positive and negative directions can be handled without limiting resulting from near-zero voltage across the transistor at maximum output current or near-zero current through the transistor at maximum output voltage. This is most critical in class B push-pull amplifiers where first one transistor stage conducts current to the load during part of one cycle and then the other stage conducts during the other part. Similar considerations also apply for distortion reduction considerations.

Limiting conditions also constitute the maximum capabilities of transistors under worst-case conditions of supply voltage, load impedance, drive signal, and temperature consistent with safe operation. In no case should the maximum voltage across a transistor ever be exceeded.

Switching and Inductive-Load Ratings

When using transistors for driving relays, deflection yokes of cathode ray tubes, or any other inductive or resonant load, current in the inductor will tend to flow in the same direction, even if interrupted by the transistor. The resultant voltage spike caused by the collapse of the magnetic field can destroy the switching device unless it is designed to handle the energy of these voltage excursions. The manufacturers of power semiconductors have special transistor types and application information relating to inductive switching circuits. In many cases, the use of protection diodes are sufficient.

Transistors are often used to switch currents into a resistive load. The various junction capacitances are voltage-dependent in the same manner as the capacitance of tuning diodes that have maximum capacitance at forward voltages, becoming less at zero voltage and lowest at reverse voltages. These capacitances and the various resistances combine into the switching delay times for turn-on and turn-off functions. If the transistor is prevented from being saturated when turned on, shorter delay times will occur for nonsaturated switching than for saturated switching. These delay times are of importance in the design of switching amplifiers or D/A converters.

Noise

Every resistor creates noise with equal and constant energy for each hertz of bandwidth, regardless of frequency. A useful number to remember is that a 1000-Ω resistor at room temperature has an open-circuit output noise voltage of 4 nanovolts per root-hertz. This converts to 40 nV in a 100-Hz bandwidth or 400 μV in a 10-kHz bandwidth.

Bipolar transistors also create noise in their input and output circuits, and every resistor in the circuit also contributes its own noise energy. The noise of a transistor is effectively created in its input junction, and all transistor noise ratings are referred to it.

In an ideal bipolar transistor, the voltage noise at the base is created by an equivalent resistor that has a value of twice the transistor input conductance at its emitter terminal, and the current noise is created by a resistor that has the value of twice the transistor input conductance at its input terminal. This means that the current noise energy is less at the base terminal of a common-emitter stage by the current gain of the transistor when compared to the current noise at the input of a grounded-base stage.

The highest signal-to-noise ratio in an amplifier can be achieved when the resistance of the signal source is equal to the ratio of amplifier input noise voltage and input noise current, and the reactive impedances have been tuned to zero. Audio frequency amplifiers usually cannot be tuned, and minimum noise may be achieved by matching transformers or by bias current adjustment of the input transistor. With low source impedances, the optimum may not be reached economically, and the equipment must then be designed to have an acceptable input noise voltage.

Practical transistors are not ideal from the standpoint of noise performance. All transistors show a voltage and current noise energy that increases inversely with frequency. At a *corner frequency* this noise will become independent of frequency. Very low noise transistors may have a corner frequency as low as a few hertz, and ordinary high-frequency devices may have a corner frequency well above the audio frequency range. Transistor noise may also be degraded by operating a transistor at more than a few percent of its maximum current rating. Poor transistor design or manufacturing techniques

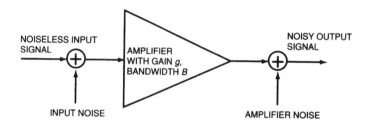

Figure 1.25 A block diagram modeling how noise is introduced to a signal during amplification. (*From* [3]. *Used with permission.*)

can result in transistors that exhibit "popcorn" noise, so named after the audible characteristics of a random low-level switching effect.

The noise level produced by thermal noise sources is not necessarily large, however, because signal power may also be low it is usually necessary to amplify the source signal. Because noise is combined with the source signal, both are then amplified, with more noise added at each successive stage of amplification. Noise can, thus, become a noticeable phenomenon (Figure 1.25).

1.7.3 Field-Effect Transistors

Field-effect transistors (FETs) have a conducting channel terminated by source and drain electrodes and a gate terminal that effectively widens or narrows the channel by the electric field between the gate and each portion of the channel. No gate current is required for steady-state control.

Current flow in the channel is by majority carriers only, analogous to current flow in a resistor. The onset of conduction is not limited by diffusion speeds but by the electric field accelerating the charged electrons.

The input impedance of an FET is a capacitance. Because of this, electrostatic charges during handling may reach high voltages that are capable of breaking down gate insulation.

FETs for common applications use silicon as the semiconducting material. Field-effect transistors are made both in p-channel and n-channel configurations. An n-channel FET has a positive drain voltage with respect to the source voltage, and a positive increase in gate voltage causes an increase in channel current. Reverse polarities exist for p-channel devices. (See Figure 1.26).

An n-channel FET has a drain voltage that is normally positive, and a positive increase in gate-to-source voltage increases drain current and *transconductance*. In single-gate field-effect transistors, drain and source terminals may often be interchanged without affecting circuit performance; however, power handling and other factors may be different. Such an interchange is not possible when two FETs are interconnected internally to form a dual-gate cascode-connected FET, or matched pairs, or when channel

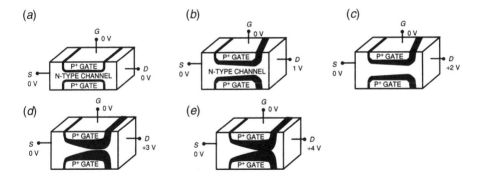

Figure 1.26 Junction FET (JFET) operational characteristics: (*a*) uniform channel from drain to source, (*b*) depletion region wider at the drain end, (*c*) depletion region significantly wider at the drain, (*d*) channel near pinchoff, (*e*) channel at pinchoff. (From [3]. Used with permission.)

conductance is controlled by gates on two sides of the channel as in insulated-gate FETs.

FET Impedance and Gain

The input impedance of a field-effect transistor is usually quite high, and is primarily capacitive. The input capacitance consists of the gate-source capacitance in parallel with the gate-drain capacitance multiplied by the stage gain + 1, assuming the FET has its source at ac-ground potential.

The output impedance of a common-source FET is also primarily capacitive as long as the drain voltage is above a critical value, which, for a junction-gate FET, is equal to the sum of the *pinch-off voltage* and gate-bias voltage. When the pinch-off voltage is applied between the gate and source terminals, the drain current is nearly shut off (the channel is pinched off). Actual FETs have a high drain resistance in parallel with this capacitance. At low drain voltages near zero volts, the drain impedance of an ideal FET is a resistor reciprocal in value to the transconductance of the FET in series with the residual end resistances between the source and drain terminals and the conducting FET channel. This permits an FET to be used as a variable resistor in circuits controlling analog signals.

At drain voltages between zero and the critical voltage, the drain current will increase with both increasing drain voltage and increasing gate voltage. This factor will cause increased saturation voltages in power amplifier circuits when compared to circuits with bipolar transistors.

Table 1.11 summarizes the basic FET amplifier configurations.

Table 1.11 Basic Field-Effect Transistor Amplifier Configurations and Operating Characteristics (*From* [4]. *Used with permission.*)

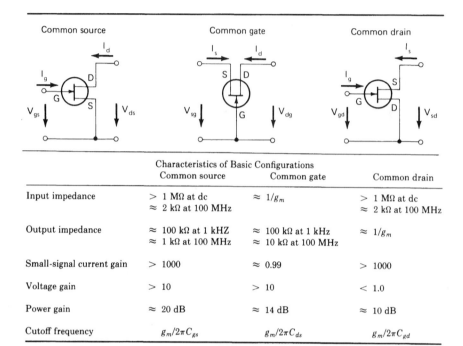

| | Characteristics of Basic Configurations | | |
	Common source	Common gate	Common drain
Input impedance	> 1 MΩ at dc ≈ 2 kΩ at 100 MHz	≈ $1/g_m$	> 1 MΩ at dc ≈ 2 kΩ at 100 MHz
Output impedance	≈ 100 kΩ at 1 kHZ ≈ 1 kΩ at 100 MHz	≈ 100 kΩ at 1 kHz ≈ 10 kΩ at 100 MHz	≈ $1/g_m$
Small-signal current gain	> 1000	≈ 0.99	> 1000
Voltage gain	> 10	> 10	< 1.0
Power gain	≈ 20 dB	≈ 14 dB	≈ 10 dB
Cutoff frequency	$g_m/2\pi C_{gs}$	$g_m/2\pi C_{ds}$	$g_m/2\pi C_{gd}$

1.7.4 Integrated Circuits

An *integrated circuit* (IC) is a combination of circuit elements that are interconnected and formed on and in a continuous substrate material. Usually, an integrated circuit is monolithic and formed by steps that produce semiconductor elements along with resistors and capacitors. A hybrid integrated circuit contains silicon chips along with circuit elements partially formed on the substrate.

The circuit elements formed in integrated circuits are more closely matched to each other than separately selected components, and these elements are in intimate thermal contact with each other. The circuit configurations used in integrated circuits take advantage of this matching and thermal coupling.

Digital Integrated Circuits

The basis of digital circuits is the logic gate that produces a high (or 1) or low (or 0) logic-level output with the proper combination of logic-level inputs. A number of these gates are combined to form a digital circuit that is part of the hardware of computers or controllers of equipment or other circuits. A digital circuit may be extremely complex, containing up to more than 1,000,000 gates.

Bipolar and field-effect transistors are the active elements of digital integrated circuits, divided into families such as *transistor-transistor logic* (TTL), *high-speed complementary metal-oxide-gate semiconductor* (HCMOS), and many others. Special families include memories, microprocessors, and interface circuits between transmission lines and logic circuits. Thousands of digital integrated circuit types in tens of families have been produced.

Linear Integrated Circuits

Linear integrated circuits are designed to process linear signals in their entirety or in part, as opposed to *digital circuits* that process logic signals only. Major classes of linear integrated circuits include operational amplifiers, voltage regulators, digital-to-analog and analog-to-digital circuits, circuits for consumer electronic equipment and communications equipment, power control circuits, and others not as easily classified.

1.8　References

1. Fink, Donald G., and Don Christiansen (eds.), *Electronic Engineers' Handbook*, McGraw-Hill, New York, N.Y., 1982.
2. Chowdhury, Badrul, "Power Distribution and Control," in *The Electronics Handbook*, Jerry C. Whitaker (ed.), pp. 1003, CRC Press, Boca Raton, FL, 1996.
3. Whitaker, Jerry C., (ed.), *The Electronics Handbook*, CRC Press, Boca Raton, FL, 1996.
4. Rhode, U., J. Whitaker, and T. Bucher, *Communications Receivers*, 2nd ed., McGraw-Hill, New York, N.Y., 1996.

1.9　Bibliography

Benson, K. Blair, and Jerry C. Whitaker, *Television and Audio Handbook for Technicians and Engineers*, McGraw-Hill, New York, N.Y., 1990.

Benson, K. Blair, *Audio Engineering Handbook*, McGraw-Hill, New York, N.Y., 1988.

Whitaker, Jerry C., and K. Blair Benson (eds), *Standard Handbook of Video and Television Engineering*, McGraw-Hill, New York, N.Y., 2000.

Whitaker, Jerry C., *Television Engineers' Field Manual*, McGraw-Hill, New York, N.Y., 2000.

2

Modulation Systems

2.1 Introduction

The primary purpose of most communications and signaling systems is to transfer information from one location to another. The message signals used in communication and control systems usually must be limited in frequency to provide for efficient transfer. This frequency may range from a few hertz for control systems to a few megahertz for video signals to many megahertz for multiplexed data signals (Table 2.1). To facilitate efficient and controlled distribution of these components, an *encoder* generally is required between the source and the transmission channel. The encoder acts to *modulate* the signal, producing at its output the *modulated waveform*. Modulation is a process whereby the characteristics of a wave (the *carrier*) are varied in accordance with a message signal, the modulating waveform. Frequency translation is usually a by-product of this process. Modulation may be continuous, where the modulated wave is always present, or pulsed, where no signal is present between pulses.

There are a number of reasons for producing modulated waves, including:

- *Frequency translation.* The modulation process provides a vehicle to perform the necessary frequency translation required for distribution of information. An input signal may be translated to its assigned frequency band for transmission or radiation.

- *Signal processing.* It is often easier to amplify or process a signal in one frequency range as opposed to another.

- *Antenna efficiency.* Generally speaking, for an antenna to be efficient, it must be large compared with the signal wavelength. Frequency translation provided by modulation allows antenna gain and beamwidth to become part of the system design considerations. The use of higher frequencies permits antenna structures of reasonable size and cost.

Table 2.1 Classification of Wireless Radio Frequency Channels by Frequently Bands, Typical Uses, and Wave Propagation Modes (*From* [1]. *Used with permission.*)

Frequency Band	Name	Typical Uses	Propagation Mode
3–30 kHz	Very low frequency (VLF)	Navigation, sonar	Ground waves
30–300 kHz	Low frequency (LF)	Navigation, telephony, telegraphy	Ground waves
0.3–3 MHz	Medium frequency (MF)	AM Broadcasting, amateur, and CB radio	Sky waves and ground waves
3–30 MHz	High frequency (HF)	Mobile, amateur, CB, and military radio	Sky waves
30–300 MHz	Very high frequency (VHF)	VHF TV, FM Broadcasting Police, air traffic control	Sky waves, tropospheric waves
0.3–3 GHz	Ultra high frequency (UHF)	UHF TV, radar, satellite communication	Tropospheric waves, space waves
3–30 GHz	Super high frequency (SHF)	Space and satellite, radar microwave relay	Direct waves, ionospheric penetration waves
30–300 GHz	Extra high frequency (EHF)	Experimental, radar radio astronomy	Direct waves, ionospheric penetration waves

- *Bandwidth modification.* The modulation process permits the bandwidth of the input signal to be increased or decreased as required by the application. Bandwidth reduction permits more efficient use of the spectrum, at the cost of signal fidelity. Increased bandwidth, on the other hand, provides increased immunity to transmission channel disturbances.

- *Signal multiplexing.* In a given transmission system, it may be necessary or desirable to combine several different signals into one baseband waveform for distribution. Modulation provides the vehicle for such *multiplexing.* Various modulation schemes allow separate signals to be combined at the transmission end and separated (*demultiplexed*) at the receiving end. Multiplexing may be accomplished by using, among other systems, *frequency-domain multiplexing* (FDM) or *time-domain multiplexing* (TDM).

Modulation of a signal does not come without the possible introduction of undesirable attributes. Bandwidth restriction or the addition of noise or other disturbances are the two primary problems faced by the transmission system designer.

2.1.1 Principles of Resonance

All RF generations rely on the principles of resonance for operation. Three basic systems exist:

- Series resonance circuits

- Parallel resonance circuits

- Cavity resonators

Series Resonant Circuits

When a constant voltage of varying frequency is applied to a circuit consisting of an inductance, capacitance, and resistance (all in series), the current that flows depends upon frequency in the manner shown in Figure 2.1. At low frequencies, the capacitive reactance of the circuit is large and the inductive reactance is small, so that most of the voltage drop is across the capacitor, while the current is small and leads the applied voltage by nearly 90°. At high frequencies, the inductive reactance is large and the capacitive reactance is low, resulting in a small current that lags nearly 90° behind the applied voltage; most of the voltage drop is across the inductance. Between these two extremes is the *resonant frequency*, at which the capacitive and inductive reactances are equal and, consequently, neutralize each other, leaving only the resistance of the circuit to oppose the flow of current. The current at this resonant frequency is, accordingly, equal to the applied voltage divided by the circuit resistance, and it is very large if the resistance is low.

The characteristics of a series resonant circuit depend primarily upon the ratio of inductive reactance ωL to circuit resistance R, known as the circuit Q:

(a)

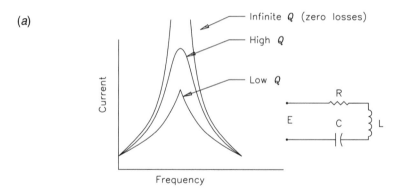

(b)

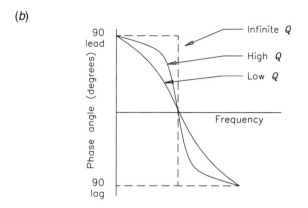

Figure 2.1 Characteristics of a series resonant circuit as a function of frequency for a constant applied voltage and different circuit Qs: (a) magnitude, (b) phase angle.

$$Q = \frac{\omega L}{R} \tag{2.1}$$

The circuit Q also may be defined by:

$$Q = 2\pi \left(\frac{E_s}{E_d} \right) \tag{2.2}$$

Where:
E_s = energy stored in the circuit
E_d = energy dissipated in the circuit during one cycle

Most of the loss in a resonant circuit is the result of coil resistance; the losses in a properly constructed capacitor are usually small in comparison with those of the coil.

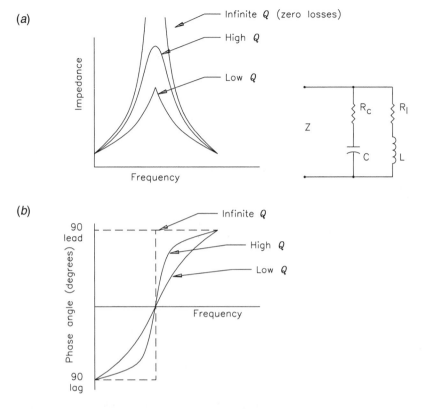

Figure 2.2 Characteristics of a parallel resonant circuit as a function of frequency for different circuit Qs: (a) magnitude, (b) phase angle.

The general effect of different circuit resistances (different values of Q) is shown in Figure 2.1. As illustrated, when the frequency differs appreciably from the resonant frequency, the actual current is practically independent of circuit resistance and is nearly the current that would be obtained with no losses. On the other hand, the current at the resonant frequency is determined solely by the resistance. The effect of increasing the resistance of a series circuit is, accordingly, to flatten the resonance curve by reducing the current at resonance. This broadens the top of the curve, giving a more uniform current over a band of frequencies near the resonant point. This broadening is achieved, however, by reducing the selectivity of the tuned circuit.

Parallel Resonant Circuits

A parallel circuit consisting of an inductance branch in parallel with a capacitance branch offers an impedance of the character shown in Figure 2.2. At low frequencies, the inductive branch draws a large lagging current while the leading current of the capacitive branch is small, resulting in a large lagging line current and a low lagging circuit impedance. At high frequencies, the inductance has a high reactance compared with the capacitance, resulting in a large leading line current and a corresponding low circuit impedance that is leading in phase. Between these two extremes is a frequency at which the lagging current taken by the inductive branch and the leading current entering the capacitive branch are equal. Being 180° out of phase, they neutralize, leaving only a small resultant in-phase current flowing in the line; the impedance of the parallel circuit is, therefore, high.

The effect of circuit resistance on the impedance of the parallel circuit is similar to the influence that resistance has on the current flowing in a series resonant circuit, as is evident when Figures 2.1 and 2.2 are compared. Increasing the resistance of a parallel circuit lowers and flattens the peak of the impedance curve without appreciably altering the sides, which are relatively independent of the circuit resistance.

The resonant frequency F_o of a parallel circuit can be taken as the same frequency at which the same circuit is in series resonance:

$$F_0 = \frac{1}{2\pi \sqrt{LC}}$$
(2.3)

Where:
L = inductance in the circuit
C = capacitance in the circuit

When the circuit Q is large, the frequencies corresponding to the maximum impedance of the circuit and to unity power factor of this impedance coincide, for all practical purposes, with the resonant frequency defined in this way. When the circuit Q is low, however, this rule does not necessarily apply.

Cavity Resonators

Any space completely enclosed with conducting walls may contain oscillating electromagnetic fields. Such a space also exhibits certain resonant frequencies when excited by electrical oscillations. Resonators of this type, commonly termed *cavity resonators*, find extensive use as resonant circuits at very high frequencies and above. For such applications, cavity resonators have a number of advantages, including:

- Mechanical and electrical simplicity

- High Q

- Stable parameters over a wide range of operating conditions

If desired, a cavity resonator can be configured to develop high shunt impedance.

The simplest cavity resonators are sections of waveguides short-circuited at each end and $\lambda_g/2$ wavelengths long, where λ_g is the guide wavelength. This arrangement results in a resonance analogous to that of a ½-wavelength transmission line short-circuited at the receiving end.

Any particular cavity is resonant at a number of frequencies, corresponding to different possible field configurations that exist within the enclosure. The resonance having the longest wavelength (lowest frequency) is termed the *dominant* or *fundamental* resonance. In the case of cavities that are resonant sections of cylindrical or rectangular waveguides, most of the possible resonances correspond to various modes that exist in the corresponding waveguides.

The resonant wavelength is proportional in all cases to the size of the resonator. If the dimensions are doubled, the wavelength corresponding to resonance will likewise be doubled. This fact simplifies the construction of resonators of unusual shapes whose proper dimensions cannot be calculated easily.

The resonant frequency of a cavity can be changed through one or more of the following actions:

- Altering the mechanical dimensions of the cavity

- Coupling reactance into the cavity

- Inserting a copper paddle inside the cavity and moving it to achieve the desired resonant frequency

Small changes in mechanical dimensions can be achieved by flexing walls, but large changes typically require some type of sliding member. Reactance can be coupled into the resonator through a coupling loop, thus affecting the resonant frequency. A copper paddle placed inside the resonator affects the normal distribution of flux and tends to raise the resonant frequency by an amount determined by the orientation of the paddle. This technique is similar to the way in which copper can be used to produce small variations in the inductance of a coil.

Coupling to a cavity resonator can be achieved by means of a coupling loop or a coupling electrode. Magnetic coupling is achieved by means of a small coupling loop oriented so as to enclose magnetic flux lines existing in the desired mode of operation. A current passed through such a loop then excites oscillations of this mode. Conversely, oscillations existing in the resonator induce a voltage in such a coupling loop. The combination of a coupling loop and cavity resonator is equivalent to the ordinary coupled circuit shown in Figure 2.3. The magnitude of magnetic coupling can be readily controlled by rotating the loop. The coupling is reduced to zero when the plane of the loop is parallel to the magnetic flux.

2.1.2 Operating Class

Amplifier stage operating efficiency is a key element in the design and application of an electronic device or system. As the power level of an amplifier generator increases, the overall efficiency of the system becomes more important. Increased efficiency translates into lower operating costs and, usually, improved reliability of the system. Put another way, for a given device dissipation, greater operating efficiency translates into higher power output. The operating mode of the final stage, or stages, is the primary determining element in the maximum possible efficiency of the system.

(a)

(b)

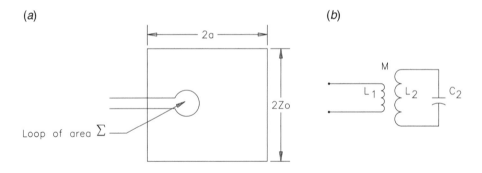

Figure 2.3 Cylindrical resonator incorporating a coupling loop: (a) orientation of loop with respect to cavity, (b) equivalent coupled circuit.

An amplifying stage is classified by its individual *class of operation*. Four primary class divisions apply to transistors and vacuum tube devices:

- *Class A*—A mode wherein the power amplifying device is operated over its linear transfer characteristic. This mode provides the lowest waveform distortion, but also the lowest efficiency. The basic operating efficiency of a class A stage is 50 percent. Class A amplifiers exhibit low intermodulation distortion, making them well suited to linear RF amplifier applications.

- *Class B*—A mode wherein the power amplifying device is operated just outside its linear transfer characteristic. This mode provides improved efficiency at the expense of some waveform distortion. *Class AB* is a variation on class B operation. The transfer characteristic for an amplifying device operating in this mode is, predictably, between class A and class B.

- *Class C*—A mode wherein the power amplifying device is operated significantly outside its linear transfer characteristic, resulting in a pulsed output waveform. High efficiency (up to 90 percent) can be realized with class C operation, but significant distortion of the waveform will occur. Class C is used extensively as an efficient RF power generator.

- *Class D*—A mode that essentially results in a switched device state. The power amplifying device is either *on* or *off*. This is the most efficient mode of operation. It is also the mode that produces the greatest waveform distortion.

The angle of current flow determines the class of operation for a power amplifying device. Typically, the following generalizations regarding conduction angle apply:

- Class A: 360°

- Class AB: between 180 and 360°

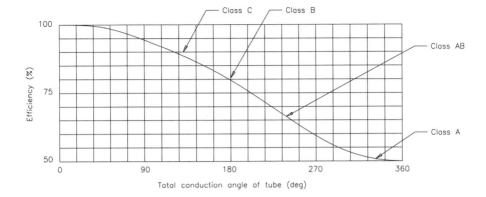

Figure 2.4 Operating efficiency as a function of conduction angle for an amplifier with a tuned load.

- Class B: 180°

- Class C: less than 180°

Subscripts also may be used to denote grid current flow for vacuum tube application. The subscript "1" means that no grid current flows in the stage; the subscript "2" denotes grid current flow. Figure 2.4 charts operating efficiency as a function of conduction angle for an RF amplifier.

2.2 Amplitude Modulation

In the simplest form of amplitude modulation, an analog carrier is controlled by an analog modulating signal. The desired result is an RF waveform whose amplitude is varied by the magnitude of the applied modulating signal and at a rate equal to the frequency of the applied signal. The resulting waveform consists of a carrier wave plus two additional signals:

- An upper-sideband signal, which is equal in frequency to the carrier *plus* the frequency of the modulating signal

- A lower-sideband signal, which is equal in frequency to the carrier *minus* the frequency of the modulating signal

This type of modulation system is referred to as *double-sideband amplitude modulation* (DSAM).

The radio carrier wave signal onto which the analog amplitude variations are to be impressed is expressed as:

$$e(t) = A E_c \cos(\omega_c t) \tag{2.4}$$

Where:

$e(t)$ = instantaneous amplitude of carrier wave as a function of time (t)
A = a factor of amplitude modulation of the carrier wave
ω = angular frequency of carrier wave (radians per second)
E_c = peak amplitude of carrier wave

If A is a constant, the peak amplitude of the carrier wave is constant, and no modulation exists. Periodic modulation of the carrier wave results if the amplitude of A is caused to vary with respect to time, as in the case of a sinusoidal wave:

$$A = 1 + \left(\frac{E_m}{E_c}\right) \cos(\omega_m t) \tag{2.5}$$

Where:

E_m/E_c = the ratio of modulation amplitude to carrier amplitude

The foregoing relationship leads to:

$$e(t) = E_c \left[1 + \left(\frac{E_m}{E_c}\right) \cos(\omega_m t) \cos(\omega_c t) \right] \tag{2.6}$$

This is the basic equation for periodic (sinusoidal) amplitude modulation. When all multiplications and a simple trigonometric identity are performed, the result is:

$$e(t) = E_c \cos(\omega_c t) + \left(\frac{M}{2}\right) \cos(\omega_c t + \omega_m t) + \left(\frac{M}{2}\right) \cos(\omega_c t - \omega_m t) \tag{2.7}$$

Where:

M = the amplitude modulation factor (E_m/E_c)

Amplitude modulation is, essentially, a multiplication process in which the time functions that describe the modulating signal and the carrier are multiplied to produce a modulated wave containing *intelligence* (information or data of some kind). The frequency components of the modulating signal are translated in this process to occupy a different position in the spectrum.

The bandwidth of an AM transmission is determined by the modulating frequency. The bandwidth required for full-fidelity reproduction in a receiver is equal to twice the applied modulating frequency.

The magnitude of the upper sideband and lower sideband will not normally exceed 50 percent of the carrier amplitude during modulation. This results in an upper-sideband power of one-fourth the carrier power. The same power exists in the lower sideband. As a result, up to one-half of the actual carrier power appears additionally in the sum of the sidebands of the modulated signal. A representation of the AM carrier and its sidebands is shown in Figure 2.5. The actual occupied bandwidth, assuming pure sinusoidal modulating signals and no distortion during the modulation process, is equal to twice the frequency of the modulating signal.

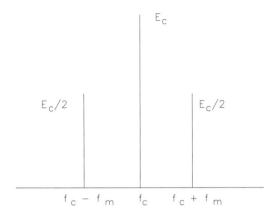

Figure 2.5 Frequency-domain representation of an amplitude-modulated signal at 100 percent modulation. E_c = carrier power, F_c = frequency of the carrier, and F_m = frequency of the modulating signal.

The extent of the amplitude variations in a modulated wave is expressed in terms of the *degree of modulation* or *percentage of modulation*. For sinusoidal variation, the degree of modulation m is determined from the following:

$$m = \frac{E_{avg} - E_{min}}{E_{avg}} \qquad (2.8)$$

Where:
E_{avg} = average envelope amplitude
E_{min} = minimum envelope amplitude

Full (100 percent) modulation occurs when the peak value of the modulated envelope reaches twice the value of the unmodulated carrier, and the minimum value of the envelope is zero. The envelope of a modulated AM signal in the time domain is shown in Figure 2.6.

When the envelope variation is not sinusoidal, it is necessary to define the degree of modulation separately for the peaks and troughs of the envelope:

$$m_{pp} = \frac{E_{max} - E_{avg}}{E_{avg}} \times 100 \qquad (2.9)$$

$$m_{np} = \frac{E_{avg} - E_{min}}{E_{avg}} \times 100 \qquad (2.10)$$

Where:

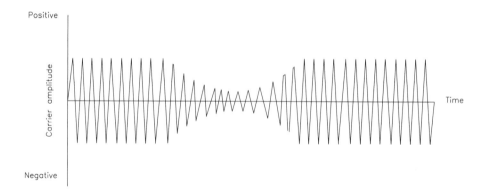

Figure 2.6 Time-domain representation of an amplitude-modulated signal. Modulation at 100 percent is defined as the point at which the peak of the waveform reaches twice the carrier level, and the minimum point of the waveform is zero.

m_{pp} = positive peak modulation (percent)
E_{max} = peak value of modulation envelope
m_{np} = negative peak modulation (percent)
E_{avg} = average envelope amplitude
E_{min} = minimum envelope amplitude

When modulation exceeds 100 percent on the negative swing of the carrier, spurious signals are emitted. It is possible to modulate an AM carrier asymmetrically; that is, to restrict modulation in the negative direction to 100 percent, but to allow modulation in the positive direction to exceed 100 percent without a significant loss of fidelity. In fact, many modulating signals normally exhibit asymmetry, most notably human speech waveforms.

The carrier wave represents the average amplitude of the envelope and, because it is the same regardless of the presence or absence of modulation, the carrier transmits no information. The information is carried by the sideband frequencies. The amplitude of the modulated envelope may be expressed as follows [5]:

$$E = E_0 + E_1 \sin(2\pi f_1 t + \Phi_1) + E_2 \sin(2\pi f_2 t + \Phi_2) \qquad (2.11)$$

Where:
E = envelope amplitude
E_0 = carrier wave crest value (volts)
E_1 = 2 × first sideband crest amplitude (volts)
f_1 = frequency difference between the carrier and the first upper/lower sidebands
E_2 = 2 × second sideband crest amplitude (volts)
f_2 = frequency difference between the carrier and the second upper/lower sidebands
Φ_1 = phase of the first sideband component

Φ_2 = phase of the second sideband component

A number of variations on the basic AM method exist. Table 2.2 lists the advantages and disadvantages of the more common techniques.

2.3   Frequency Modulation

Frequency modulation is a technique whereby the phase angle or phase shift of a carrier is varied by an applied modulating signal. The *magnitude* of frequency change of the carrier is a direct function of the *magnitude* of the modulating signal. The *rate* at which the frequency of the carrier is changed is a direct function of the *frequency* of the modulating signal. In FM modulation, multiple pairs of sidebands are produced. The actual number of sidebands that make up the modulated wave is determined by the *modulation index* (MI) of the system.

2.3.1   Modulation Index

The modulation index is a function of the frequency deviation of the system and the applied modulating signal:

$$MI = \frac{F_d}{M_f} \tag{2.12}$$

Where:
MI = the modulation index
F_d = frequency deviation
M_f = modulating frequency

The higher the MI, the more sidebands produced. It follows that the higher the modulating frequency for a given deviation, the fewer number of sidebands produced, but the greater their spacing.

To determine the frequency spectrum of a transmitted FM waveform, it is necessary to compute a Fourier series or Fourier expansion to show the actual signal components involved. This work is difficult for a waveform of this type, because the integrals that must be performed in the Fourier expansion or Fourier series are not easily solved. The result, however, is that the integral produces a particular class of solution that is identified as the *Bessel function*, illustrated in Figure 2.7.

The carrier amplitude and phase, plus the sidebands, can be expressed mathematically by making the modulation index the argument of a simplified Bessel function. The general expression is given from the following equations:

RF output voltage = $E_1 = E_c + S_{1u} - S_{1l} + S_{2u} - S_{2l} + S_{3u} - S_{3l} + S_{nu} - S_{nl}$

Carrier amplitude = $E_c = A \left[J_0(M) \sin \omega c(t) \right]$

Table 2.2 Comparison of Amplitude Modulation Techniques (*From* [1]. *Used with permission.*)

Modulation Scheme	Advantages	Disadvantages	Comments
DSB-SC	Good power efficiency. Good low-frequency response.	More difficult to generate than DSB+C. Detection requires coherent local oscillator, pilot, or phase-locked loop (PLL). Poor spectrum efficiency.	
DSB+C (AM)	Easier to generate than DSB-SC, especially at high-power levels. Inexpensive receivers using envelope detection.	Poor power efficiency. Poor Spectrum efficiency. Poor low-frequency response. Exhibits threshold effect in noise.	Used in commercial AM.
SSB-SC	Excellent spectrum efficiency.	Complex transmitter design. Complex receiver design (same as DSB-SC). Poor low-frequency response.	Used in military communication systems, and to multiplex multiple phone calls onto long-haul microwave links.
SSB+C	Good spectrum efficiency. Low receiver complexity.	Poor power efficiency. Complex transmitters. Poor low-frequency response. Poor noise performance.	
VSB-SC	Good spectrum efficiency. Excellent low-frequency response. Transmitter easier to build than for SSB.	Complex receivers (same as DSB-SC).	
VSB+C	Good spectrum efficiency. Good low-frequency response. Inexpensive receivers using envelope detection.	Poor power efficiency. Poor performance in noise.	Used in commercial TV
QAM	Good low-frequency response. Good spectrum efficiency.	Complex receivers. Sensitive to frequency and phase errors.	Two SSB signals may be preferable.

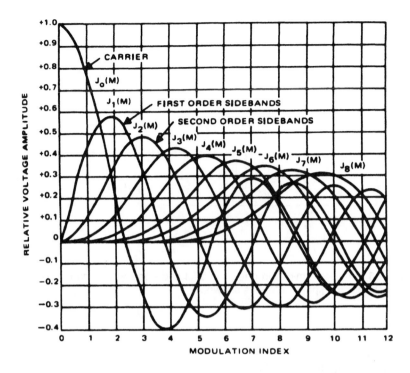

Figure 2.7 Plot of Bessel functions of the first kind as a function of modulation index.

First-order upper sideband = $S_{1u} = J_1(M) \sin(\omega c + \omega m)t$

First-order lower sideband = $S_{1l} = J_1(M) \sin(\omega c - \omega m)t$

Second-order upper sideband = $S_{2u} = J_2(M) \sin(\omega c + 2\omega m)t$

Second-order lower sideband = $S_{2l} = J_2(M) \sin(\omega c - 2\omega m)t$

Third-order upper sideband = $S_{3u} = J_3(M) \sin(\omega c + 3\omega m)t$

Third-order lower sideband = $S_{3l} = J_3(M) \sin(\omega c - 3\omega m)t$

Nth-order upper sideband = $S_{nu} = J_n(M) \sin(\omega c + n\omega m)t$

Nth-order lower sideband = $S_{nl} = J_n(M) \sin(\omega c - n\omega m)t$

Where:
A = the unmodulated carrier amplitude constant

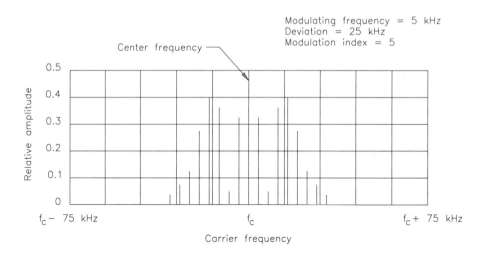

Figure 2.8 RF spectrum of a frequency-modulated signal with a modulation index of 5 and other operating parameters as shown.

J_0 = modulated carrier amplitude
$J_1, J_2, J_3...J_n$ = amplitudes of the nth-order sidebands
M = modulation index
$\omega_c = 2\pi F_c$, the carrier frequency
$\omega_m = 2\pi F_m$, the modulating frequency

Further supporting mathematics will show that an FM signal using the modulation indices that occur in a wideband system will have a multitude of sidebands. From the purist point of view, *all* sidebands would have to be transmitted, received, and demodulated to reconstruct the modulating signal with complete accuracy. In practice, however, the channel bandwidths permitted FM systems usually are sufficient to reconstruct the modulating signal with little discernible loss in fidelity, or at least an acceptable loss in fidelity.

Figure 2.8 illustrates the frequency components present for a modulation index of 5. Figure 2.9 shows the components for an index of 15. Note that the number of significant sideband components becomes quite large with a high MI. This simple representation of a single-tone frequency-modulated spectrum is useful for understanding the general nature of FM, and for making tests and measurements. When typical modulation signals are applied, however, many more sideband components are generated. These components vary to the extent that sideband energy becomes distributed over the entire occupied bandwidth, rather than appearing at discrete frequencies.

Although complex modulation of an FM carrier greatly increases the number of frequency components present in the frequency-modulated wave, it does not, in general, widen the frequency band occupied by the energy of the wave. To a first approximation,

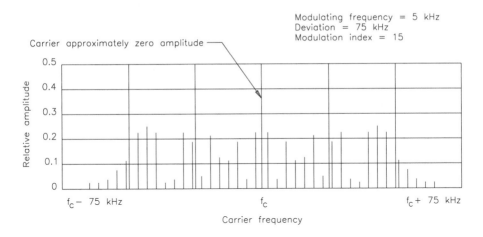

Figure 2.9 RF spectrum of a frequency-modulated signal with a modulation index of 15 and operating parameters as shown.

this band is still roughly twice the sum of the maximum frequency deviation at the peak of the modulation cycle plus the highest modulating frequency involved.

FM is not a simple frequency translation, as with AM, but involves the generation of entirely new frequency components. In general, the new spectrum is much wider than the original modulating signal. This greater bandwidth may be used to improve the *signal-to-noise ratio* (S/N) of the transmission system. FM thereby makes it possible to exchange bandwidth for S/N enhancement.

The power in an FM system is constant throughout the modulation process. The output power is increased in the amplitude modulation system by the modulation process, but the FM system simply distributes the power throughout the various frequency components that are produced by modulation. During modulation, a wideband FM system does not have a high amount of energy present in the carrier. Most of the energy will be found in the sum of the sidebands.

The constant-amplitude characteristic of FM greatly assists in capitalizing on the low noise advantage of FM reception. Upon being received and amplified, the FM signal normally is clipped to eliminate all amplitude variations above a certain threshold. This removes noise picked up by the receiver as a result of man-made or atmospheric signals. It is not possible (generally speaking) for these random noise sources to change the frequency of the desired signal; they can affect only its amplitude. The use of *hard limiting* in the receiver will strip off such interference.

2.3.2 Phase Modulation

In a phase modulation (PM) system, intelligence is conveyed by varying the phase of the RF wave. Phase modulation is similar in many respects to frequency modulation, except in the interpretation of the modulation index. In the case of PM, the modulation index depends only on the amplitude of the modulation; MI is independent of the frequency of the modulating signal. It is apparent, therefore, that the phase-modulated wave contains the same sideband components as the FM wave and, if the modulation indices in the two cases are the same, the relative amplitudes of these different components also will be the same.

The modulation parameters of a PM system relate as follows:

$$\Delta f = m_p \times f_m \tag{2.13}$$

Where:
Δf = frequency deviation of the carrier
m_p = phase shift of the carrier
f_m = modulating frequency

In a phase-modulated wave, the phase shift m_p is independent of the modulating frequency; the frequency deviation Δf is proportional to the modulating frequency. In contrast, with a frequency-modulated wave, the frequency deviation is independent of modulating frequency. Therefore, a frequency-modulated wave can be obtained from a phase modulator by making the modulating voltage applied to the phase modulator inversely proportional to frequency. This can be readily achieved in hardware.

2.4　Pulse Modulation

The growth of digital processing and communications has led to the development of modulation systems tailor-made for high-speed, spectrum-efficient transmission. In a *pulse modulation* system, the unmodulated carrier usually consists of a series of recurrent pulses. Information is conveyed by modulating some parameter of the pulses, such as amplitude, duration, time of occurrence, or shape. Pulse modulation is based on the *sampling principle*, which states that a message waveform with a spectrum of finite width can be recovered from a set of discrete samples if the sampling rate is higher than twice the highest sampled frequency (the Nyquist criteria). The samples of the input signal are used to modulate some characteristic of the carrier pulses.

2.4.1　Digital Modulation Systems

Because of the nature of digital signals (on or off), it follows that the amplitude of the signal in a pulse modulation system should be one of two heights (present or absent/positive or negative) for maximum efficiency. Noise immunity is a significant advantage of such a system. It is necessary for the receiving system to detect only the presence or absence (or polarity) of each transmitted pulse to allow complete reconstruction of the original intelligence. The pulse shape and noise level have minimal effect (to a point). Furthermore, if the waveform is to be transmitted over long distances, it is possible to regenerate the original signal exactly for retransmission to the next relay point. This feature is in striking contrast to analog modulation systems in which each modulation step introduces some amount of noise and signal corruption.

In any practical digital data system, some corruption of the intelligence is likely to occur over a sufficiently large span of time. Data encoding and manipulation schemes have been developed to detect and correct or conceal such errors. The addition of error-correction features comes at the expense of increased system overhead and (usually) slightly lower intelligence throughput.

2.4.2 Pulse Amplitude Modulation

Pulse amplitude modulation (PAM) is one of the simplest forms of data modulation. PAM departs from conventional modulation systems in that the carrier exists as a series of pulses, rather than as a continuous waveform. The amplitude of the pulse train is modified in accordance with the applied modulating signal to convey intelligence, as illustrated in Figure 2.10. There are two primary forms of PAM sampling:

- *Natural sampling* (or *top sampling*), where the modulated pulses follow the amplitude variation of the sampled time function during the sampling interval.

- *Instantaneous sampling* (or *square-topped sampling*), where the amplitude of the pulses is determined by the instantaneous value of the sampled time function corresponding to a single instant of the sampling interval. This "single instant" may be the center or edge of the sampling interval.

There are two common methods of generating a PAM signal:

- Variation of the amplitude of a pulse sequence about a fixed nonzero value (or *pedestal*). This approach constitutes double-sideband amplitude modulation.

- Double-polarity modulated pulses with no pedestal. This approach constitutes double-sideband suppressed carrier modulation.

2.4.3 Pulse Time Modulation

A number of modulating schemes have been developed to take advantage of the noise immunity afforded by a constant amplitude modulating system. *Pulse time modulation* (PTM) is one of those systems. In a PTM system, instantaneous samples of the intelligence are used to vary the time of occurrence of some parameter of the pulsed carrier. Subsets of the PTM process include:

- *Pulse duration modulation* (PDM), where the time of occurrence of either the leading or trailing edge of each pulse (or both pulses) is varied from its unmodulated position by samples of the input modulating waveform. PDM also may be described as *pulse length* or *pulse width* modulation (PWM).

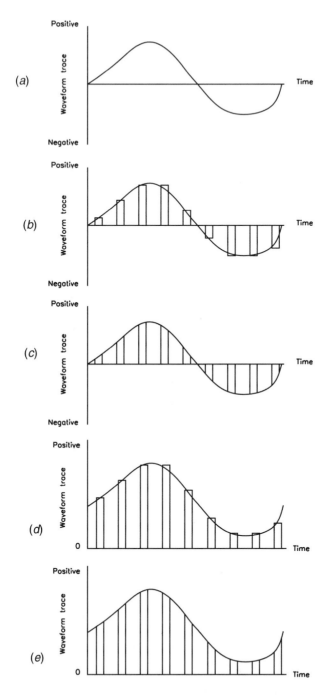

Figure 2.10 Pulse amplitude modulation waveforms: (*a*) modulating signal; (*b*) square-topped sampling, bipolar pulse train; (*c*) topped sampling, bipolar pulse train; (*d*) square-topped sampling, unipolar pulse train; (*e*) top sampling, unipolar pulse train.

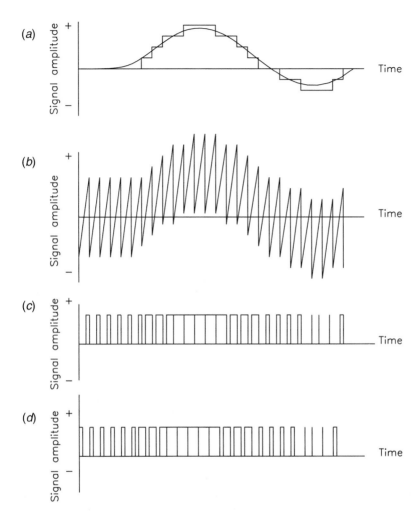

Figure 2.11 Pulse time modulation waveforms: (*a*) modulating signal and sample-and-hold (S/H) waveforms, (*b*) sawtooth waveform added to S/H, (*c*) leading-edge PTM, (*d*) trailing-edge PTM.

- *Pulse position modulation* (PPM), where samples of the modulating input signal are used to vary the position in time of pulses, relative to the unmodulated waveform. Several types of pulse time modulation waveforms are shown in Figure 2.11.

- *Pulse frequency modulation* (PFM), where samples of the input signal are used to modulate the frequency of a series of carrier pulses. The PFM process is illustrated in Figure 2.12.

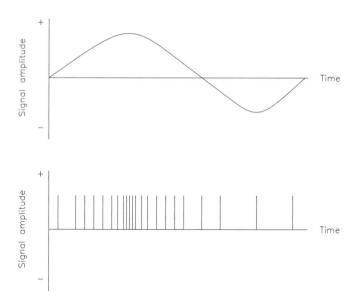

Figure 2.12 Pulse frequency modulation.

It should be emphasized that all of the pulse modulation systems discussed thus far may be used with both analog and digital input signals. Conversion is required for either signal into a form that can be accepted by the pulse modulator.

2.4.4 Pulse Code Modulation

The pulse modulation systems discussed previously are *unencoded* systems. *Pulse code modulation* (PCM) is a scheme wherein the input signal is *quantized* into discrete steps and then sampled at regular intervals (as in conventional pulse modulation). In the *quantization* process, the input signal is sampled to produce a code representing the instantaneous value of the input within a predetermined range of values. Figure 2.13 illustrates the concept. Only certain discrete levels are allowed in the quantization process. The code is then transmitted over the communications system as a pattern of pulses.

Quantization inherently introduces an initial error in the amplitude of the samples taken. This *quantization error* is reduced as the number of quantization steps is increased. In system design, tradeoffs must be made regarding low quantization error, hardware complexity, and occupied bandwidth. The greater the number of quantization steps, the wider the bandwidth required to transmit the intelligence or, in the case of some signal sources, the slower the intelligence must be transmitted.

In the classic design of a PCM encoder, the quantization steps are equal. The quantization error (or *quantization noise*) usually can be reduced, however, through the use of nonuniform spacing of levels. Smaller quantization steps are provided for

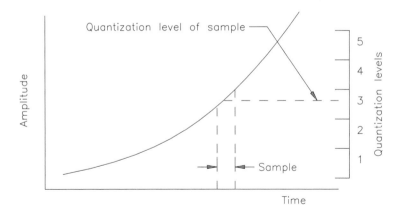

Figure 2.13 The quantization process.

weaker signals, and larger steps are provided near the peak of large signals. Quantization noise is reduced by providing an encoder that is matched to the *level distribution* (*probability density*) of the input signal.

Nonuniform quantization typically is realized in an encoder through processing of the input (analog) signal to compress it to match the desired nonuniformity. After compression, the signal is fed to a uniform quantization stage.

2.4.5 Delta Modulation

Delta modulation (DM) is a coding system that measures changes in the direction of the input waveform, rather than the instantaneous value of the wave itself. Figure 2.14 illustrates the concept. The clock rate is assumed to be constant. Transmitted pulses from the pulse generator are positive if the signal is changing in a positive direction; they are negative if the signal is changing in a negative direction.

As with the PCM encoding system, quantization noise is a parameter of concern for DM. Quantization noise can be reduced by increasing the sampling frequency (the pulse generator frequency). The DM system has no fixed maximum (or minimum) signal amplitude. The limiting factor is the slope of the sampled signal, which must not change by more than one level or step during each pulse interval.

2.4.6 Digital Coding Systems

A number of methods exist to transmit digital signals over long distances in analog transmission channels. Some of the more common systems include:

- *Binary on-off keying* (BOOK), a method by which a high-frequency sinusoidal signal is switched on and off corresponding to 1 and 0 (on and off) periods in the

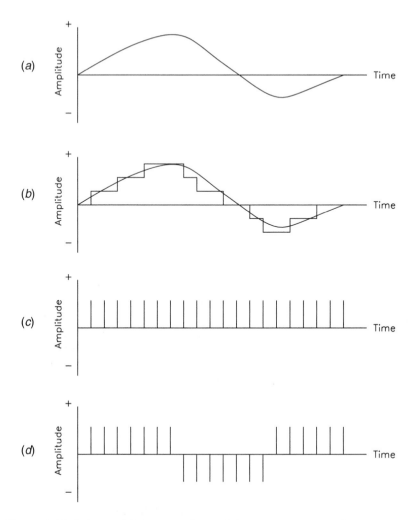

Figure 2.14 Delta modulation waveforms: (*a*) modulating signal, (*b*) quantized modulating signal, (*c*) pulse train, (*d*) resulting delta modulation waveform.

input digital data stream. In practice, the transmitted sinusoidal waveform does not start or stop abruptly, but follows a predefined ramp up or down.

- *Binary frequency-shift keying* (BFSK), a modulation method in which a continuous wave is transmitted that is shifted between two frequencies, representing 1s and 0s in the input data stream. The BFSK signal may be generated by switching between two oscillators (set to different operating frequencies) or by applying a binary baseband signal to the input of a voltage-controlled oscillator (VCO). The transmitted signals often are referred to as a *mark* (binary digit 1) or a *space* (binary digit 0). Figure 2.15 illustrates the transmitted waveform of a BFSK system.

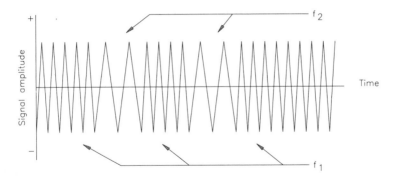

Figure 2.15 Binary FSK waveform.

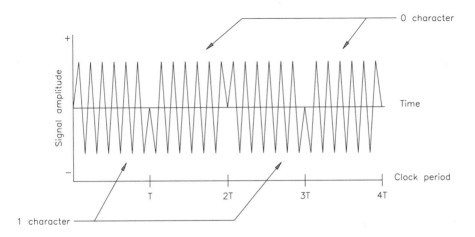

Figure 2.16 Binary PSK waveform.

- *Binary phase-shift keying* (BPSK), a modulating method in which the phase of the transmitted wave is shifted 180° in synchronism with the input digital signal. The phase of the RF carrier is shifted by $\pi/2$ radians or $-\pi/2$ radians, depending upon whether the data bit is a 0 or a 1. Figure 2.16 shows the BPSK transmitted waveform.

- *Quadriphase-shift keying* (QPSK), a modulation scheme similar to BPSK except that quaternary modulation is employed, rather than binary modulation. QPSK requires half the bandwidth of BPSK for the same transmitted data rate.

2.4.7 Baseband Digital Pulse Modulation

After the input samples have been quantized, they are transmitted through a channel, received, and converted back to their approximate original form [2]. The format (modulation scheme) applied to the quantized samples is determined by a number of factors, not the least of which is the channel through which the signal passes. A number of different formats are possible and practical.

Several common digital modulation formats are shown in Figure 2.17. The first (a) is referred to as *non-return-to-zero* (NRZ) polar because the waveform does not return to zero during each signaling interval, but switches from $+V$ to $-V$, or vice versa, at the end of each signaling interval (NRZ unipolar uses the levels V and 0). On the other hand, the *unipolar return-to-zero* (RZ) format, shown in (b) returns to zero in each signaling interval. Because bandwidth is inversely proportional to pulse duration, it is apparent that RZ requires twice the bandwidth that NRZ does. Also, RZ has a nonzero dc component, whereas NRZ does not necessarily have a nonzero component (unless there are more 1s than 0s or vice versa). An advantage of RZ over NRZ is that a pulse transition is guaranteed in each signaling interval, whereas this is not the case for NRZ. Thus, in cases where there are long strings of 1s or 0s, it may be difficult to synchronize the receiver to the start and stop times of each pulse in NRZ-based systems. A very important modulation format from the standpoint of synchronization considerations is NRZ-*mark*, also known as *differential encoding*, where an initial reference bit is chosen and a subsequent 1 is encoded as a change from the reference and a 0 is encoded as no change. After the initial reference bit, the current bit serves as a reference for the next bit, and so on. An example of this modulation format is shown in (c).

Manchester is another baseband data modulation format that guarantees a transition in each signaling interval and does not have a dc component. Also known as *biphase* or *split phase*, this scheme is illustrated in (d). The format is produced by *OR*ing the data clock with an NRZ-formatted signal. The result is a + to − transition for a logic 1, and a − to + zero crossing for a logic 0.

A number of other data formats have been proposed and employed in the past, but further discussion is beyond the scope of this chapter.

2.5 Spread Spectrum

As the name implies, a *spread spectrum* system requires a frequency range substantially greater than the basic information-bearing signal. Spread spectrum systems have some or all of the following properties:

- Low interference to other communications systems

- Ability to reject high levels of external interference

- Immunity to jamming by hostile forces

- Provision for secure communications paths

- Operability over multiple RF paths

Spread spectrum systems operate with an entirely different set of requirements than transmission systems discussed previously. Conventional modulation methods are designed to provide for the easiest possible reception and demodulation of the transmitted

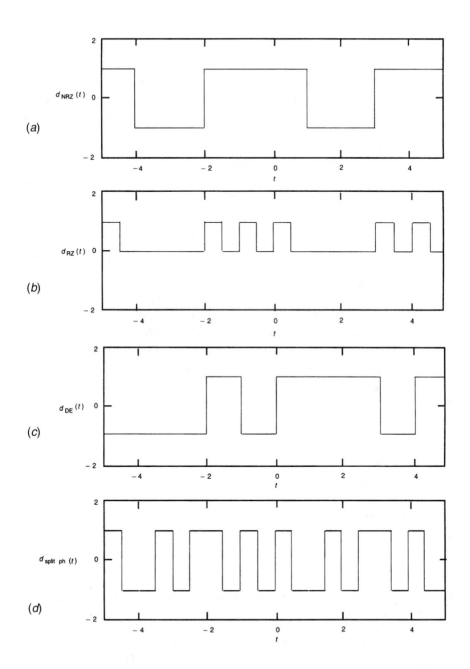

Figure 2.17 Various baseband modulation formats: (*a*) non-return-to zero, (*b*) unipolar return-to-zero, (*c*) differential encoded (NRZ-mark), (*d*) split phase. (*From* [8]. *Used with permission.*)

intelligence. The goals of spread spectrum systems, on the other hand, are secure and reliable communications that cannot be intercepted by unauthorized persons. The most common modulating and encoding techniques used in spread spectrum communications include:

- *Frequency hopping*, where a random or *pseudorandom* number (PN) sequence is used to change the carrier frequency of the transmitter. This approach has two basic variations: *slow frequency hopping*, where the hopping rate is smaller than the data rate, and *fast frequency hopping*, where the hopping rate is larger than the data rate. In a fast frequency-hopping system, the transmission of a single piece of data occupies more than one frequency. Frequency-hopping systems permit multiple-access capability to a given band of frequencies because each transmitted signal occupies only a fraction of the total transmitted bandwidth.

- *Time hopping*, where a PN sequence is used to switch the position of a message-carrying pulse within a series of frames.

- *Message corruption*, where a PN sequence is added to the message before modulation.

- *Chirp spread spectrum*, where linear frequency modulation of the main carrier is used to spread the transmitted spectrum. This technique is commonly used in radar and also has been applied to communications systems.

In a spread spectrum system, the signal power is divided over a large bandwidth. The signal, therefore, has a small average power in any single narrowband slot. This means that a spread spectrum system can share a given frequency band with one or more narrowband systems. Furthermore, because of the low energy in any particular band, detection or interception of the transmission is difficult.

2.6   References

1. Kubichek, Robert, "Amplitude Modulation," in *The Electronics Handbook*, Jerry C. Whitaker (ed.), CRC Press, Boca Raton, FL, pp. 1175–1187, 1996.
2. Ziemer, Rodger E., "Pulse Modulation," in *The Electronics Handbook*, Jerry C. Whitaker (ed.), CRC Press, Boca Raton, FL, pp. 1201–1212, 1996.

2.7   Bibliography

Benson, K. B., and Jerry. C. Whitaker (eds.), *Television Engineering Handbook*, McGraw-Hill, New York, 1986.

Benson, K. B., and Jerry. C. Whitaker, *Television and Audio Handbook for Technicians and Engineers*, McGraw-Hill, New York, 1989.

Crutchfield, E. B. (ed.), *NAB Engineering Handbook*, 8th Ed., National Association of Broadcasters, Washington, DC, 1991.

Fink, D., and D. Christiansen (eds.), *Electronics Engineers' Handbook*, 3rd Ed., McGraw-Hill, New York, 1989.

Fink, D., and D. Christiansen (eds.), *Electronics Engineers' Handbook*, 2nd Ed., McGraw-Hill, New York, 1982.

Hulick, Timothy P., "Using Tetrodes for High Power UHF," *Proceedings of the Society of Broadcast Engineers*, Vol. 4, SBE, Indianapolis, IN, pp. 52-57, 1989.

Jordan, Edward C., *Reference Data for Engineers: Radio, Electronics, Computer and Communications*, 7th Ed., Howard W. Sams, Indianapolis, IN, 1985.

Laboratory Staff, *The Care and Feeding of Power Grid Tubes*, Varian Eimac, San Carlos, CA, 1984.

Mendenhall, G. N., "Fine Tuning FM Final Stages," *Broadcast Engineering*, Intertec Publishing, Overland Park, KS, May 1987.

Whitaker, Jerry. C., *Maintaining Electronic Systems*, CRC Press, Boca Raton, FL, 1992.

Whitaker, Jerry C., *Power Vacuum Tubes Handbook*, 2nd ed., CRC Press, Boca Raton, FL, 1999.

Whitaker, Jerry. C., *Radio Frequency Transmission Systems: Design and Operation*, McGraw-Hill, New York, 1991.

3

Analog and Digital Circuits

3.1 Introduction

Amplifiers are the functional building blocks of electronic systems, and each of these building blocks typically contains several amplifier stages coupled together. An amplifier may contain its own power supply or require one or more external sources of power. The active component of each amplifier stage is usually a transistor or an FET. Other amplifying components, such as vacuum tubes, can also be used in amplifier circuits if the operating power and/or frequency of the application demands it.

3.2 Single-Stage Transistor/FET Amplifier

The single-stage amplifier can best be described using a single transistor or FET connected as a *common-emitter* or *common-source* amplifier, using an npn transistor (Figure 3.1a) or an n-channel FET (Figure 3.1b) and treating pnp transistors or p-channel FET circuits by simply reversing the current flow and the polarity of the voltages.

At zero frequency (dc) and at low frequencies, the transistor or FET amplifier stage requires an input voltage E_1 equal to the sum of the input voltages of the device (the transistor V_{be} or FET V_{gs}) and the voltage across the resistance R_e or R_s between the common node (ground) and the emitter or source terminal. The input current I_1 to the amplifier stage is equal to the sum of the current through the external resistor connected between ground and the base or gate and the base current I_b or gate current I_g drawn by the device. In most FET circuits, the gate current may be so small that it can be neglected, while in transistor circuits the base current I_b is equal to the collector current I_c divided by the current gain beta of the transistor. The input resistance R_1 to the amplifier stage is equal to the ratio of input voltage E_1 to input current I_1.

The input voltage and the input resistance of an amplifier stage increases as the value of the emitter or source resistor becomes larger.

The output voltage E_2 of the amplifier stage, operating without any external load, is equal to the difference of supply voltage V+ and the product of collector or drain load resistor R_1 and collector current I_c or drain current I_d. An external load will cause the device to draw an additional current I_2, which increases the device output current.

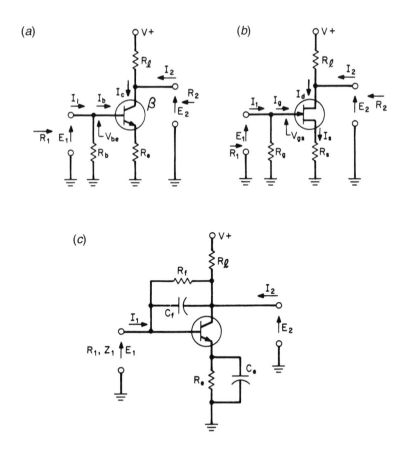

Figure 3.1 Single-stage amplifier circuits: (a) common-emitter NPN, (b) common-source n-channel FET, (c) single-stage with current and voltage feedback.

As long as the collector-to-emitter voltage is larger than the saturation voltage of the transistor, collector current will be nearly independent of supply voltage. Similarly, the drain current of an FET will be nearly independent of drain-to-source voltage as long as this voltage is greater than an equivalent saturation voltage. This saturation voltage is approximately equal to the difference between gate-to-source voltage and *pinch-off* voltage, the latter being the bias voltage that causes nearly zero drain current. In some FET data sheets, the pinch-off voltage is referred to as the *threshold voltage*. At lower supply voltages, the collector or drain current will become less until it reaches zero, when the drain-to-source voltage is zero or the collector-to-emitter voltage has a very small reverse value.

The output resistance R_2 of a transistor or FET amplifier stage is—in effect—the parallel combination of the collector or drain load resistance and the series connection of two resistors, consisting of R_e or R_s, and the ratio of collector-to-emitter voltage and collector current or the equivalent drain-to-source voltage and drain current. In actual devices, an additional resistor, the relatively large output resistance of the device, is connected in parallel with the output resistance of the amplifier stage.

The collector current of a single-stage transistor amplifier is equal to the base current multiplied by the current gain of the transistor. Because the current gain of a transistor may be specified as tightly as a two-to-one range at one value of collector current, or it may have just a minimum value, knowledge of the input current is usually not quite sufficient to specify the output current of a transistor.

3.2.1 Impedance and Gain

The input impedance is the ratio of input voltage to input current, and the output impedance is the ratio of output voltage to output current. As the input current increases, the output current into the external output load resistor will increase by the current amplification factor of the stage. The output voltage will decrease because the increased current flows from the collector or drain voltage supply source into the collector or drain of the device. Therefore, the voltage amplification is a negative number having the magnitude of the ratio of output voltage change to input voltage change.

The magnitude of voltage amplification is often calculated as the product of transconductance G_m of the device and the load resistance value. This can be done as long as the emitter or source resistance is zero or the resistor is bypassed with a capacitor that effectively acts as a short circuit for all signal changes of interest but allows the desired bias currents to flow through the resistor. In a bipolar transistor, the transconductance is approximately equal to the emitter current multiplied by 39, which is the charge of a single electron divided by the product of Boltzmann's constant and absolute temperature in degrees Kelvin. In a field-effect transistor, this value will be less and usually proportional to the input-bias voltage, with reference to the pinch-off voltage.

The power gain of the device is the ratio of output power to input power, often expressed in decibels. Voltage gain or current gain can be stated in decibels but must be so marked.

The resistor in series with the emitter or source causes negative feedback of most of the output current, which reduces the voltage gain of the single amplifier stage and raises its input impedance (Figure 3.1c). When this resistor R_e is bypassed with a capacitor C_e, the amplification factor will be high at high frequencies and will be reduced by approximately 3 dB at the frequency where the impedance of capacitor C_e is equal to the emitter or source input impedance of the device, which in turn is approximately equal to the inverse of the transconductance G_m of the device (Figure 3.2a). The gain of the stage will be approximately 3 dB higher than the dc gain at the frequency where the impedance of the capacitor is equal to the emitter or source resistor. These simplifications hold in cases where the product of transconductance and resistance values are much larger than 1.

A portion of the output voltage may also be fed back to the input, which is the base or gate terminal. This resistor R_r will lower the input impedance of the single amplifier

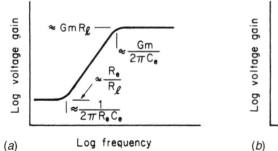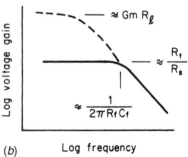

Figure 3.2 Feedback amplifier voltage gains: (*a*) current feedback, (*b*) voltage feedback.

stage, reduce current amplification, reduce output impedance of the stage, and act as a supply voltage source for the base or gate. This method is used when the source of input signals, and internal resistance R_s, is coupled with a capacitor to the base or gate and a group of devices with a spread of current gains, transconductances, or pinch-off voltages must operate with similar amplification in the same circuit. If the feedback element is also a capacitor C_f, high-frequency current amplification of the stage will be reduced by approximately 3 dB when the impedance of the capacitor is equal to the feedback resistor R_f and voltage gain of the stage is high (Figure 3.2*b*). At still higher frequencies, amplification will decrease at the rate of 6 dB per octave of frequency. It should be noted that the base-collector or gate-drain capacitance of the device has the same effect of limiting high-frequency amplification of the stage; however, this capacitance becomes larger as the collector-base or drain-gate voltage decreases.

Feedback of the output voltage through an impedance lowers the input impedance of an amplifier stage. Voltage amplification of the stage will be affected only as this lowered input impedance loads the source of input voltage. If the source of input voltage has a finite source impedance and the amplifier stage has very high voltage amplification and reversed phase, the effective amplification for this stage will approach the ratio of feedback impedance to source impedance and also have reversed phase.

3.2.2 Common-Base or Common-Gate Connection

For the common-base or common-gate case, voltage amplification is the same as in the common-emitter or common-source connection; however, the input impedance is approximately the inverse of the transconductance of the device. (See Figure 3.3*a*.) As a benefit, high-frequency amplification will be less affected because of the relatively lower emitter-collector or source-drain capacitance and the relatively low input impedance. This is the reason why the *cascade connection* (Figure 3.3*b*) of a com-

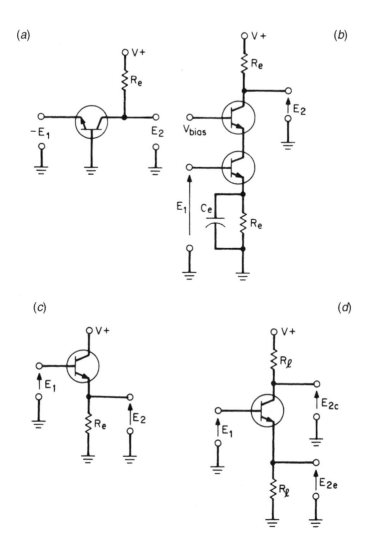

Figure 3.3 Transistor amplifier circuits: (a) common-base NPN, (b) cascode NPN, (c) common-collector NPN emitter follower, (d) split-load phase inverter.

mon-emitter amplifier stage driving a common-base amplifier stage exhibits nearly the dc amplification of a common-emitter stage with the wide bandwidth of a common-base stage. Another advantage of the common-base or common-gate amplifier stage is stable amplification at very high frequencies and ease of matching to RF transmission-line impedances, usually 50 to 75 Ω.

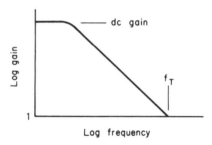

Figure 3.4 Amplitude-frequency response of a common-emitter or common-source amplifier.

3.2.3 Common-Collector or Common-Drain Connection

The voltage gain of a transistor or FET is slightly below 1.0 for the common-collector or common-drain configuration. However, the input impedance of a transistor so connected will be equal to the value of the load impedance multiplied by the current gain of the device plus the inverse of the transconductance of the device (Figure 3.3c). Similarly, the output impedance of the stage will be the impedance of the source of signals divided by the current gain of the transistor plus the inverse of the transconductance of the device.

When identical resistors are connected between the collector or drain and the supply voltage and the emitter or source and ground, an increase in base or gate voltage will result in an increase of emitter or source voltage that is nearly equal to the decrease in collector or drain voltage. This type of connection is known as the *split-load phase inverter*, useful for driving push-pull amplifiers, although the output impedances at the two output terminals are unequal (Figure 3.3d).

The current gain of a transistor decreases at high frequencies as the emitter-base capacitance shunts a portion of the transconductance, thereby reducing current gain until it reaches a value of 1 at the transition frequency of the transistor (Figure 3.4). From this figure it can be seen that the output impedance of an emitter-follower or common-collector stage will increase with frequency, having the effect of an inductive source impedance when the input source to the stage is resistive. If the source impedance is inductive, as it might be with cascaded-emitter followers, the output impedance of such a combination can be a negative value at certain high frequencies and be a possible cause of amplifier oscillation. Similar considerations also apply to common-drain FET stages.

3.2.4 Bias and Large Signals

When large signals have to be handled by a single-stage amplifier, distortion of the signals introduced by the amplifier itself must be considered. Although feedback can reduce distortion, it is necessary to ensure that each stage of amplification operates in

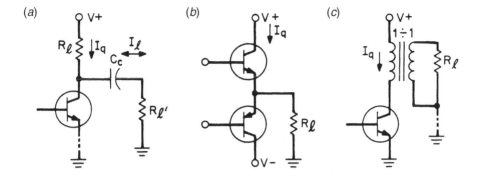

Figure 3.5 Output load-coupling circuits: (*a*) ac-coupled, (*b*) series-parallel ac, push-pull half-bridge, (*c*) single-ended transformer-coupled.

a region where normal signals will not cause the amplifier stage to operate with nearly zero voltage drop across the device or to operate the device with nearly zero current during any portion of the cycle of the signal. Although described primarily with respect to a single-device-amplifier stage, the same holds true for any amplifier stage with multiple devices, except that here at least one device must be able to control current flow in the load without being *saturated* (nearly zero voltage drop) or *cut off* (nearly zero current).

If the single-device-amplifier load consists of the collector or drain load resistor only, the best operating point should be chosen so that in the absence of a signal, one-half of the supply voltage appears as a *quiescent voltage* across the load resistor R_l. If an additional resistive load R_l is connected to the output through a coupling capacitor C_c (Figure 3.5*a*), the maximum peak load current I_l in one direction is equal to the difference between quiescent current I_l of the stage and the current that would flow if the collector resistor and the external load resistor were connected in series across the supply voltage. In the other direction, the maximum load current is limited by the quiescent voltage across the device divided by the load resistance. The quiescent current flows in the absence of an alternating signal and is caused by bias voltage or current only. Because most audio frequency signals (and others, depending upon the application) have positive and negative peak excursions of equal probability, it is usually advisable to have the two peak currents be equal. This can be accomplished by increasing the quiescent current as the external load resistance decreases. Video signals, on the other hand, are typically unidirectional in nature.

When several devices contribute current into an external load resistor (Figure 3.5*b*), one useful strategy is to set bias currents so that the sum of all transconductances remains as constant as practical, which means a design for minimum distortion. This operating point for one device is near one-quarter the peak device current for push-pull FET stages and at a lesser value for bipolar push-pull amplifiers.

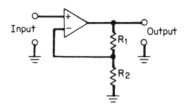

Figure 3.6 Operational amplifier with unbalanced input and output signals and a fixed level of feedback to set the voltage gain V_g, which is equal to $(1 + R)/R$.

When the load resistance is coupled to the single-device-amplifier stage with a transformer (Figure 3.5c), the optimum bias current should be nearly equal to the peak current that would flow through the load impedance at the transformer with a voltage drop equal to the supply voltage.

3.3   Operational Amplifiers

An operational amplifier is a circuit (device) with a pair of differential input terminals that have very high gain to the output for differential signals of opposite phase at each input and relatively low gain for common-mode signals that have the same phase at each input (see Figure 3.6). An external feedback network between the output and the minus (–) input and ground or signal, sets the circuit gain, with the plus (+) input at signal or ground level. Most operational amplifiers require a positive and a negative power supply voltage. One to eight operational amplifiers may be contained on one substrate mounted in a plastic, ceramic, or hermetically sealed metal-can package. Operational amplifiers may require external capacitors for circuit stability or may be internally compensated. Input stages may be field-effect transistors for high input impedance or bipolar transistors for low-offset voltage and low-voltage noise. Available types of operational amplifiers number in the hundreds. Precision operational amplifiers generally have more tightly controlled specifications than general-purpose types. Table 3.1 details the most common application and their functional parameters

The input-bias current of an operational amplifier is the average current drawn by each of the two inputs, + and –, from the input and feedback circuits. Any difference in dc resistance between the circuits seen by the two inputs multiplied by the input-bias current will be amplified by the circuit gain and become an *output-offset voltage*. The *input-offset current* is the difference in bias current drawn by the two inputs, which when multiplied by the sum of the total dc resistance in the input and feedback circuits and the circuit gain, becomes an additional output-offset voltage. The *input-offset voltage* is the internal difference in bias voltage within the operational amplifier, which when multiplied by the circuit gain, becomes an additional output-offset voltage. If the normal input voltage is zero, the open-circuit output voltage is the sum of the three offset voltages.

Table 3.1 Common Op-Amp Circuits (*From* [1]. *Used with permission.*)

No.	Type of Circuit	Schematic	Circuit Gain or Variable of Interest	Input Resistance or Input Currents or Voltages	Special Requirements or Remarks
1	Noninverting amplifier		$\dfrac{v_O}{v_i} = 1 + \dfrac{R_2}{R_1}$	$R_{\text{in}} \approx \infty$ (ideally)	
2	Buffer		$\dfrac{v_O}{v_i} = 1$	$R_{\text{in}} \approx \infty$ (ideally)	Special case of circuit 1
3	Difference amplifier		$v_O = \dfrac{R_2}{R_1}(v_2 - v_1)$	$i_1 = \dfrac{v_1 - v_2\dfrac{R_b}{(R_a + R_b)}}{R_1}$ $i_2 = \dfrac{v_2}{R_a + R_b}$	$\dfrac{R_1}{R_2} = \dfrac{R_a}{R_b}$
4	Adder		$v_O = -\left\{ v_1\dfrac{R_f}{R_1} + v_2\dfrac{R_f}{R_2} + \cdots + v_n\dfrac{R_f}{R_n} \right\}$	$i_1 = \dfrac{v_1}{R_1}$ $i_2 = \dfrac{v_2}{R_2}$ $\cdots$ $i_n = \dfrac{v_n}{R_n}$	
5	Variable gain circuit		$\dfrac{v_O}{v_i} = (2Kx - K)$ $0 \le x \le 1, \quad K > 1$	$i = \dfrac{v_i}{R_3} + \dfrac{Kv_i(1 - x)}{R}$	Potentiometer R_3 adjusts the gain over the range $-K$ to $+K$.

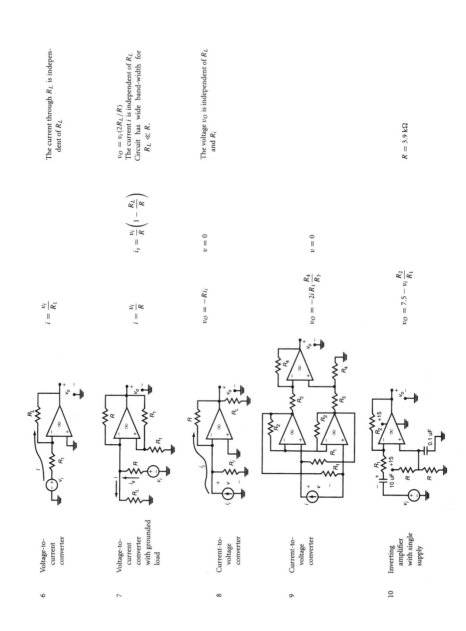

6	Voltage-to-current converter	$i = \dfrac{v_i}{R_1}$		The current through R_L is independent of R_L
7	Voltage-to-current converter with grounded load	$i = \dfrac{v_i}{R}$	$i_s = \dfrac{v_i}{R}\left(1 - \dfrac{R_L}{R}\right)$	$v_O = v_i(2R_L/R)$ The current i is independent of R_L Circuit has wide band-width for $R_L \ll R$.
8	Current-to-voltage converter	$v_O = -R i_i$	$v = 0$	The voltage v_O is independent of R_L and R_i
9	Current-to-voltage converter	$v_O = -2 i R_1 \dfrac{R_4}{R_3}$	$v = 0$	
10	Inverting amplifier with single supply	$v_O = 7.5 - v_i \dfrac{R_2}{R_1}$		$R = 3.9\ \text{k}\Omega$

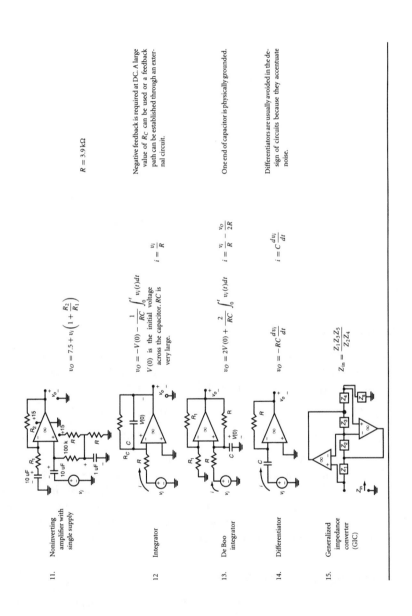

11. Noninverting amplifier with single supply

$$v_O = 7.5 + v_i \left(1 + \frac{R_2}{R_1}\right)$$

$R = 3.9\,\text{k}\Omega$

12. Integrator

$$v_O = -V(0) - \frac{1}{RC}\int_0^t v_i(t)\,dt$$

$V(0)$ is the initial voltage across the capacitor. RC is very large.

$$i = \frac{v_i}{R}$$

Negative feedback is required at DC. A large value of R_C can be used or a feedback path can be established through an external circuit.

13. De Boo integrator

$$v_O = 2V(0) + \frac{2}{RC}\int_0^t v_i(t)\,dt$$

$$i = \frac{v_i}{R} - \frac{v_O}{2R}$$

One end of capacitor is physically grounded.

14. Differentiator

$$v_O = -RC\frac{dv_i}{dt}$$

$$i = C\frac{dv_i}{dt}$$

Differentiators are usually avoided in the design of circuits because they accentuate noise.

15. Generalized impedance converter (GIC)

$$Z_{in} = \frac{Z_1 Z_3 Z_5}{Z_2 Z_4}$$

Table 3.2 Comparison of Counting in the Decimal, Binary, and Octal Systems

Decimal	Binary	Octal
0	0	0
1	1	1
2	10	2
3	11	3
4	100	4
5	101	5
6	110	6
7	111	7
8	1000	10
9	1001	11
10	1010	12
11	1011	13
12	1100	14
13	1101	15
14	1110	16
15	1111	17

3.4 Digital Circuits

Digital signals differ from analog in that only two steady-state levels are used for the storage, processing, and/or transmission of information. The definition of a digital transmission format requires specification of the following parameters:

- The type of information corresponding to each of the binary levels

- The frequency or rate at which the information is transmitted as a bilevel signal

The digital coding of signals for most applications uses a scheme of binary numbers in which only two digits, 0 and 1, are used. This is called a *base*, or *radix*, of 2. It is of interest that systems of other bases are used for some more complex mathematical applications, the principal ones being *octal* (8) and *hexadecimal* (16). Table 3.2 compares the decimal, binary, and octal counting systems. Note that numbers in the decimal system are equal to the number of items counted, if used for a tabulation.

3.4.1 Analog-to-Digital (A/D) Conversion

Because the inputs and outputs of devices that interact with humans usually deal in analog values, the inputs must be represented as numbered sequences corresponding to the analog levels of the signal. This is accomplished by sampling the signal levels and assigning a binary code number to each of the samples. The rate of sampling must

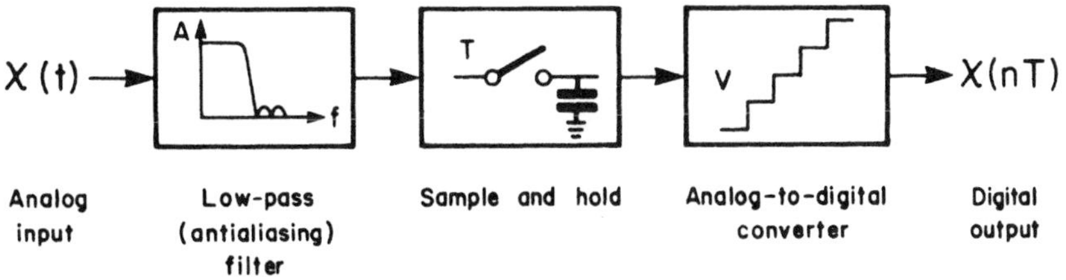

$X(t) \rightarrow$ [Low-pass filter] $\rightarrow$ [Sample and hold] $\rightarrow$ [Analog-to-digital converter] $\rightarrow X(nT)$

Analog input — Low-pass (antialiasing) filter — Sample and hold — Analog-to-digital converter — Digital output

Figure 3.7 Analog-to-digital converter block diagram.

be substantially higher than the highest signal frequency in order to cover the bandwidth of the signal and to avoid spurious patterns (*aliasing*) generated by the interaction between the sampling signal and the higher signal frequencies. A simplified block diagram of an A/D converter (ADC) is shown in Figure 3.7. The *Nyquist law* for digital coding dictates that the sample rate must be at least twice the cutoff frequency of the signal of interest to avoid these effects.

The sampling rate, even in analog sampling systems, is crucial. Figure 3.8*a* shows the spectral consequence of a sampling rate that is too low for the input bandwidth; Figure 3.8*b* shows the result of a rate equal to the theoretical minimum value, which is impractical; and Figure 3.8*c* shows typical practice. The input spectrum must be limited by a low-pass filter to greatly attenuate frequencies near one-half the sampling rate and above. The higher the sampling rate, the easier and simpler the design of the input filter becomes. An excessively high sampling rate, however, is wasteful of transmission bandwidth and storage capacity, while a low but adequate rate complicates the design and increases the cost of input and output analog filters.

3.4.2 Digital-to-Analog (D/A) Conversion

The digital-to-analog converter (DAC) is, in principle, quite simple. The digital stream of binary pulses is decoded into discrete, sequentially timed signals corresponding to the original sampling in the A/D. The output is an analog signal of varying levels. The time duration of each level is equal to the width of the sample taken in the A/D conversion process. The analog signal is separated from the sampling components by a low-pass filter. Figure 3.9 shows a simplified block diagram of a D/A. The deglitching sample-and-hold circuits in the center block set up the analog levels from the digital decoding and remove the unwanted high-frequency sampling components.

Each digital number is converted to a corresponding voltage and stored until the next number is converted. Figure 3.10 shows the resulting spectrum. The energy surrounding the sampling frequency must be removed, and an output low-pass filter is used to accomplish that task. One cost-effective technique used in compact disk players and other applications is called *oversampling*. A new sampling rate is selected that is a

(a)

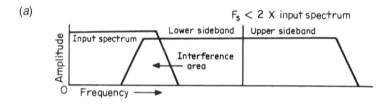

(b)

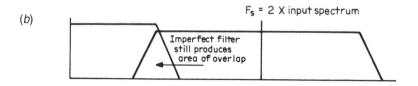

(c)

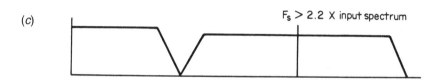

Figure 3.8 Relationship between sampling rate and bandwidth: (a) a sampling rate too low for the input spectrum, (b) the theoretical minimum sampling rate (F_s), which requires a theoretically perfect filter, (c) a practical sampling rate using a practical input filter.

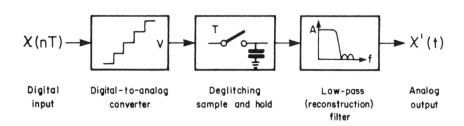

Figure 3.9 Digital-to-analog converter block diagram.

whole multiple of the input sampling rate. The new rate is typically two or four times the old rate. Every second or fourth sample is filled with the input value, while the others are set to zero. The result is passed through a digital filter that distributes the energy in the real samples among the empty ones and itself. The resulting spectrum (for a 4×

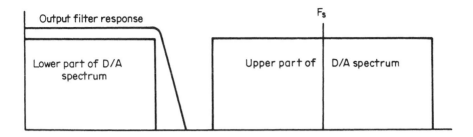

Figure 3.10 Output filter response requirements for a common D/A converter.

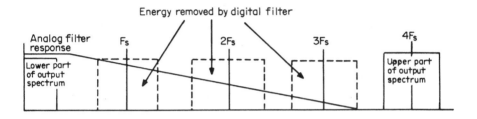

Figure 3.11 The filtering benefits of oversampling.

oversampling system) is shown in Figure 3.11. The energy around the 4× sample frequency must be removed, which can be done simply because it is so distant from the upper band edge. The response of the output filter is chiefly determined by the digital processing and is therefore very stable with age, in contrast to a strictly analog filter, whose component values are susceptible to drift with age and other variables.

3.4.3 Combinational Logic

When the inputs to a logic circuit have only one meaning for each, the circuit is said to be *combinational*. These devices tend to have names reflecting the function they will perform, such as AND, OR, exclusive OR, latch, flip-flop, counter, and gate. Logic circuits are usually documented through the use of schematic diagrams. For simple devices, the shape of the symbol tells the function it performs, while the presence of small bubbles at the points of connection tell whether that point is high or low when the function is being performed. More complicated functions are shown as rectangular boxes. Figure 3.12 shows a collection of common logic symbols.

The clocking input to memory devices and counters is indicated by a small triangle at (usually) the inside left edge of the box. If the device is a transparent latch, the output follows the input while the clock input is active, and the output is "frozen" when the

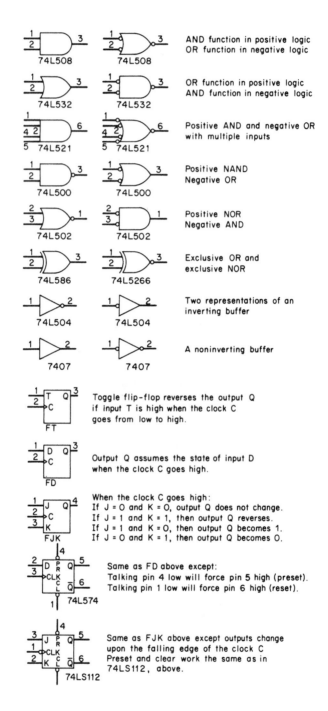

Figure 3.12 Symbols used in digital system block diagrams.

clock becomes inactive. A flip-flop, on the other hand, is an *edge-triggered* device. The output is allowed to change only upon a transition of the clock input from low to high (no bubble) or high to low (bubble present).

Three types of flip-flops are shown in Figure 3.12:

- A *T* (toggle) flip-flop, which will reverse its output state when clocked while the T input is active.

- A *D* flip-flop, which will allow the output to assume the state of the D input when clocked.

- A *J-K* flip-flop. If both J and K inputs are inactive, the output does not change when clocked. If both are active, the output will toggle as in T. If J and K are different, the output will assume the state of the J input when clocked, similar to the D case.

Flip-flops, latches, and counters are often supplied with additional inputs used to force the output to a known state. An *active set* input will force the output into the active state, while a *reset input* will force the output into the inactive state. Counters also have inputs to force the output states; there are two types:

- Asynchronous, in which the function (preset or clear) is performed immediately

- Synchronous, in which the action occurs on the next clock transition

Usually, if both preset and clear are applied at once, the clear function outranks the preset function. Figure 3.13 shows some common logic stages and their truth tables. These gates and a few simple rules of *Boolean algebra*, the basics of which are shown in Table 3.3, facilitate the design of very complex circuits.

3.4.4 Boolean Algebra

Boolean algebra provides a means to analyze and design binary systems. It is based on the seven postulates given in Table 3.4. All other Boolean relationships are derived from these postulates. The OR and AND operations are normally designated by the arithmetic operator symbols + and • and are referred to as *sum* and *product* operators in basic digital logic literature. A set of theorems derived from the postulates, given in Table 3.5, facilitate the development of more complex logic systems.

3.4.5 Logic Device Families

Resistor-transistor-logic (RTL) is mostly of historic interest only. It used a 3.6-V positive power supply, and was essentially incompatible with the logic families that came later. The packages were round with a circular array of wires (not pins) for circuit board mounting. Inputs were applied to the base of a transistor, and the transistor was turned on directly by the input signal if it was high. An open input could usually be considered as an "off" or "0."

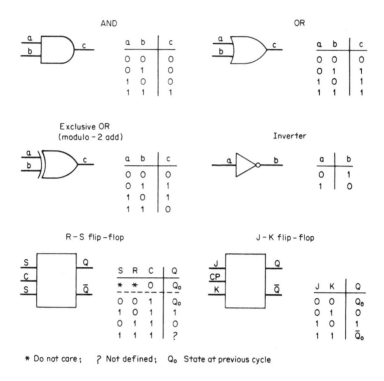

Figure 3.13 Basic logic circuits and truth tables.

Diode-Transistor Logic (DTL)

RTL was followed by the popular DTL, mounted in a DIP (dual in-line package). It had 14 or 16 stiff pins arranged in two parallel rows 0.3 in apart with the pins on 0.1-in centers. For simple devices, such as a two-input NAND gate, four gates were packaged into one DIP. The stiff pins made possible the use of sockets. An internal resistor attached to the positive 5.0-V supply turned on the input transistor. Input signals were applied through diodes such that if an input signal were low, it pulled down the resistor's current, and the transistor turned off. It is important to remember that a disconnected DTL or TTL input is a logic high. The DTL output circuit was pulled low by a transistor and pulled up to +5 V by an internal resistor. As a result, fall times were faster than rise times.

Transistor-Transistor Logic (TTL)

TTL, like DTL, supplies its own turn-on current but uses a transistor instead of a resistor. The inputs do not use diodes but instead use multiple emitters on an input transistor. The output is pulled down by one transistor and pulled up by another. There are

Table 3.3 Fundamental Rules of Boolean Algebra

Rules and relations	
	$0 + 0 = 0$
	$0 \cdot 0 = 0$
	$1 + 1 = 1$
	$1 \cdot 1 = 1$
	$0 \cdot 1 = 0$
	$0 + 1 = 1$
	$\bar{0} = 1$
	$\bar{1} = 0$
	$a \cdot a = a$
	$a + a = a$
	$\bar{a} \cdot a = 0$
	$\bar{a} + a = 1$
	$0 + a = a$
	$0 \cdot a = 0$
	$1 + a = 1$
	$1 \cdot a = a$
	$a + a \cdot b = a$
	$a \cdot (a + b) = a$
	$a + a \cdot b = a + b$
Commutative law	$a + b = b + a$
	$a \cdot b = b \cdot a$
Associative law	$(a + b) + c = a + (b + c)$
	$(a \cdot b) \cdot c = a \cdot (b \cdot c)$
Distributive law	$a \cdot (b + c) = a \cdot b + a \cdot c$
DeMorgan's rules	$\overline{(a \cdot b)} = \bar{a} + \bar{b}$
	$\overline{(a + b)} = \bar{a} \cdot \bar{b}$

† • denotes AND, + denotes OR.

a considerable number of family variations on this basic design. For example, the 7400 device (a two-input NAND gate) has the following common variations:

- 7400—the prototype

- 74L00—a low-power version, but with relatively slow switching speed

- 74S00—(Schottky) fast but power-hungry

- 74LS00—low power and relatively slow speed

- 74AS00—advanced Schottky

- 74ALS00—similar to LS, but with improved performance

- 74F00—F for fast

All variants can be used in the presence of the others, but doing so complicates the design rules that determine how many inputs can be driven by one output. The dividing

Table 3.4 Boolean Postulates (*From* [1]. *Used with permission.*)

Postulate	Name	Meaning	Forms	
			(a)	(b)
			OR	**AND**
1	Definition	$\exists$ a set $\{K\} = \{a, b, \ldots\}$ of two or more elements and two binary operators	$+$	$\bullet$
			$\vee$	$\wedge$
		$\exists \{K\} = \{a, b, a + b, a \cdot b, \ldots\}$	$\cup$	$\cap$
2	Substitution Law	$\text{expression}_1 = \text{expression}_2$ If one replaced by the other does not alter the value		
3	Identity Element	$\exists$ identity elements for each operator	$a + 0 = a$	$a \bullet 1 = a$
4	Commutativity	For every a and b in K	$a + b = b + a$	$a \bullet b = b \bullet a$
5	Associativity	For every a, b, and c in K	$a + (b + c) = (a + b) + c$	$a \cdot (b \cdot c) = (a \cdot b) \cdot c$
6	Distributivity	For every a, b, and c in K	$a + (b \cdot c) = (a + b) \cdot (a + c)$	$a \cdot (b + c) = (a \cdot b) + (a \cdot c)$
7	Complement	For every a in K $\exists$ a complement in K $\ni$	$a + \bar{a} = 1$	$a \bullet \bar{a} = 0$

Table 3.5 Boolean Theorems (*From* [1]. *Used with permission.*)

Theorem		Forms (a)	(b)
8	Idempotency	$a + a = a$	$a \cdot a = a$
9	Complement Theorem	$a + 1 = 1$	$a \cdot 0 = 0$
10	Absorption	$a + ab = a$	$a(a + b) = a$
11	Extra Element Elimination	$a + \bar{a}b = a + b$	$a(\bar{a} + b) = ab$
12	De Morgan's Theorem	$\overline{a + b} = \bar{a} \cdot \bar{b}$	$\overline{ab} = \bar{a} + \bar{b}$
13	Concensus	$ab + \bar{a}c + bc = ab + \bar{a}c$	$(a + b)(\bar{a} + c)(b + c) = (a + b)(\bar{a} + c)$
14	Complement Theorem 2	$ab + a\bar{b} = a$	$(a + b)(a + \bar{b}) = a$
15	Concensus 2	$ab + a\bar{b}c = ab + ac$	$(a + b)(a + \bar{b} + c) = (a + b)(a + c)$
16	Concensus 3	$ab + \bar{a}c = (a + c)(\bar{a} + b)$	$(a + b)(\bar{a} + c) = ac + \bar{a}b$

line between an input high and an input low in this example is about 1.8 V. A high output is guaranteed to be 2.4 V or greater, while an output low will be 0.8 V or less.

NMOS and PMOS

Metal-oxide semiconductor (MOS) logic devices use field-effect transistors as the switching elements. The initial letter tells whether the device uses – or p-type dopant on the silicon. At low frequencies, MOS devices are very frugal in power consumption. Early MOSs were fairly slow, but smaller conductor sizes reduced on-chip capacitance and therefore charging time.

Complementary MOS (CMOS)

A very popular logic family, CMOS devices uses both p- and n-type transistors. At direct current, input currents are almost zero. Output current rises with frequency because the output circuit must charge and discharge the capacitance of the inputs it is driving. Early CMOS devices were fairly slow when powered with a 5-V supply, but performance improved when powered at 10 or 15 V. Modern microscopic geometry produces CMOS parts that challenge TTL speeds while using less power.

The input decision level of a CMOS device is nominally midway between the positive supply and ground. The logic state of an open input is indeterminate. It can and will wander around depending on which of the two input transistors is leaking the most. Unused inputs must be returned either to ground or the supply rail. CMOS outputs, unlike TTL, are very close to ground when low and very close to the supply rail when high. CMOS can drive TTL inputs, however, in a 5-V environment, the CMOS decision level of 2.5 V is too close to the TTL guaranteed output high for reliable operation. The solution is an external pull-up resistor between the output pin of the TTL part and the supply rail.

Early CMOS devices had their own numbering system (beginning at 4000) that was totally different from the one used for TTL parts. Improvements in speed and other performance metrics spawned subfamilies that tended toward a return to the use of the 7400 convention; for example, 74HC00 is a high-speed CMOS part.

Emitter-Coupled Logic (ECL)

ECL has almost nothing in common with the families previously discussed. Inputs and outputs are push-pull. The supply voltage is negative with respect to ground at –5.2 V. Certain advantages accrue from this configuration:

- Because of the push-pull input-output, inverters are not needed. To invert, simply reverse the two connections.

- The differential-amplifier construction of ECL input and output stages causes the total current through the device to be almost constant.

- The output voltage swing is small and, from a crosstalk standpoint, is opposed by the complementary output.

- Driving a balanced transmission line does not require a line-driver because an ECL output (with some resistors) is a line-driver.

Because the transistors in ECL are never saturated, they operate at maximum speed. Early ECL was power-hungry, but newer ECL gate-array products are available that will toggle well into the gigahertz range without running hot.

3.4.6 Scaling of Digital Circuit Packages

The term small-scale integration (SSI) includes those packages containing, for example, a collection of four gates, a 4-bit counter, a 4-bit adder, and any other item of less than about 100 gate equivalents. Large-scale integration (LSI) describes more complex circuitry, such as an asynchronous bit-serial transmitter-receiver, or a DMA (direct memory access) controller, involving a few thousand gate equivalents. Very large scale integration (VLSI) represents tens of thousands of gate equivalents or more, such as a microprocessor or a graphics controllers. LSI and VLSI devices are typically packaged in a larger version of the DIP package, usually with the two rows spaced 0.6 in or more, and having 24 to 68 pins or more.

Many devices are available in dual in-line packages designed to be soldered to the surface of a circuit board rather than using holes in the circuit board. The pin spacing is 0.05 in or less. The *leadless chip carrier* is another surface-mount device with contact spacing of 0.05 in or less, and an equal number of contacts along each edge of the square. Well over 100 contacts can be accommodated in such packages. Sockets are available for these packages, but once the package is installed, a special tool is required to extract it. Yet another large-scale package is called the *pin-grid array*, with pins protruding from the bottom surface of a flat, square package in a row-and-column "bed-of-nails" array. The pin spacing is 0.1 in or less. For this device, more than 200 pins may be incorporated. Extraction tools are available for these packages as well.

3.4.7 Representation of Numbers and Numerals

A single bit, terminal, or flip-flop in a binary system can have only two states. When a single bit is used to describe numerals, by convention those two numerals are 0 and 1. A group of bits, however, can describe a larger range of numbers. Conventional groupings are identified in the following sections.

Nibble

A *nibble* is a group of 4 bits. It is customary to show the binary representation with the *least significant bit* (LSB) on the right. The LSB has a decimal value of 1 or 0. The next most significant bit has a value of 2 or 0, and the next, 4 or 0, and the *most significant bit* (MSB), 8 or 0. The nibble can describe any value from binary 0000 (= 0 decimal) and 1111 (= 8 + 4 + 2 + 1 = 15 decimal), inclusive. The 16 characters used to signify the 16 values of a nibble are the ordinary numerals 0 through 9, followed by the letters of the alphabet A through F. The 4-bit "digit" is a *hexadecimal representation*.

Octal, an earlier numbering scheme, used groupings of 3 bits to describe the numerals 0 through 7. Used extensively by the Digital Equipment Corporation, it has fallen out of use, but is still included in some figures for reference.

Byte

A *byte* is a collection of 8 bits, or 2 nibbles. It can represent numbers (a number is a collection of numerals) in two ways:

- Two hexadecimal digits, the least significant representing the number of 1s, and the most significant the number of 16s. The total range of values is 0 through 255 (FF).

- Two decimal digits, the least significant representing the number of 1s, and limited to the range of numerals 0 through 9, and the most significant representing the number of 10s, again limited to the range 0 through 9.

The use of 4 bits to represent decimal numbers is called *binary-coded decimal* (BCD). The use of a byte to store two numerals is called *packed* BCD. The least significant nibble is limited to the range of 0 through 9, as is the upper nibble, thus representing 00 through 90. The maximum value of the byte is 99.

Word

A *word*, usually a multiple of 8 bits, is the largest array of bits that can be handled by a system in one action of its logic. In most personal computers, a word is 16 or 32 bits. Larger workstations use words of 32 and 64 bits in length. In all cases, the written and electronically mapped representation of the numeric value of the word is either hexadecimal or packed BCD.

Negative Numbers

When a byte or word is used to describe a *signed number* (one that may be less than zero), it is customary for the most significant bit to represent the sign of the number, 0 meaning positive and 1 negative. This representation is known as *two's complement*. To negate (make negative) a number, simply show the number in binary, make all the zeros into 1s, and all the 1s into zeros, and then add 1.

Floating Point

In engineering work, the range of numerical values is tremendous and can easily overflow the range of values offered by 64-bit (and smaller) systems. Where the accuracy of a computation can be tolerably expressed as a percentage of the input values and the result, *floating-point calculation* is used. One or two bytes are used to express the *characteristic* (a power of 10 by which to multiply everything), and the rest are used to express the *mantissa* (that fractional power of 10 to be multiplied by). This is commonly referred to as *engineering notation*. (See Table 3.6.)

Table 3.6 Number and Letter Representations

Decimal	Binary	Octal	Hexadecimal
0	0	0	0
1	1	1	1
2	10	2	2
3	11	3	3
4	100	4	4
5	101	5	5
6	110	6	6
7	111	7	7
8	1000	10	8
9	1001	11	9
10	1010	12	A
11	1011	13	B
12	1100	14	C
13	1101	15	D
14	1110	16	E
15	1111	17	F
81	01010001	121	51
250	11111010	372	FA
+127	01111111 (signed)	177	7F
−1	11111111 (signed)	377	FF
−128	10000000 (signed)	200	80

Compare

A *comparison* involves negating one of the two numbers being compared, then adding them and testing the result. If the test shows zero, the two numbers are equal. If not, the test reveals which of the two is greater than or less than the other, and the appropriate bits in the status register are set.

Jump

The orderly progression of the program counter may be interrupted and instructions fetched from a new location in memory, usually based upon a test or a comparison. For example, "If the result is zero, jump to location X and begin execution there; if the result is positive, jump to Y and begin execution there; else keep on counting." This ability is probably the most powerful asset of a computer because it permits logic-based branching of a program.

3.4.8 Errors in Digital Systems

When a digital signal is transmitted through a noisy path, errors can occur. Early methods to deal with this problem included generating one or more digital words, using *check sums*, *cyclic redundancy checks*, and similar error-coding schemes, and appending the result at the end of a block of transmitted data. Upon reception, the same arithmetic was used to generate the same results, which were compared to the data appended to the transmission. If they were identical, it was unlikely that an error had occurred. If they differed, an error was assumed to have occurred, and a retransmission was requested. Such methods, thus, performed only error detection. In the case of many digital transmission systems, however, retransmission is not possible and methods must be employed that not only detect but correct errors.

Error Detection and Correction

Given a string of 8-bit bytes, additional bytes can be generated using *Galois field arithmetic* and appended to the end of the string. The length of the string and the appended bytes must be 256 or less, since 8 bits can have no more than 256 different states. If 2 bytes are generated, upon reconstruction 2 *syndrome* (symptom) bytes are generated. If they are zero, there was likely no error. If they are nonzero, then after arithmetic processing, 1 byte "points" to the location of the damaged byte in the string, while the other contains the 8-bit *error pattern*. The error pattern is used in a bit-wise exclusive OR function upon the offending byte, thus reversing the damaged bits and correcting the byte. With 4 check bytes, 2 flawed bytes can be pinpointed and corrected; with 6, 3 can be treated; and so on. If the number of bytes in the string is significantly less than 256, for example, 64, the error-detection function becomes more robust because, if the error pointer points to a nonexistent byte, it may be assumed that the error-detection system itself made a mistake.

Errors in digital recorders, for example, fall into two classes: random errors brought on by thermal random noise in the reproduce circuitry, and dropouts (long strings of lost signal resulting from tape imperfections. The error detection and correction system of digital recorders is designed to cope with both types of errors. Figure 3.14 shows how data can be arranged in rows and columns, with separate check bytes generated for each row and each column in a two-dimensional array. The data is recorded (and reproduced) in row order. In the example given in the figure, it can be seen that a long interruption of signal will disrupt every tenth byte. The row corrector cannot cope with this, but it is likely that the column corrector can because it "sees" the burst error as being spread out over a large number of columns.

The column corrector, if taken alone, can correct $N/2$ errors, where N is the number of check bytes. Given knowledge of which rows are uncorrectable by the row corrector, then N errors can be corrected. Generally, the row (or "inner") corrector acts on errors caused by random noise, while the column (or "outer") corrector takes care of burst errors.

Generally, error detection and correction schemes have the following characteristics:

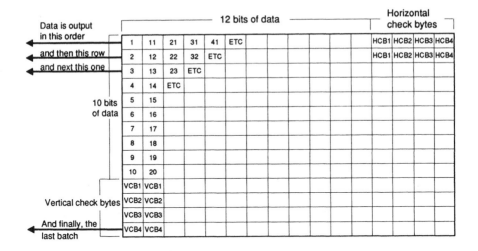

Figure 3.14 An example of row and column two-dimensional error-detection coding.

- Up to a threshold error rate, all errors are corrected.

- If the error rate is greater than the above first threshold, the system will flag the blocks of data it is unable to correct. This allows other circuits to attempt to conceal the error.

- Above an even higher error rate, the system will occasionally fail and either stop producing output data entirely, or simply pass along the data, correcting what it can and letting the rest pass through.

Error Concealment

When the error-correction system is overloaded and error-ridden samples are identified, it is typical practice in communications applications to calculate an estimation of the bad sample. In video applications, for example, samples that are visually nearby and that are not corrupted can be used to calculate an estimate of the damaged sample. The estimate is then substituted for the unusable sample. In the recording or transmission process, the video data samples are scrambled in a way that maximizes the chance that a damaged sample will be surrounded by good ones.

In the case of audio, the samples can be scrambled such that failure of the correction system is most likely to result in every alternate sample being in error. Replacement of a damaged audio sample can then consist of summing the previous (good) sample and the following (good) sample and dividing by 2. If the error rate becomes unreasonable, then the last good sample is simply repeated, or "held."

Video error concealment is roughly 10 times more effective than audio conceal-ment, due in large part to differences in the way the eye and ear interpret and process in-put information.

3.5 References

1. Whitaker, Jerry C., (ed.), *The Electronics Handbook*, CRC Press, Boca Raton, FL, 1996.

3.6 Bibliography

Benson, K. Blair (ed.), *Audio Engineering Handbook*, McGraw-Hill, New York, N.Y., 1988.

Boyer, Robert, JeanLuc Grimaldi, Jacques Oyaux, and Jacques Vallee, "Serial Inter-face Within the Digital Studio," 1. *Soc. Motion Pict. Telev.*, November 1984.

Busby, E. Stanley, "Digital Fundamentals," in *Television and Audio Handbook for Technicians and Engineers*, K. Blair Benson and Jerry C. Whitaker (eds.), McGraw-Hill, New York, N.Y., 1990.

EBU: Publication Tech 3247.E, Technical Centre of the EBU, Brussels, 1985.

Fink, Donald (ed.), *Electronics Engineers' Handbook*, McGraw-Hill, New York, N.Y., 1982.

Texas Instruments, *2-mm CMOS Standard Cell Data Book*, Chapter 8, Texas Instru-ments, Dallas, TX, 1986.

Whitaker, Jerry C., and K. Blair Benson (eds.), *Standard Handbook of Video and Tele-vision Engineering*, third ed., McGraw-Hill, New York, N.Y., 2000.

Whitaker, Jerry C., (ed.), *Video and Television Engineer's Field Manual*, McGraw-Hill, New York, N.Y., 2000.

4

Systems Engineering

4.1   Introduction

When the owner or company executive of a technical facility decides to proceed with new construction or the renovation of a technical plant, this person will probably enlist the help of a qualified architect and general contractor. However, another important team member should be included—the *system engineer*. This person may already be a member of the staff, an outside engineering consultant, or a system integration contractor. Without a system engineer, design and development will not proceed as quickly, and a greater chance exists for serious, costly miscalculations.

The system engineer plays a major role in developing a successful facility plan. A good system engineer has a wealth of technical information. With it, this person can expedite the planning and design process considerably, and can help keep overall facility costs down. Many architectural firms cannot provide technical equipment planning on an appropriate level for technical facilities. And, on an industrial or commercial level, they probably cannot provide anything beyond cursory services, unless they bring in an outside system engineering consultant.

Researching a manufacturer's literature to determine the heat output or power consumption of one piece of equipment, for example, can be a time-consuming task. If an architect has no previous experience with the equipment, this approach can cost time and money. Most owners delegate technical facility planning to their operations and maintenance engineering personnel. This is unfortunate because even with their hands-on experience and knowledge about the equipment, they often find themselves unable to devote the necessary time to the project while they're handling normal duties. And, no matter how knowledgeable these technical people are, they often lack the necessary engineering skills and design experience of a system engineer. There is no substitute for a qualified, experienced system engineer.

4.2   The System Engineer

As an electrical engineering graduate, the system engineer has the training and knowledge required for designing electronic systems. This person has hands-on experience with electronic equipment assembly and testing techniques. With this knowledge, the system engineer can avoid the pitfalls that are often encountered by those with less education and/or experience.

An experienced system engineer is already familiar with design techniques, drawings, and specifications. This person also shares the architect's skills in reading drawings and visualizing real, 3-D environments. A familiarity with facility layout requirements enables the system engineer to recognize and avoid costly problems. It can be disappointing, for example, to see the space set aside for future expansion being consumed by the last-minute addition of more equipment racks.

Experienced system engineers, because they are familiar with proper construction and wiring techniques, can also assure quality workmanship on the job. They can catch and correct work that has been done improperly. In other words, an experienced system engineer is management's eyes and ears.

4.2.1 Outside Engineering Contractor

When the size of a project warrants the employment of a system engineer, the company's owner or its executives should approve the expense. The alternative, if frequent changes to a facility are not anticipated, is to go outside of the company and retain the services of a system design engineering consultant or a system integration contractor. There are many circumstances in which consulting system engineers represent the most cost-effective solution to engineering needs. Consultants contribute specialized expertise and experience to projects when these resources are not available within a firm.

Consultants can be used for peak-period or unique projects. They can help equalize the work load of permanent employees. Consultants can provide impartial analysis and can make valuable contributions regarding problems, products, and plans. Because they are independent business people, they know the value of good judgment and unbiased opinion. Consultants can be effective as catalysts for innovation and change when fresh thinking is needed on a project or in a department. Consultants can provide formal and informal training at all company levels. They can be especially useful during expansion, by recommending and training new full-time employees.

When near-term business situations are uncertain, consultants can perform the work without long-term commitments. As independent contractors rather than employees, consultants offer specialized services or skills to clients for a fee. They are reimbursed either on a fixed-fee basis, a fixed fee determined by a percentage of the equipment cost, or for time and expenses. Consultants provide their clients with expertise and specialized knowledge that are usually not available within the client's organization. Because they are exposed to a variety of situations, they have developed experience through the application of many successful approaches to business and technical challenges. Consultants are project- or task-oriented and can devote the necessary time to accomplish an assignment independent of other responsibilities.

Consultants solve problems. They are experts. They are frequently asked to implement solutions—to function as a *system integrator.*

System integration contractors provide turnkey systems design, fabrication, and installation. They can often provide the same skills and services rendered by professional consulting engineers. They can handle all phases of a project, from a project's conception through completion. As the equipment supplier, training and warranty support are also offered as a part of their contract. Typically, system integrators are reimbursed either on a fixed-fee basis or by a percentage of the equipment cost. Although a fixed-fee arrangement is usually best for the owner, because larger system integration contractors are also equipment vendors, they may prefer to charge a commission for larger projects based on equipment costs, rather than charging an engineering fee. These companies will probably specify equipment made by manufacturers with whom they have a dealer agreement. They may also specify products that provide them the greatest profit margin.

If competitive bids must be solicited for a project, an independent consultant should usually be retained to prepare the specification documents and a *request for proposal* (RFP). This is done to avoid any advantage or bias that may be built into bid documents prepared by a system integrator. Upon the owner's acceptance of the bid specification, an RFP can be sent to several system integrators. This process also provides checks on the specifications and may introduce some alternative ideas.

Whatever the arrangement, a qualified system engineer can help contain equipment costs, meet construction schedules, and ensure that a technical facility will perform as required.

4.2.2 Design Development

System design is executed in a series of steps that lead to an operational unit. Appropriate research and preliminary design work is completed in the first phase of the project—the *design development* phase. This phase is designed to fully delineate all project requirements and to identify any constraints. Based on initial ideas and information, the design concepts are modified until all parties are satisfied and approval is given for the final design work to begin. The first objective of this phase is to answer the following questions:

- What are the functional requirements of the product?

- What are the physical requirements of the product?

- What are the performance requirements of the product?

- Are there any constraints limiting design decisions?

- Will existing equipment be used, and is it acceptable?

- Will this be a new facility or a renovation?

- Will this be a retrofit or upgrade to an existing system?

- Will this be a stand-alone system?

The equipment and functional requirements of each of the major technical areas are identified by the engineer as this person works closely with the owner's representatives. With facility renovation, the system engineer's first step is to analyze existing equipment. This person will visit the site to gather detailed information about the existing facility. The system engineer, usually confronted with a mixture of acceptable and unacceptable equipment, must sort out the equipment that meets current standards and determine which items should be replaced. After soliciting input from the facility's technical personnel, the system engineer then develops a list of needed equipment.

One of the system engineer's most important contributions is the ability to identify and meet the owner's needs, and to do so within the project's budget. Based on the owner's initial concepts and any subsequent equipment utilization research conducted by the system engineer, the desired capabilities are identified as precisely as possible. Design parameters and objectives are defined and reviewed. Functional efficiency is maximized to allow operation by a minimum number of people. Future needs are also investigated at this time, and future technical system expansion is considered.

After management approves the equipment list, preliminary system plans are drawn up for review and further development. If architectural drawings of the facility are available, they can be used as a starting point for laying out an equipment floor plan. The system engineer uses this floor plan to be certain adequate space is provided for present and future equipment, and adequate clearance is furnished for maintenance and convenient operation. Equipment identification is then added to the architect's drawings.

Documentation should include, but not be limited to, a list of major equipment:

- Equipment prices
- Technical system functional block diagrams
- Custom item descriptions
- Rack and console elevations
- Equipment floor plans

The preliminary drawings and other supporting documents are prepared to record design decisions and to illustrate the design concepts to the owner and/or facility manager. Renderings, scale models, or full-size mockups may also be needed to better illustrate, clarify, or test design ideas.

Ideas and concepts must be exchanged and understood by all concerned parties. Good communication skills are essential. The bulk of the creative work is carried out in the design development phase. The physical layout—the look and feel—and the functionality of the facility will all have been decided and agreed upon by the completion of this phase. If the design concepts appear feasible, and the cost is within the anticipated budget, management can authorize work to proceed on the final detailed design.

4.2.3 Level of Detail

With the research and preliminary design development completed, the design details must be concluded. The system engineer prepares complete detailed documentation and specifications necessary for the fabrication and installation of the technical systems. These include all major and minor components. Drawings must show the final configuration and the relationship of each component to other elements of the system. The drawings will also show how these components interface with other building services, such as air conditioning and electrical power. This documentation must communicate the design requirements to the other design professionals, including the construction and installation contractors.

In this phase, the system engineer develops final, detailed flow diagrams that show the interconnection of all equipment. Cable interconnection information for each type of signal is taken from the flow diagrams and recorded on the cable schedule. Cable paths are measured and timing calculations performed. *Timed cable lengths* (used for video and other special services) are entered onto the cable schedule.

Special custom items are defined and designed. Detailed schematics and assembly diagrams are drawn. Parts lists and specifications are finalized, and all necessary details are worked out for these items. Mechanical fabrication drawings are prepared for consoles and other custom-built cabinetry.

The system engineer provides the architect with layouts of cable runs and connections. Such detailed documentation simplifies equipment installation and facilitates future changes in the system. During preparation of final construction documents, the architect and the system engineer can confirm the layout of the technical equipment wire ways, including access to flooring, conduits, trenches, and overhead wire trays.

Dimensioned floor plans and elevation drawings are required to show placement of equipment, lighting, electrical cable ways, ducts, conduits, and HVAC ductwork. Requirements for special construction, electrical, lighting, HVAC, finishes, and acoustical treatments must be prepared and submitted to the architect for inclusion in the architectural drawings and specifications. This type of information, along with cooling and electrical power requirements, also must be provided to the mechanical and electrical engineering consultants (if used on the project) so they can begin their design calculations.

Equipment heat loads are calculated and submitted to the HVAC consultant. Steps are taken when locating equipment to avoid any excessive heat buildup within the equipment enclosures, while maintaining a comfortable environment for the operators.

Electrical power loads are calculated and submitted to the electrical consultant. Also, steps are taken to provide for sufficient power and proper phase balance.

4.2.4 Management Support

The system engineer can assist in ordering equipment and helping to coordinate the move to a new or renovated facility. This is critical if a lot of existing equipment is being relocated. With new equipment, the facility owner will find the system engineer's knowledge of prices, features, and delivery times a valuable asset. A good system engineer will make sure that equipment arrives in ample time to allow for sufficient testing and installation. A good working relationship with equipment manufacturers helps guarantee their support of and speedy response to the owner's needs.

The system engineer can also provide engineering management support during planning, construction, installation, and testing to help qualify and select contractors, resolve problems, explain design requirements, and assure quality workmanship by the contractors and the technical staff.

The procedures described in this section outline an ideal scenario. Management may often try to bypass many of the foregoing steps to save money. This, they reason, will eliminate unnecessary engineering costs and allow construction to begin immediately. By using in-house personnel, a small company may attempt to handle the job without professional help. With inadequate design detail and planning, which can result from using unqualified people, the job of setting technical standards and making the system work then defaults to the construction contractors, in-house technical staff, or the installation contractor. This can result in costly and uncoordinated work-arounds and, of course, delays and added costs during construction, installation, and testing. This also makes the project less manageable and less likely to be completed successfully.

The size of a technical facility can vary from a small, one-room operation to a large, multi-million-dollar plant or network. Management should recruit a qualified system engineer for projects that involve large amounts of money and other resources.

4.3   The Project Team

The persons who plan and carry out a project compose the *project team*. The project team's makeup will vary depending on the size of the company and the complexity of the project. Management is responsible for providing the necessary human resources to complete the project.

4.3.1 Executive Management

The *executive manager* is the person who can authorize the project's undertaking. This person can allocate funds and delegate authority to others to accomplish this task. Motivation and commitment are important aspects for accomplishing the goals of the organization. The ultimate responsibility for a project's success lies with the executive manager. This person's job is to complete tasks through others by assigning group responsibilities, coordinating activities among groups, and resolving group conflicts. The executive manager establishes policy, provides broad guidelines, approves the project master plan, resolves conflicts, and assures project compliance with commitments.

Executive management delegates the project management functions and assigns authority to qualified professionals, allocates a capital budget for the project, supports the project team, and establishes and maintains a healthy relationship with project team members.

Management is responsible for providing clear information and goals—up front—based upon management's needs and initial research. Before initiating a project, the company executive should be familiar with daily facility operation and should analyze how the company works, how the people do their jobs, and what tools are needed to accomplish the work. An executive should consider certain points before initiating a project:

- What is the current capital budget for equipment?
- Why does the staff currently use specific equipment?
- What function of the equipment is the weakest within the organization?
- What functions are needed, but cannot be accomplished, with current equipment?
- Is the staff satisfied with current hardware?
- Are there any reliability problems or functional weaknesses?
- What is the maintenance budget, and is it expected to remain steady?
- How soon must the changes be implemented?
- What is expected from the project team?

Only after answering the appropriate questions will the executive manager be ready to bring in expert project management and engineering assistance. Unless the manager has made a systematic effort to evaluate all of the obvious points about the facility requirements, the not-so-obvious points may be overlooked. Overall requirements must be divided into their component parts. Do not try to tackle ideas that have too many branches. Keep the planning as basic as possible. If the company executive does not attempt to investigate the needs and problems of a facility thoroughly before consulting experts, the expert advice will be shallow and incomplete, no matter how good the engineer.

Engineers work with the information they are given. They put together plans, recommendations, budgets, schedules, purchases, hardware, and installation specifications based upon the information they receive from interviewing management and staff. If the management and staff have failed to go through the planning, reflection, and refinement cycle before those interviews, the company will probably waste time and money.

4.3.2 Project Manager

Project management is an outgrowth of the need to accomplish large, complex projects in the shortest possible time, within the anticipated cost, and with the required performance and reliability. Project management is based on the realization that modern organizations may be so complex that they preclude effective management using traditional organizational structures and relationships. Project management can be applied to any undertaking that has a specific objective.

The *project manager* has the authority to carry out a project and has been given the right to direct the efforts of the project team members. The project manager gains power from the acceptance and respect that is provided by superiors and subordinates, and has the power to act and is committed to group goals.

The project manager is responsible for the successful completion of the project, on schedule, and within budget. This person will use whatever resources are necessary to accomplish the goal in the most efficient manner. The project manager provides project schedule, and financial and technical requirement direction. This person also evaluates and reports on project performance. This requires planning, organizing, staffing, directing, and controlling all aspects of the project.

In this leadership role, the project manager is required to perform many tasks:

- Assemble the project organization.
- Develop the project plan.
- Publish the project plan.
- Set measurable and attainable project objectives.
- Set attainable performance standards.
- Determine which scheduling tools (PERT, CPM, and/or GANTT) are right for the project.
- Use the scheduling tools, and develop and coordinate the project plan. This includes the budget, resources, and the project schedule.
- Develop the project schedule.
- Develop the project budget.
- Manage the budget.
- Recruit personnel for the project.
- Select subcontractors.
- Assign work, responsibility, and authority so team members can make maximum use of their abilities.
- Estimate, allocate, coordinate, and control project resources.
- Deal with specifications and resource needs that are unrealistic.
- Determine the right level of administrative and computer support.
- Train project members to fulfill their duties and responsibilities.
- Supervise project members, giving them day-to-day instructions, guidance, and discipline, as required, to fulfill their duties and responsibilities.
- Design and implement reporting and briefing information systems or documents that respond to project needs.
- Control the project.

Some basic project management practices can improve the chances for success. Consider the following:

- Secure the necessary commitments from top management to make the project a success.

- Establish an action plan that will be easily adopted by management.

- Use a work breakdown structure that is comprehensive and easy-to-use.

- Establish accounting practices that help, not hinder, successful project completion.

- Prepare project team job descriptions properly up front to eliminate conflict later.

- Select project team members appropriately the first time.

After the project is underway, follow these steps:

- Manage the project, but make the oversight reasonable and predictable.

- Persuade team members to accept and participate in the plans.

- Motivate project team members for best performance.

- Coordinate activities so they are carried out in relation to their importance, with a minimum of conflict.

- Monitor and minimize interdepartmental conflicts.

- Get the most out of project meetings without wasting the team's productive time. Develop an agenda for each meeting, and start on time. Conduct one piece of business at a time. Assign responsibilities where appropriate. Agree on follow-up and accountability dates. Indicate the next step for the group. Set the time and place for the next meeting. Then, end on time.

- Spot problems and take corrective action before it is too late.

- Discover the strengths and weaknesses in project team members, and manage them to obtain desired results.

- Help team members solve their own problems.

- Exchange information with subordinates, associates, superiors, and others about plans, progress, and problems.

- Make the best of available resources.

- Measure project performance.

- Determine, through formal and informal reports, the degree to which progress is being made.

- Determine causes of and possible ways to act upon significant deviations from planned performance.

- Take action to correct an unfavorable trend, or to take advantage of an unusually favorable trend.

- Look for areas where improvements can be made.

- Develop more effective and economical methods of managing.

- Remain flexible.
- Avoid "activity traps."
- Practice effective time management.

When dealing with subordinates, employees should:

- Know what they are supposed to do, preferably in terms of an end product.
- Have a clear understanding of what their authority is, and its limits.
- Know what their relationship with other people is.
- Know what constitutes a job well done in terms of specific results.
- Know when and what they are doing exceptionally well.
- Be shown concrete evidence that there are rewards for work well done and *exceptionally* well done.
- Know where and when they are falling short of expectations.
- Be informed of what can and should be done to correct unsatisfactory results.
- Feel that their superior has an interest in them individually.
- Feel that their superior believes in them and is enthusiastic for them to succeed and progress.

By fostering a good relationship with associates, managers will have less difficulty communicating with them. The fastest, most effective communication takes place among people with common viewpoints.

4.3.3 Engineering Manager

The *engineering manager* in a technical facility usually manages the technical staff, which may be made up of graduate engineers and technicians. The engineering manager is committed to technical quality and the functional integrity of the facility.

If a company has no project manager, the engineering manager may assume this role.

4.3.4 System Engineer

The term *system engineer* means different things to different people. The system engineer provides the employer with the experience gained from many successful approaches to technical problems developed through hands-on exposure to a variety of situations. This person is a professional with knowledge and experience, possessing skills in one or more specialized and learned fields. The system engineer is an expert in a given field, highly trained in analyzing problems and developing solutions that satisfy management objectives.

Education in electronics theory is a prerequisite for designing systems that employ electronic components. As a graduate engineer, the system engineer has the education required to design electronic facilities correctly. Knowledge of testing techniques and theory enables this person to specify system components and performance, and to measure the results. Drafting and writing skills permit efficient preparation of the necessary documentation needed to communicate the design to the technicians and contractors who will have to build and install the system.

Training in personnel relations, a part of the engineering curriculum, helps the system engineer deal with subordinates and management. A good system engineer has a wealth of technical information that can be used to speed up the design process and help in making cost-effective decisions. If the system engineer does not have the needed information, this person knows where to find it.

The system engineer performs the following functions:

- Receives input from management and staff.

- Researches the project and develops a workable design.

- Solves technical problems related to the design and integration of the system into a facility.

- Concentrates on results and focuses work according to the employer's objectives.

The degree to which these objectives are achieved is an important measure of the system engineer's contribution. In some cases, the system engineer may have to assume the responsibilities of planning and managing a project.

The system engineer's duties will vary, depending on the size of the project and the management organization. Aside from designing the system, this person has to answer questions and solve problems that may arise during hardware fabrication and installation. The system engineer must also monitor installation quality and workmanship. Hardware and software will have to be tested and calibrated upon completion, which is also a concern of the system engineer. Depending on the complexity of the new installation, the system engineer also may have to provide orientation and operating instructions to the users.

Other key members of the project team include the following:

- Architect—responsible for design of the structure.

- Mechanical engineer—responsible for HVAC and other mechanical designs.

- Structural engineer—responsible for concrete and steel structures.

- Construction contractors—responsible for executing the plans developed by the architect, mechanical engineer, and structural engineer.

Small in-house projects can be completed on an informal basis. This is probably the normal routine for uncomplicated projects. In a large facility project, however, the system engineer's involvement usually begins with preliminary planning and continues through fabrication, installation, and testing. A project's scope will determine the work that will be required of the system engineer. The scope is an outline of the endeavors to which pursuits, activities, and interests will be confined. Consequently, the scope of the project must be formulated and agreed upon by the project participants. The extent and the limits of the work also must be determined. In this case, the intent of the scope is to fully delineate the work to be carried out by the project's system engineer. Subjects to be considered include:

- What work is to be done by the system engineer.

- What the results of the work will be.

- What the end product of the work will be.

4.4 Budget Requirements Analysis

The need for a project may originate with management, operations staff, technicians, or engineers. In any case, some sort of logical reasoning or a specific production requirement will justify the need. On small projects, such as the addition of one piece of equipment, money must only be available for the purchase and installation costs. When the need justifies a large project, the final cost is not always immediately apparent. The project must be analyzed by dividing it into its constituent parts or elements:

- Equipment and parts

- Materials

- Resources, including money and time needed for project completion

An executive summary or capital project budget request, which contains a detailed breakdown of these elements, can provide the information management needs to determine the return on investment, and to make an informed decision on whether or not to authorize the project.

A capital project budget request, which contains the minimum information, may consist of the following items:

- **Project name**—a name that describes the result of the project, such as "control room upgrade."

- **Project number** (if required). A large organization that does many projects will use some kind of project numbering system, or it may use a budget code assigned by the accounting department.

- **Project description**—a brief description of what the project will accomplish, such as "design the technical system upgrade for the renovation of production control room 2."

- **Initiation date**—the date the request will be submitted.

- **Completion date**—the date the project will be completed.

- **Justification**—the reason the project is needed.

- **Material cost breakdown**—a list of equipment, parts, and materials required for construction, fabrication, and equipment installation.

- **Total material cost.**

- **Labor cost breakdown**—a list of personnel required to complete the project, their hourly pay rates, the number of hours they will spend on the project, and the total cost for each person.

- **Total project cost**—the sum of material and labor costs.

- **Payment schedule**—an estimation of individual amounts that will be paid out during the course of the project, and the approximate dates that each will be payable.

- Preparer's name and the date prepared.

- Approval signature(s) and date(s) approved.

More detailed analysis, such as return on investment, can be carried out by an engineer. Financial analysis, however, should be left to the accountants, who have access to company financial data.

4.4.1 Feasibility Study and Technology Assessment

In cases where an attempt must be made to implement new technology, and where a determination must be made as to whether certain equipment can perform a desired function, a feasibility study should be conducted. The system engineer may be called upon to assess the state of the art in order to develop a new application. In addition to a capital project budget request, an executive summary or a more detailed report of evaluation test results may be required to help management make its decision.

4.4.2 Project Tracking and Control

A project team member may be selected by the project manager to report the status of work during the course of the project. A standardized *project status report* form can provide consistent and complete information to the project manager. The purpose is to supply information to the project manager regarding work completed and money spent on resources and materials.

A project status report containing minimum information should contain the following items:

- Project number (if required)

- Date prepared

- Project name

- Project description

- Start date

- Completion date (the date this part of the project was completed)

- Total material cost

- Labor cost breakdown

- Preparer's name

4.4.3 Change Order

After all or part of a project design has been approved and money has been allocated, any changes may increase or decrease the cost. Several factors can affect the cost:

- Material

- Resources, such as labor and special tools or construction equipment

- Costs incurred because of manufacturing or construction delays

Management should know about such changes, and will want to control them. For this reason, a method of reporting changes to management and soliciting its approval should be instituted. The best way to do this is with a *change order request* or *change order*. A change order includes a brief description and reason for the change and a summary of the effect it will have on costs and the project schedule.

Management will exercise its authority to approve or disapprove each change, based upon its understanding of the cost and benefits and the perceived need for the modification of the original plan. Therefore, the system engineer should provide as much information and explanation as may be necessary to make the change clear and understandable to management.

A change order form, containing the minimum information, should contain the following items:

- Project number

- Date prepared

- Project name

- Labor cost breakdown

- Preparer's name

- Description of the change

- Reason for the change

- Equipment and materials to be added or deleted

- Material costs or savings

- Labor costs or savings

- Total cost of this change (increase or decrease)

4.5   Electronic System Design

Performance standards and specifications must be established in advance for a technical facility project. This will set the performance level of equipment that is acceptable for the system and affect the size of the budget. Signal quality, stability, reliability, and accuracy are examples of the kinds of parameters that must be specified. Access and processor speeds are important parameters when dealing with computer-driven products. The system engineer must confirm whether selected equipment conforms to the standards.

At this point, it must be determined what functions each component in the system will be required to fulfill, and how each will perform with other components in the system. The management and operation staff usually know what they would like the system to do, and how they can best accomplish the task. They should select equipment that they think will do the job. With a familiarity of the capabilities of different equipment, the system engineer should be able to contribute to this function/definition stage. Following is a list of questions that must be answered:

- What functions must be available to the operators?

- What functions are secondary and, therefore, not necessary?

- What level of automation should be required to perform a function?

- How accessible should the controls be?

Over-engineering or over-design must be avoided. Such serious and costly mistakes are often made by engineers and company staff when planning technical system requirements. A staff member may, for example, ask for a feature or capability without fully understanding its complexity or the additional cost it may impose. Other portions of the system may have to be compromised to implement the additional feature. An experienced system engineer will be able to spot this and determine whether the trade-offs and added engineering time and cost are really justified.

When existing equipment is used, an inventory list should be made. This is the preliminary part of a final equipment list. Normally, when confronted with a mixture of acceptable and unacceptable equipment, the system engineer must determine what meets current standards and what should be replaced. Then, after soliciting input from facility technical personnel, the system engineer develops a summary of equipment needs, including future acquisitions. One of the system engineer's most important contributions is the ability to identify and meet these needs within the facility budget.

A list of major equipment is then prepared. The system engineer selects equipment based on experience with the products and on owner preferences. Existing equipment is often reused. A number of considerations are discussed with the facility owner to determine the best product selection. Some major points include:

- Budget restrictions

- Space limitations

- Performance requirements

- Ease of operation

- Flexibility

- Functions and features

- Past performance history

- Manufacturer support

The system engineer's goal is to choose and install equipment that will meet the project's functional requirements efficiently and economically. Simplified block diagrams of the video, audio, control, data, RF, and communication systems are drawn and then discussed with the owner and presented for approval.

4.5.1 Developing a Flow Diagram

The flow diagram is a schematic drawing used to show the interconnections among all equipment that will be installed. It differs from a block diagram because it contains much more detail. Every wire and cable must be included on these drawings. See Figure 4.1 for a typical flow diagram of a video production facility.

The starting point for preparing a flow diagram can vary depending on the information available from the design development phase of the project, and on the similarity of the project to previous projects. If a similar system has been designed previously, the diagrams from that project can be modified to include the equipment and functionality required for the new system. New equipment models can be shown on the diagram in place of their counterparts, and minor wiring changes can be made to reflect the new equipment connections and changes in functional requirements. This method is efficient and easy to complete.

If the facility requirements do not fit any previously completed design, the block diagram and equipment list are used as a starting point. Essentially, the block diagram is expanded and details are added to show all of the equipment and interconnections, and to show any details necessary to describe the installation and wiring completely.

An additional design feature that is desirable for specific applications is the capability to disconnect a rack assembly easily from the system and relocate it. This would be used if a system was pre-built at a system integration facility and later moved and installed at the client's site. With this type of situation, the interconnecting cable harnessing scheme must be well planned and identified on the drawings and cable schedules.

4.5.2 Estimating Cable Lengths

Cable lengths are calculated using dimensions taken from the floor plans and rack elevations and should be included on the cable schedule. The quantity of each cable type can then be estimated for pricing and purchasing. A typical cable schedule database printout is shown in Figure 4.2.

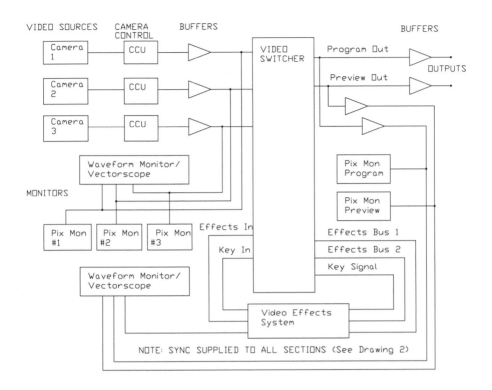

Figure 4.1 Example flow diagram of a video production facility.

4.5.3 Signal Timing Considerations

For certain signal paths, the length of the cable connecting two pieces of equipment may critically affect the timing relationship of that signal as it relates to others in the system. Calculate these critical cable lengths and include them on the cable schedule.

Electrical signals travel through cable at a velocity determined by the physical properties of the cable. Using the published value for the velocity of propagation, calculate the amount of delay in a given length of cable. The velocity for video cables may vary from 66 percent to 78 percent, depending on the manufacturer. The delay may also be determined experimentally by simply measuring the amount of delay produced in a given length of cable being used.

When building a facility that has critical timing requirements, keep cables as short as possible to minimize signal attenuation and crosstalk. This requires keeping interconnected equipment as close together as possible. It is best to locate all of the distribution equipment in the same or adjacent racks. Because most video cabling among distribution elements must be timed or of matching lengths, short cables make the job manageable, and, at the same time, cable costs are kept low.

WIRE NO.	DESCRIPTION	CONN	SOURCE	DESTINATION	CONN	CABLE TYPE		CABLE TIMELENGTH	NO.OF WIRES	COLOR CODE
1031	BB1	BNC	DAO1 OUT 1	CCU1 GENLOCK IN	BNC	RK7560		26'	1	YEL
1032	BB1	BNC	DAO1 OUT 2	CCU2 GENLOCK IN	BNC	RK7560		26'	1	YEL
1033	BB1	BNC	DAO1 OUT 3	CCU3 GENLOCK IN	BNC	RK7560		26'	1	YEL
1034	BB1	BNC	DAO1 OUT 4	CCU4 GENLOCK IN	BNC	RK7560		26'	1	YEL
1035	BB	BNC	DAO2 IN LOOP	DA03 IN	BNC	RK7560		1'	1	YEL
1036	BB1	BNC	DAO2 OUT 1	P1 REF VID IN	BNC	RK7560		21'	1	YEL
1037	BB1	BNC	DAO2 OUT 2	P2 REF VID IN	BNC	RK7560		25'	1	YEL
1038	BB1	BNC	DAO2 OUT 3	P3 REF VID IN	BNC	RK7560		27'	1	YEL
1039	BB1	BNC	DAO2 OUT 4	P4 REF VID IN	BNC	RK7560		29'	1	YEL
1040	BB	BNC	DAO3 IN LOOP	DA04 IN	BNC	RK7560		1'	1	YEL
1041	BB1	BNC	DAO3 OUT 1	P5 EXT REF IN	BNC	RK7560		29'	1	YEL
1042	BB1	BNC	DAO3 OUT 2	P6 REF VID IN (FUTURE)	BNC	RK7560				
1043	BB1	BNC	DAO3 OUT 3	R1 REF VID IN	BNC	RK7560		29'	1	YEL
1044	BB1	BNC	DAO3 OUT 4	VTR H TBC COMP VID IN	BNC	RK7560		34'	1	YEL
1045	BB	BNC	DAO4 IN LOOP	DA05 IN	BNC	RK7560		1'	1	YEL
1046	BB1	BNC	DAO4 OUT 1	VTR I TBC GENLOCK IN	BNC	RK7560		34'	1	YEL
1047	BB1	BNC	DAO4 OUT 2	CG1 GEN LOCK VID IN	BNC	MINI		47'	10	1
1048	BB1	BNC	DAO4 OUT 3	CG2 ENC PGM IN	BNC	MINI		39'	10	1
1049	BB1	BNC	DAO4 OUT 4	CG3 GEN LOCK VID IN	BNC	RK7560		10'6"	1	YEL
1050	BB	BNC	DAO5 IN LOOP	DA51 IN	BNC	RK7560		6'	1	YEL
1051	BB1	BNC	DAO5 OUT 1	DVE VBS/S IN	BNC	RK7560		8'9"	1	YEL
1052	BB1	BNC	DAO5 OUT 2	TBC GENLOCK IN	BNC	RK7560		5'6"	1	YEL
1053	BB1	BNC	DAO5 OUT 3	TCG 2 VID IN	BNC	RK7560		5'5"	1	YEL
1054	BB1	BNC	DAO5 OUT 4	CG3 CC SYNC IN	BNC	RK7560				WHT
1055	TBC ADV SYNC OUT	BNC	VTR H TBC ADV SYNC OUT	VTR H SYNC IN	BNC	RK7560		4'6"	1	WHT
1056	TBC ADV SYNC OUT	BNC	VTR I TBC ADV SYNC OUT	VTR I VID IN 1	BNC	RK7560		4'6"	1	WHT
1057	SC TO VCR	BNC	TBC VTR SC OUT	VTR I SC IN	BNC	VP618PE		4'6"	1	BLK
1058	BB1	BNC	P5 REF VID IN LOOP OUT	P5 TBC COMP VID IN	BNC	RK7560		6'	1	YEL
1059										
1060	BB2	BNC	DA51 OUT 1	TEST SRCE SWR IN 1	BNC	RK7560	E	7'6"	1	YEL
1061	BB2	BNC	DA51 OUT 2	PS EXT REF IN	BNC	RK7560	G	25'7"	1	YEL

Figure 4.2 Wiring database printout for a portion of the facility illustrated in Figure 4.1.

Cable Loss and Equalization

The high frequency response of a cable decreases with increasing frequency. The loss can be compensated for by using an equalizing amplifier with a response curve that complements the cable loss. For video applications, a typical distribution amplifier (DA) has six outputs isolated from one another by fan-out resistors. Because the equalization is adjusted to produce a flat response at the end of a length of a specific type of cable, all of the cables being driven by the amplifier must be the same type and length.

4.6 Facility Design

The best way to design a facility is to begin with the architectural drawings of the existing building or planned construction. If architectural drawings are not available, it is necessary to have the architect prepare them. For small renovation projects, the system engineer may prepare the needed drawings to plan equipment layout.

Before any details are confirmed, a site visit should be made to record and confirm building space dimensions, clearances, and access to building services. Also, existing rack and console dimensions and locations should be measured. If the site is some distance away, photograph important elements, such as existing construction details or current equipment configurations, to reduce the need to travel back to the site.

4.6.1 Preliminary Space Planning

Whether the project involves new construction or renovation of an existing building, current facilities and equipment are reviewed to determine a starting point for the planning process. Building and room layouts are determined by studying each function and its relationship to all others. Functional requirements of each operational department are assessed to determine the gross space requirements of areas to be expanded or renovated. Key facility personnel are interviewed to determine past experiences, future trends, operational requirements for immediate use, and future needs of the facility. This should include the number of present employees and those anticipated in the future.

Environmental factors, such as noise, vibration, RF interference, power line interference, temperature, and humidity also must be considered. Accessibility to utilities, such as communications, power, air supply, fuel, and water, must be calculated. Air conditioning is a major concern in all large facilities that employ a lot of equipment or lighting.

After management approves the equipment list, a rough schematic layout is prepared in conjunction with the architect's preliminary drawings. The system engineer examines this layout to be certain that it provides adequate space for present and future equipment and for maintenance and operation. Equipment identification is then added to the architect's schematic, and the procedure continues to the design-development phase. An example of an architectural floor plan is shown in Figure 4.3. Equipment placement in rack assemblies is illustrated in Figure 4.4.

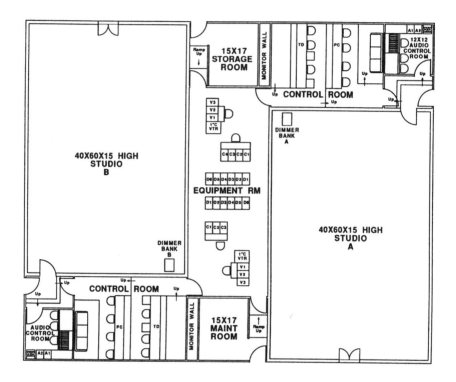

Figure 4.3 Architectural floor plan of a new facility.

Design renderings (drawings or paintings created by an artist or drafter to show a realistic flat or perspective view of a design) are then produced. Full-color 3-D models can be generated by a computer for viewing from different perspectives. The printout of any view can be used as the rendering.

A color and materials presentation board is usually prepared for review by decision makers. The presentation may include the following:

- Artist or computer renderings

- Color chips

- Wood types

- Work surface laminates

- Metal samples

- Samples of carpeting, furniture fabrics, and wall coverings

Several different combinations may be prepared. The samples and renderings are attached to a board or heavy paper stock for easy presentation.

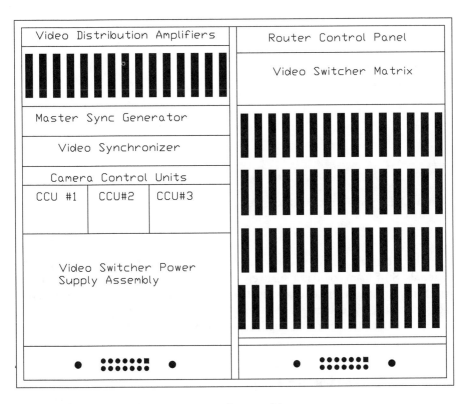

Figure 4.4 Equipment placement in a rack assembly.

4.6.2 Design Models and Mockups

When a drawing cannot be interpreted easily by the owner and/or staff, a scale model or full-size mockup of the facility (or portions of it) can be constructed. This will help familiarize them with the design, allowing them to make decisions and changes. Models can also be used to present a design concept to company executives. Models can provide a cost-effective way to evaluate new ideas. Inexpensive materials can be formed to represent racks, consoles, or equipment. For example, the top and four side views of an enclosure can be drawn or plotted at a reduced scale on stiff paper so that the drawings touch at adjoining surfaces. When cut out, they are glued together to form a 3-D model. Flaps added to the drawing make it easier to join the surfaces. The more detail provided in the mockup, the better.

Blocks of wood can also be cut to the shape of the equipment being modeled. Cut-out drawings of the equipment features are pasted on the block's surfaces to add realism. Plastic scale models of structural components, piping, furniture, and other elements are available from model manufacturers and can be used to enhance the presentation.

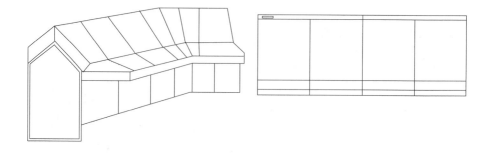

Figure 4.5 Equipment rack and console mockup for planning purposes.

Full-scale mockups, like models, can be built using any combination of construction materials. Stiff foam board is a relatively easy and inexpensive material to use to prepare full-size models. Pieces can be cut to any shape and joined to form 3-D models of racks, consoles, and equipment. (See Figure 4.5.) Actual-size drawings of equipment outlines, or more detailed representations, can then be pasted in place on the surfaces of the mockup.

4.6.3 Construction Considerations

Demolition and construction of existing structures may have to be specified by the system engineer. Electrical power, lighting, and air conditioning requirements must be identified and layout drawings prepared for use by the electrical and mechanical engineering consultants and the architect.

During preparation of final construction documents, the architect and the system engineer can confirm the layout of technical equipment wire ways, including access to flooring, conduits, trenches, and overhead raceways. At this point, the system engineer also provides layouts of cable runs and connections. This makes equipment installation and future changes much easier. An overhead cable routing plan is shown in Figure 4.6.

When it is necessary to install coaxial cables in conduit, follow National Electrical Code (NEC) requirements for conduit fill and the number of pull boxes. More pull boxes or larger conduit is required in conduit runs that have many bends. Specify direct-burial-type cable when the conduit or cable trays are underground, and where there is a possibility of standing water. Conduit and cable trays should be designed to accommodate the minimum bend radius requirements for the cables being used. The recommended minimum bend radius for coaxial cable, for a single permanent bend, is 10 times the cable diameter.

Debur and remove all sharp edges and splinters from installed conduit. Remove construction debris from inside the lines to prevent damage to the cable jacket during pulling. Cover openings to the conduit to prevent contamination or damage from other con-

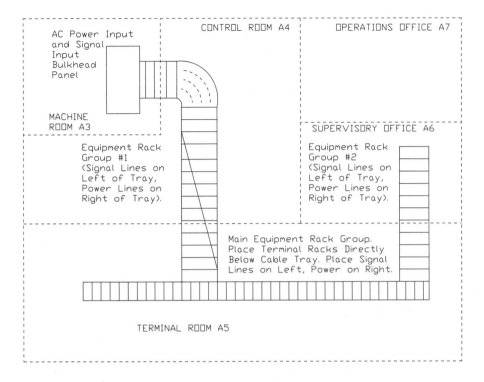

Figure 4.6 Cable wire routing plan for a new facility.

struction activities. If the cable is damaged during pulling, moisture could enter the cable.

4.6.4 Component Selection and Installation

Equipment selection is normally based on the function it will perform. User input about operational ease and flexibility of certain models is also important. However, to ensure that the most cost-effective choice is made, consider certain technical issues before making a decision. The system engineer should research, test (when required), and provide the technical input needed for selecting hardware and software.

Equipment features and functional capabilities are probably the main concerns of the users and management. Technical performance data and specifications are important considerations that should be contributed by the system engineer during the selection process. For a piece of equipment to qualify, technical specifications must be checked to ensure that they meet set standards for the overall system. Newly introduced products should be tested and compared before a decision is made. The experienced system engineer can test and measure the equipment to evaluate its performance.

A simple visual inspection inside a piece of equipment by an experienced technician or engineer can uncover possible weaknesses, design flaws, and problem areas that may affect reliability or make maintenance difficult. It is therefore advisable to request a sample of the equipment for evaluation before committing to its use.

The availability of replacement parts is another important consideration when specifying products. A business that depends on its equipment functioning to specifications requires that the service technician be able to make repairs when needed in a timely manner. Learn about manufacturer replacement parts policies and their reliability. When possible, select equipment that uses standard off-the-shelf components that are available from multiple sources. Avoid equipment that incorporates custom components that are available only from the equipment manufacturer. This will make it easier to acquire replacements, and the cost of the parts will—most likely—be less.

When possible, specify products manufactured by the same company. Avoid mixing brands. Maintenance technicians will more easily become familiar with equipment maintenance, and experience gained while repairing one piece of hardware can be directly applied to another of the same model. Service manuals published by the same manufacturer will be similar and therefore easier to understand and use to locate the information or diagram needed for a repair.

Commonality of replacement parts will keep the parts inventory requirements and the inventory cost low. Because the technical staff will be dealing with the manufacturer on a regular basis, familiarity with the company's representatives makes it easier to get technical support quickly.

Sometimes components are selected that are not really compatible, such as differing signal levels or impedances. The responsibility then falls on the system engineer to devise a fix to make that component compatible with the rest of the system. The component may have been originally selected because of its low price, but additional components, engineering, and labor costs often offset the expected savings. Extra wiring and components can also clutter the equipment enclosure, hampering access to the equipment inside. Nonstandard mounting facilities on equipment can add unnecessary cost and can result in a less than elegant solution.

4.7 Technical Documentation

Engineering documentation describes the practices and procedures used within the industry to specify a design and communicate the design requirements to technicians and contractors. Documentation preparation should include, but not be limited to, the generation of technical system flow diagrams, material and parts lists, custom item fabrication drawings, and rack and console elevations. The required documents include the following:

- Documentation schedule

- Signal flow diagram

- Equipment schedule

- Cable schedule

- Patch panel assignment schedule

- Rack elevation drawing

- Construction detail drawing

- Console fabrication mechanical drawing

- Duct and conduit layout drawing

- Single-line electrical flow diagram

4.7.1 Documentation Tracking

The documentation schedule provides a means of keeping track of the project's paperwork. During engineering design, drawings are reviewed, and changes are made. A system for efficiently handling changes is essential, especially on big projects that require a large amount of documentation.

Completed drawings are submitted for management approval. A set of originals is signed by the engineers and managers who are authorized to check the drawings for correctness and to approve the plans.

4.7.2 Symbols

Because there are only limited informal industry standards for the design of electronic component symbols to represent equipment and other elements in a system, custom symbols are usually created by the designer. Each organization develops its own symbols. The symbols that exist apply to component-level devices, such as integrated circuits, resistors, and diodes. Some common symbols apply to system-level components, such as amplifiers and speakers. Figure 4.7 shows some of the more common component-level symbols currently used in electronics.

The proliferation of manufacturers and equipment types makes it impractical to develop a complete library, but, by following basic rules for symbol design, new component symbols can be produced easily as they are added to the system.

For small systems built with a few simple components, all of the input and output signals can be included on one symbol. However, when the system uses complex equipment with many inputs and outputs with different types of signals, it is usually necessary to draw different diagrams for each type of signal. For this reason, each component requires a set of symbols, with a separate symbol assigned for each signal type, showing its inputs and outputs only. For example, a videotape recorder (VTR) will require a set of symbols for audio, video, sync, time code, and control signal.

If abbreviations are used, be consistent from one drawing to the next, and develop a dictionary of abbreviations for the drawing set. Include the dictionary with the documentation.

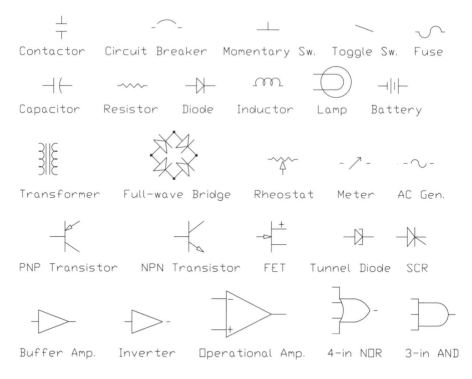

Figure 4.7 Schematic representations of common electrical components and devices.

4.7.3 Cross-Referencing Documentation

In order to tie all of the documentation together and to enable fabricators and install-ers to understand the relationships between the drawings, the documents should in-clude reference designations common to the project. That way, items on one type of document can be located on another type. For example, the location of a piece of equipment can be indicated at its symbol on the flow diagram so that the technician can identify it on the rack elevation drawing and in the actual rack.

A flow diagram is used by the installation technicians to assemble and wire the sys-tem components together. All necessary information must be included to avoid confu-sion and delays. When designing a symbol to represent a component in a flow diagram, include all of the necessary information to identify, locate, and wire that component into the system. The information should include the following:

- Generic description of the component or its abbreviation. When no abbreviation exists, create one. Include it in the project manual reference section and in the notes on the drawing.

- Model number of the component.

- Manufacturer of the component.

- All input and output connections with their respective name and/or number.

4.7.4 Specifications

Specifications are a compilation of knowledge about how something should be done. An engineer condenses years of personal experience, and that of others, into the specification. The more detailed the specification, the higher the probability that the job will be done right.

The *project manual* is the document where specifications and other printed project documentation is compiled.

4.7.5 Working with the Contractors

The system engineer must provide support and guidance to contractors during the procurement, construction, installation, testing, and acceptance phases of a project. The system engineer can assist in ordering equipment and can help coordinate the move to a new or renovated facility. This can be critical if a great deal of existing equipment is being relocated. In the case of new equipment, the system engineer's knowledge of prices, features, and delivery times is invaluable to the facility owner.

The steps to assure quality workmanship from contractors on a job include the following:

- Clarify details.

- Clarify misunderstandings about the requirements.

- Resolve problems that may arise.

- Educate contractors about the requirements of the project.

- Assure that the work conforms to the specifications.

- Evaluate and approve change requests.

- Provide technical support to the contractors when needed.

4.7.6 Computer-Based Tools

Technology is evolving so rapidly that it takes a great deal of time just to keep up with the changes. Competition forces change and improvements that would otherwise not take place at such a rapid pace. In this environment, engineering skills must be augmented with tools to speed the design process. Many of the tasks required of the system engineer can be accelerated and the resulting documentation enhanced with the aid of computer programs. Computer aided design (CAD) tools include application software from simple word processing and spreadsheet programs to complex simulation, 3-D graphic modeling, and artificial intelligence. Software is commonly used in new construction and renovations to perform the following tasks:

- Document tracking
- Documentation preparation
- Correspondence
- Report generation
- Technical manual publication
- List management
- Mechanical design
- Electrical design
- Schematic capture

4.8   Professional Association Directory

Most professional organizations in the electronics industry have established standards or recommended practices that can serve as a guide to the system engineer planning a new facility or modifying an existing one. The following list includes the primary organizations involved in the disciplines covered in this book.

Advanced Television Systems Committee (ATSC)
    1776 K Street, N.W.
    Washington, DC 20006
    USA
    202-828-3130
American National Standards Institute (ANSI)
    655 15th Street, N.W.
    Suite 300
    Washington, DC 20005
    USA
    202-639-4090
Audio Engineering Society (AES)
    60 East 42nd Street
    New York, NY 10165
    USA
    212-661-2355
Cable Television Laboratories (CableLabs)
    1050 Walnut Street
    Suite 500
    Boulder, CO 80302
    USA
    303-939-8500
Department of Trade and Industry (DoTI)
    Radio Regulatory Division

Ashdown House
123 Victoria Street
London, SW1E 6RB
England
+44-71-215-5000

Electronic Industries Association (EIA)
Engineering Department
2001 I Street, N.W.
Washington DC 20006
USA
202-457-4971

European Broadcasting Union (EBU)
Technical Centre
Avenue Albert Lancaster, 32
B-1 180
Brussels, Belgium
+32-2-375-5990

Federal Communications Commission (FCC)
1919 M Street, N.W.
Washington DC 20554
USA
202-653-8247

Illumination Engineering Society of North America (IES)
345 East 47th Street
New York, NY 10017
USA
212-705-7926

Independent Television Commission (ITC)
70, Brompton Road
Knightsbridge
London SW3 1 EY
England
+44-71-584 7011

Institute of Electrical and Electronics Engineers (IEEE)
345 East 47th Street
New York, NY 10017
USA
212-705-7900

Institution of Electrical Engineers (IEE)
P.O. Box 96
Michael Faraday House
Six Hills Way
Stevenage, Herts. SG1 2SD
England
+44-438-313311

International Electrotechnical Commission (IEC)
3, rue de Varembe
P.O. Box 131
1211 Geneva 20
Switzerland
+41-22-34-01-50

International Society for Optical Engineering (SPIE)
P.O. Box 10
Bellingham, WA 98227-0010
USA
206-676-3290

International Telecommunications Union (ITU)
International Radio Consultative Committee
Place des Nations
1211 Geneva 20
Switzerland
+41-22-99-51-11

National Association of Broadcasters (NAB)
1771 N Street, N.W.
Washington, DC 20036
USA
202-429-5300

National Cable Television Association (NCTA)
1724 Massachusetts Avenue, N.W.
Washington, DC 20036
USA
202-775-3550

National Institute of Standards and Technology (NIST)
Department of Commerce
Gaithersburg, MD 20899
USA
202-921-1000

National Telecommunications and Information Administration (NTIA)
14th Street and Constitution Avenue, N.W.
Washington, DC 20203
USA
202-337-1551

National Transcommunications, Ltd. (NTL)
Crawley Court
Winchester
Hampshire S021 2QA
England
+44 962 823434

Society of Broadcast Engineers (SBE)
P.O. Box 20450
Indianapolis, IN 46220
USA
317-253-1640
Society of Motion Picture and Television Engineers (SMPTE)
595 West Heartsease Avenue
White Plains, NY 10607-1824
USA
914-761-1100
Telecommunications Industry Association (TIA)
1722 I Street, N.W.
Suite 440
Washington, DC 20006
USA
202-457-4936
Underwriters Laboratories (UL)
333 Pfingsten Road
Northbrook, IL 60062
USA
312-272-8800

4.9 Bibliography

DeSantis, Eugene, "Planning for Facility Construction," in *Interconnecting Electronic Systems*, Jerry C. Whitaker, Eugene DeSantis and C. Robert Paulson, CRC Press, 1992.

5

Facility Construction Issues

5.1 Introduction

There are a number of important elements that go into designing and building a successful electronics facility. Without doubt, the most critical issues include grounding, power distribution, cooling, and equipment racks. Each of these will be examined, in order.

5.2 Facility Grounding

The primary purpose of grounding electronic hardware is to prevent electrical shock. The National Electrical Code (NEC) and local building codes are designed to provide for the safety of the workplace. Local codes must always be followed. Occasionally, code sections are open for some interpretation. If in doubt, consult a field inspector.

Codes constantly are being changed or expanded, because new situations arise that were not anticipated when the codes were written. Sometimes, an interpretation will depend on whether the governing safety standard applies to building wiring or to a factory-assembled product to be installed in a building. Underwriters Laboratories (UL) and other qualified testing organizations examine products at the request and expense of manufacturers or purchasers. They "list" products if the examination reveals that the device or system presents no significant safety hazard when installed and used properly.

Municipal and county safety inspectors generally accept UL and other qualified testing laboratory certification listings as evidence that a product is safe to install. Without a listing, the end user may not be able to obtain the necessary wiring permits and inspection sign-off. On-site wiring must conform to local wiring codes. Most codes are based on the NEC. Electrical codes specify wiring materials, wiring devices, circuit protection, and wiring methods.

5.2.1 Planning the Ground System

The attention given to the design and installation of a facility ground system is a key element in the day-to-day reliability of the plant. A well-designed and -installed ground network is invisible to the engineering staff. A marginal ground system, however, can cause problems on a regular basis. Grounding schemes range from simple to complex, but any system serves three primary purposes:

- It provides for operator safety.

- It protects electronic equipment from damage caused by transient disturbances.

- It diverts stray RF energy from sensitive audio, video, control, and computer equipment.

Most engineers view grounding mainly as a method to protect equipment from damage or malfunction. However, the most important element is operator safety. The 120 or 208 Vac line current that powers most equipment can be dangerous—even deadly—if handled improperly. Grounding of equipment and structures provides protection against wiring errors or faults that could endanger human life.

Proper grounding is basic to protection against ac line disturbances. This applies whether the source of the disturbance is lightning, power-system switching activities, or faults in the distribution network. Proper grounding is also a key element in preventing RF interference in transmission or computer equipment. A facility with a poor ground system may experience RFI problems on a regular basis.

Implementing an effective ground network is not an easy task. It requires planning, quality components, and skilled installers. It is not inexpensive. However, proper grounding is an investment that will pay off in facility reliability.

A ground system consists of two key elements:

- The earth-to-grounding electrode interface outside the facility.

- The ac power and signal-wiring systems inside the facility.

5.2.2 Establishing an Earth Ground

The grounding electrode is the primary element of any ground system. The electrode can take many forms. In all cases, its purpose is to interface the electrode (a conductor) with the earth (a semiconductor). Grounding principles have been refined to a science. Still, however, many misconceptions exist. An understanding of proper grounding procedures begins with the basic earth-interface mechanism.

Grounding Interface

The grounding electrode (or ground rod) interacts with the earth to create a hemisphere-shaped volume. This is illustrated in Figure 5.1. The size of this volume is related to the size of the grounding electrode. The length of the electrode has a much greater effect than the diameter. Studies have demonstrated that the earth-to-electrode resistance from a driven ground rod increases exponentially with the distance from that rod. At a given point, the change becomes insignificant. It has been found that for maximum effectiveness of the earth-to-electrode interface, each ground rod must

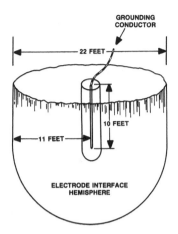

Figure 5.1 The effective earth-interface hemisphere resulting from a single driven ground rod. The 90 percent effective area of the rod extends to a radius of approximately 1.1 times the length of the rod. (*After* [1].)

have a hemisphere-shaped volume with a diameter that is approximately 2.2 times the rod's length.

The constraints of economics and available real estate place practical limitations on the installation of a ground system. It is important, however, to keep the 2.2 rule in mind, because it allows the facility design engineer to take advantage of the available resources. Figure 5.2 illustrates the effects of locating ground rods too close (less than 2.2 times the rod length). An overlap area is created that effectively wastes some of the earth-to-electrode capabilities of the two ground rods. Research has shown, for example, that two 10 ft ground rods driven only 1 ft apart provide about the same resistivity as a single 10 ft rod.

There are two schools of thought with regard to ground-rod length. The first is that extending ground-rod length beyond about 10 ft is of little value for most types of soil. The reason is presented in Figure 5.3, where ground resistance is plotted as a function of ground-rod length. Beyond 10 ft, a point of diminishing returns is reached. The second school of thought is that optimum earth-to-electrode interface is achieved with long (40 ft or greater) rods, driven to penetrate the local water table. With this type of installation, consider the difficulty in attempting to drive long ground rods. This discussion assumes that the composition of the soil around the grounding electrode is reasonably uniform. Depending on the location, however, this may not be the case.

Horizontal grounding electrodes provide essentially the same resistivity as an equivalent-length vertical electrode. As Figure 5.4 demonstrates, the difference between a 10 ft vertical and a 10 ft horizontal ground rod is negligible (275 Ω vs. 250 Ω). This comparison includes the effects of the vertical connection element from the surface of the ground to the horizontal rod. By itself, the horizontal ground rod provides an earth-interface resistivity of approximately 308 Ω when buried at a depth of 36-in.

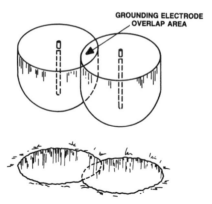

Figure 5.2 The effect of overlapping earth interface hemispheres by placing two ground rods at a spacing less than 2.2 times the length of either rod. The overlap area represents wasted earth-to-grounding electrode interface capability. (*After* [1].)

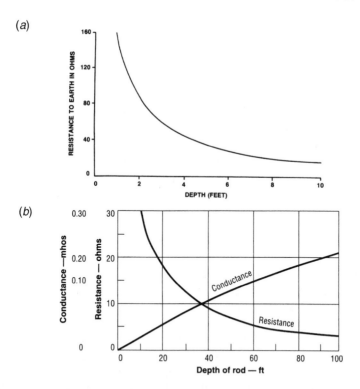

Figure 5.3 Charted grounding resistance as a function of ground-rod length: (*a*) data demonstrating that ground-rod length in excess of 10 ft produces diminishing returns (1-in.-diameter rod) [1], (*b*) data demonstrating that ground system performance continues to improve as depth increases. (*Chart* b *from* [2]. *Used with permission.*)

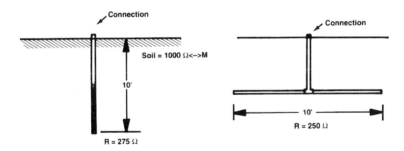

Figure 5.4 The effectiveness of vertical ground rods compared with horizontal ground rods. (*After* [1].)

Ground rods come in many sizes and lengths. The more popular sizes are 1/2-, 5/8-, 3/4-, and 1-in. The 1/2-in size is available in steel, with stainless-clad, galvanized, or copper-clad rods. All-stainless-steel rods also are available. Ground rods can be purchased in unthreaded or threaded (sectional) lengths. The sectional sizes are typically 9/16- or 1/2-in rolled threads. Couplers are made from the same materials as the rods. These couplers can be used to join 8 or 10 ft rods together. A 40 ft ground rod is driven one 10 ft section at a time.

The type and size of ground rod used is determined by how many sections are to be connected and how hard or rocky the soil is. Copper-clad 5/8-in × 10 ft rods are probably the most popular. Rod diameter has minimal effect on final ground impedance. Copper cladding is designed to prevent rust, not for better conductivity. Although the copper certainly provides a better conductor interface to earth, the steel that it covers is also an excellent conductor when compared with ground conductivity. The thickness of the cladding is important only as far as rust protection is concerned.

Soil Resistivity

Wide variations in soil resistivity can be found within a given geographic area, as documented in Table 5.1. The wide range of values results from differences in moisture content and mineral content, and from temperature.

Chemical Ground Rods

A chemically activated ground system is a common alternative to the conventional ground rod. The idea behind it is to increase the earth-to-electrode interface by conditioning the soil surrounding the rod. Experts have known for many years that the addition of ordinary table salt (NaCl) to soil will reduce the resistivity of the earth-to-ground electrode interface. With the proper soil moisture level (4 to 12 percent), *salting* can reduce soil resistivity from 10,000 Ω/m to less than 100 Ω/m. Salting the area surrounding a ground rod (or group of rods) follows a predictable

Table 5.1 Typical Resistivity of Common Soil Types

Type of soil	Resistivity in Ω/cm		
	Average	Minimum	Maximum
Filled land, ashes, salt marsh	2,400	600	7,000
Top soils, loam	4,100	340	16,000
Hybrid soils	16,000	1,000	135,000
Sand and gravel	90,000	60,000	460,000

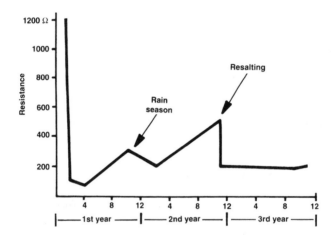

Figure 5.5 The effect of soil salting on ground-rod resistance with time. The expected resalting period, shown here as two years, varies depending on the local soil conditions and the amount of moisture present.

life-cycle pattern, which is illustrated in Figure 5.5. Subsequent salt applications are rarely as effective as the initial salting.

Various approaches have been tried over the years to solve this problem. One product is shown in Figure 5.6. This chemically activated grounding electrode consists of a 2 ½-in-diameter copper pipe filled with rock salt. Breathing holes are provided on the top of the assembly, and seepage holes are located at the bottom. The theory of operation is simple: Moisture is absorbed from the air (when available) and is then absorbed by the salt. This creates a solution that seeps out of the base of the device and conditions the soil in the immediate vicinity of the rod.

Another approach is shown in Figure 5.7. This device incorporates a number of ports (holes) in the assembly. Moisture from the soil (and rain) is absorbed through the ports. The metallic salts subsequently absorb the moisture, forming a saturated solution

(a)   (b)

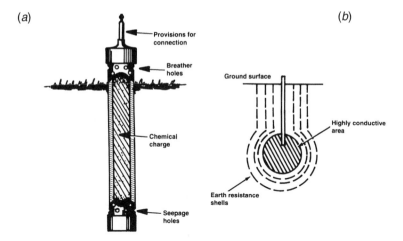

Figure 5.6 An air-breathing chemically activated ground rod: (a) breather holes at the top of the device permit moisture penetration into the chemical charge section of the rod; (b) a salt solution seeps out of the bottom of the unit to form a conductive shell. (*After* [1].)

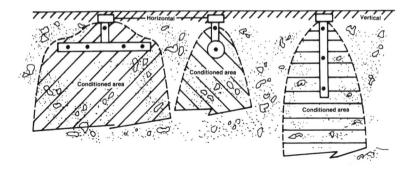

Figure 5.7 An alternative approach to the chemically activated ground rod. Multiple holes are provided on the ground-rod assembly to increase the effective earth-to-electrode interface. Note that chemical rods can be produced in a variety of configurations. (*After* [1].)

that seeps out of the ports and into the earth-to-electrode hemisphere. Tests have shown that if the moisture content is within the required range, earth resistivity can be reduced by as much as 100:1. Figure 5.8 shows the measured performance of a typical chemical ground rod in three types of soil.

Implementations of chemical ground-rod systems vary depending on the application. Figure 5.9 illustrates a counterpoise ground consisting of multiple leaching aper-

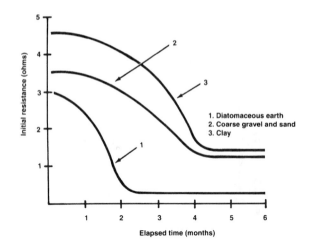

Figure 5.8 Measured performance of a chemical ground rod. (*After* [1].)

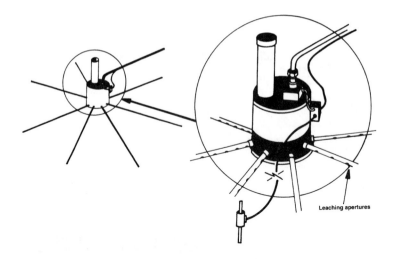

Figure 5.9 Hub and spoke counterpoise ground system.

tures connected to a central hub in a spoke fashion. The system is serviceable because additional salt compound can be added to the hub at required intervals to maintain the effectiveness of the ground. Figure 5.10 shows a counterpoise system consisting of individual chemical ground rods interconnected with radial wires that are buried below the surface.

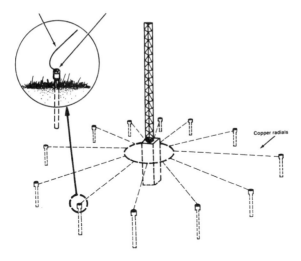

Figure 5.10 Tower grounding scheme using buried copper radials and chemical ground rods.

Ufer Ground System

Driving ground rods is not the only method of achieving a good earth-to-electrode interface. The concept of the *Ufer ground* has gained interest because of its simplicity and effectiveness. The Ufer approach (named for its developer), however, must be designed into a new structure. It cannot be added later. The Ufer ground takes advantage of the natural chemical- and water-retention properties of concrete to provide an earth ground. Concrete retains moisture for 15 to 30 days after a rain. The material has a ready supply of ions to conduct current because of its moisture-retention properties, mineral content, and inherent pH. The large mass of any concrete foundation provides a good interface to ground.

A Ufer system, in its simplest form, is made by routing a solid-copper wire (no. 4 gauge or larger) within the foundation footing forms before concrete is poured. Figure 5.11 shows one such installation. The length of the conductor run within the concrete is important. Typically, a 20 ft run (10 ft in each direction) provides a 5 Ω ground in 1,000 Ω/m soil.

As an alternative, steel reinforcement bars (rebar) can be welded together to provide a rigid, conductive structure. A ground lug is provided to tie equipment to the ground system in the foundation. The rebar must be welded, not tied, together. If it is only tied, the resulting poor connections between rods can result in arcing during a current surge. This can lead to deterioration of the concrete in the affected areas.

The design of a Ufer ground is not to be taken lightly. Improper installation can result in a ground system that is subject to problems. The grounding electrodes must be kept a minimum of 3-in from the bottom and sides of the concrete to avoid the possibil-

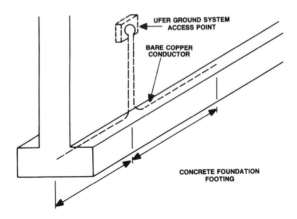

Figure 5.11 The basic concept of a Ufer ground system, which relies on the moisture-retentive properties of concrete to provide a large earth-to-electrode interface. Design of such a system is critical. Do not attempt to build a Ufer ground without the assistance of an experienced contractor.

ity of foundation damage during a large lightning strike. If an electrode is placed too near the edge of the concrete, a surge could turn the water inside the concrete to steam and break the foundation apart.

The Ufer approach also can be applied to guy-anchor points or the tower base, as illustrated in Figure 5.12. Welded rebar or ground rods sledged in place after the rebar cage is in position may be used. By protruding below the bottom concrete surface, the ground rods add to the overall electrode length to help avoid thermal effects that may crack the concrete. The maximum length necessary to avoid breaking the concrete under a lightning discharge is determined by the following:

- Type of concrete (density, resistivity, and other factors)
- Water content of the concrete
- How much of the buried concrete surface area is in contact with the ground
- Ground resistivity
- Ground water content
- Size and length of the ground rod
- Size of lightning flash

The last variable is a gamble. The 50 percent mean occurrence of lightning strikes is 18 A, but super strikes can occur that approach 100 to 200 kA.

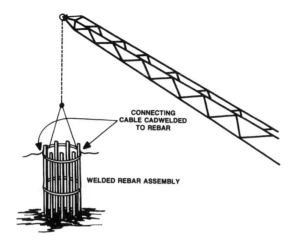

Figure 5.12 The Ufer ground system as applied to a transmission-tower base or guy-wire anchor point. When using this type of ground system, bond all rebar securely to prevent arcing in the presence of large surge currents.

Before implementing a Ufer ground system, consult a qualified contractor. Because the Ufer ground system will be the primary grounding element for the facility, it must be installed correctly.

5.2.3 Bonding Ground-System Elements

A ground system is only as good as the methods used to interconnect the component parts. Do not use soldered-only connections outside the equipment building. Crimped/brazed and *exothermic* (*Cadwelded*) connections are preferred. To make a proper bond, all metal surfaces must be cleaned, any finish removed to bare metal, and surface preparation compound applied. Protect all connections from moisture by appropriate means—usually sealing compound and heat-shrink tubing.

It is not uncommon for an untrained installer to use soft solder to connect the elements of a ground system. Such a system is doomed from the start. Soft-soldered connections cannot stand up to the acid and mechanical stress imposed by the soil. The most common method of connecting the components is silver-soldering. The process requires the use of brazing equipment, which may be unfamiliar to many system engineers. The process uses a high-temperature/high-conductivity solder to complete the bonding process. For most grounding systems, however, the best bonding approach is the Cadwelding process. (Cadweld is a registered trademark of Erico Corporation.)

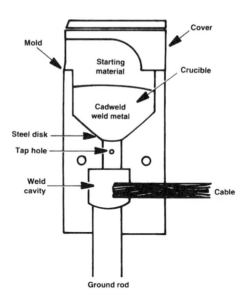

Figure 5.13 Typical Cadweld mold for connecting a cable to a ground rod.

Cadwelding

Cadwelding is the preferred method of connecting the elements of a ground system. Molten copper is used to melt connections together, forming a permanent bond. This process is particularly useful in joining dissimilar metals. In fact, if copper and galvanized cable must be joined, Cadwelding is the only acceptable means. The completed connection will not loosen or corrode and will carry as much current as the cable connected to it.

Cadwelding is accomplished by dumping powdered metals (copper oxide and aluminum) from a container into a graphite crucible and igniting the material with a flint lighter. Reduction of the copper oxide by the aluminum produces molten copper and aluminum oxide slag. The molten copper flows over the conductors, bonding them together.

Figure 5.13 shows a typical Cadweld mold. A variety of special-purpose molds are available to join different-size cables and copper strap. Figure 5.14 shows a bonding form for a copper-strap-to-ground-rod interface.

Ground-System Inductance

Conductors interconnecting sections or components of an earth ground system must be kept as short as possible to be effective. The inductance of a conductor is a major factor in its characteristic impedance to surge energy. For example, consider a no. 6 AWG copper wire 10 m long. The wire has a dc resistance of 0.013 Ω and an induc-

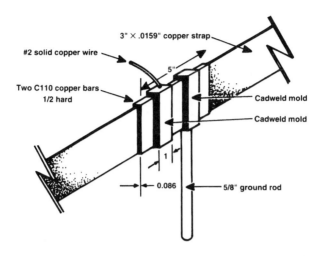

Figure 5.14 Cadweld mold for connecting a copper strap to a ground rod.

tance of 10 μH. For a 1,000 A lightning surge with a 1 μs rise time, the resistive volt-age drop will be 13 V, but the reactive voltage drop will be 10 kV. Further, any bends in the conductor will increase its inductance and further decrease the effectiveness of the wire. Bends in ground conductors should be gradual. A 90° bend is electrically equivalent to a 1/4-turn coil. The sharper the bend, the greater the inductance.

Because of the fast rise time of most lightning discharges and power-line transients, the *skin effect* plays an important role in ground-conductor selection. When planning a facility ground system, view the project from an RF standpoint.

5.2.4 Designing a Building Ground System

After determining the required grounding elements, they must be connected together in a unified system. Many different approaches can be taken, but the goal is the same: Establish a low-resistance, low-inductance path to surge energy. Figure 5.15 shows a building ground system using a combination of ground rods and buried bare-copper radial wires. This design is appropriate when the building is large or when it is located in an urban area. This approach also may be used when the facility is located in a high-rise building that requires a separate ground system. Most newer office build-ings have ground systems designed into them. If a comprehensive building ground system is provided, use it. For older structures (constructed of wood or brick), a sepa-rate ground system is required.

Figure 5.16 shows another approach in which a perimeter ground strap is buried around the building and ground rods are driven into the earth at regular intervals (2.2 times the rod length). The ground ring consists of a one-piece copper conductor that is bonded to each ground rod.

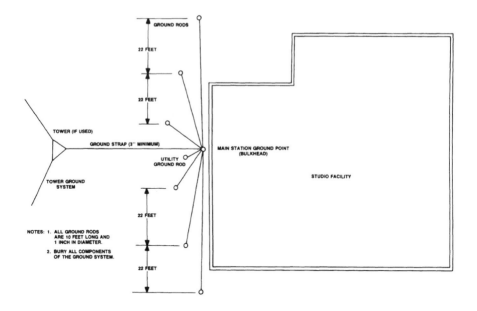

Figure 5.15 A facility ground system using the hub-and-spoke approach. The available real estate at the site will dictate the exact configuration of the ground system. If a tower is located at the site, the tower ground system is connected to the building ground, as shown.

If a transmission or microwave tower is located at the site, connect the tower ground system to the main ground point via a copper strap. The width of the strap must be at least 1 percent of the length and, in any event, not less than 3-in wide. The building ground system is not a substitute for a tower ground system, no matter the size of the tower. The two systems are treated as independent elements, except for the point at which they interconnect.

Tie the utility company power system ground rod to the main facility ground point as required by the local electrical code. Do not consider the building ground system to be a substitute for the utility company ground rod. The utility rod is important for safety reasons and must not be disconnected or moved. Do not remove any existing earth ground connections to the power line neutral connection. Doing so may violate local electrical code.

Bury all elements of the ground system to reduce the inductance of the overall network. Do not make sharp turns or bends in the interconnecting wires. Straight, direct wiring practices reduce the overall inductance of the system and increase its effectiveness in shunting fast rise-time surges to earth. Figure 5.17 illustrates the interconnection of a tower and building ground system.

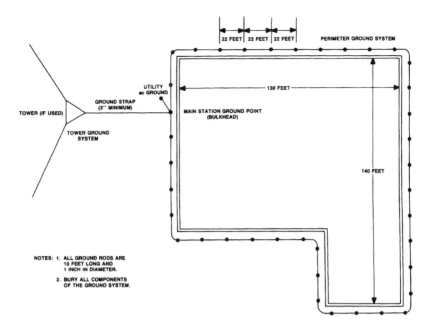

Figure 5.16 Facility ground using a perimeter ground-rod system. This approach works well for buildings with limited available real estate.

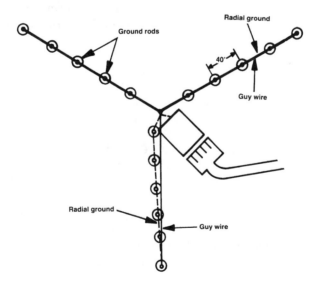

Figure 5.17 A typical guy-anchor and tower-radial grounding scheme. The radial ground is no. 6 copper wire. The ground rods are 5/8 in x 10 ft.

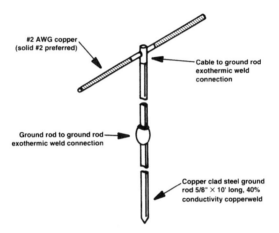

Figure 5.18 Preferred bonding method for below-grade elements of the ground system.

In most areas, soil conductivity is high enough to permit rods to be connected with no. 6 or larger bare-copper wire. In areas of sandy soil, use copper strap. A wire buried in low-conductivity, sandy soil tends to be inductive and less effective in dealing with fast-rise-time current surges. Again, make the width of the ground strap at least 1 percent of its overall length. Connect buried elements of the system, as shown in Figure 5.18.

Bulkhead Panel

The *bulkhead panel* is the cornerstone of an effective facility grounding system. The concept of the bulkhead is simple: Establish one reference point to which all cables entering and leaving the equipment building are grounded and to which all transient-suppression devices are mounted. Figure 5.19 shows a typical bulkhead installation for a communications facility. The panel size depends on the spacing, number, and dimensions of the coaxial lines, power cables, and other conduit entering or leaving the building.

To provide a weatherproof point for mounting transient-suppression devices, the bulkhead can be modified to accept a subpanel, as shown in Figure 5.20. The subpanel is attached so that it protrudes through an opening in the wall and creates a secondary plate on which transient suppressors are mounted and grounded. A typical cable/suppressor-mounting arrangement for a communications site is shown in Figure 5.21. To handle the currents that may be experienced during a lightning strike or large transient on the utility company ac line, the bottom-most subpanel flange (which joins the subpanel to the main bulkhead) must have a total surface-contact area of at least 0.75-in^2 per transient suppressor.

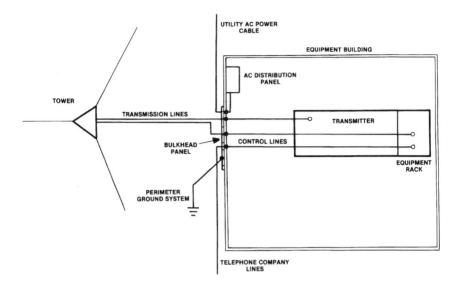

Figure 5.19 The basic design of a bulkhead panel for a facility. The bulkhead establishes the grounding reference point for the plant.

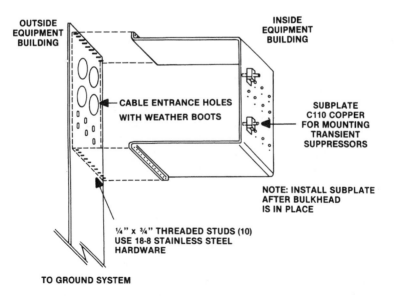

Figure 5.20 The addition of a subpanel to a bulkhead as a means of providing a mounting surface for transient-suppression components. To ensure that the bulkhead is capable of handling high surge currents, use the hardware shown.

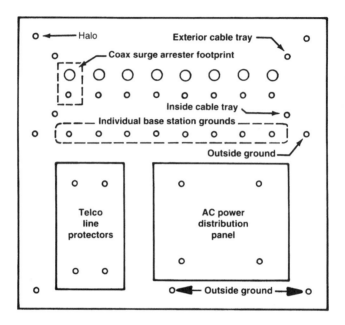

Figure 5.21 Mounting-hole layout for a communications site bulkhead subpanel.

Because the bulkhead panel carries significant current during a lightning strike or ac line disturbance, it must be constructed of heavy material. The recommended material is 1/8-in C110 (solid copper) 1/2 hard. Use 18-8 stainless-steel mounting hardware to secure the subpanel to the bulkhead.

Because the bulkhead panel establishes the central grounding point for all equipment within the building, it must be tied to a low-resistance (and low-inductance) perimeter ground system. The bulkhead establishes the main facility ground point, from which all grounds inside the building are referenced. A typical bulkhead installation for a small communications site is shown in Figure 5.22.

Bulkhead Grounding

A properly installed bulkhead panel will exhibit lower impedance and resistance to ground than any other equipment or cable-grounding point at the facility. Because the bulkhead panel will be used as the central grounding point for all of the equipment inside the building, the lower the inductance to the perimeter ground system, the better. The best arrangement is to simply extend the bulkhead panel down the outside of the building, below grade, to the perimeter ground system. This approach is illustrated in Figure 5.23.

If cables are used to ground the bulkhead panel, secure the interconnection to the outside ground system along the bottom section of the panel. Use multiple no. 1/0 or

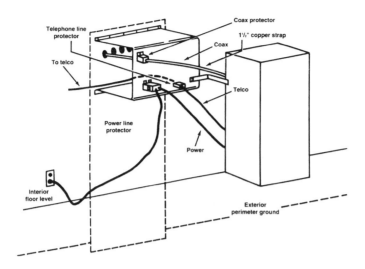

Figure 5.22 Bulkhead installation at a small communications site.

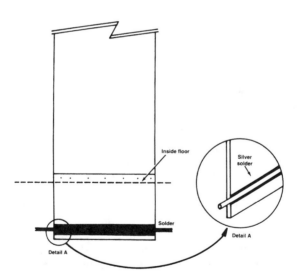

Figure 5.23 The proper way to ground a bulkhead panel and provide a low-inductance path for surge currents stripped from cables entering and leaving the facility. The panel extends along the building exterior to below grade. It is silver-soldered to a no. 2/0 copper wire that interconnects with the outside ground system.

larger copper wire or several solid-copper straps. If using strap, attach with stainless-steel hardware, and apply joint compound for aluminum bulkhead panels. Clamp and Cadweld, or silver-solder, for copper/brass panels. If no. 1/0 or larger wire is used, employ crimp lug and stainless-steel hardware. Measure the dc resistance. It should be no greater than 0.01 Ω between the ground system and the panel. Repeat this measurement on an annual basis.

5.2.5 Checklist for Proper Grounding

A methodical approach is necessary in the design of a facility ground system. Consider the following points:

1. Install a bulkhead panel to provide mechanical support, electric grounding, and lightning protection for coaxial cables, power feeds, and telephone lines entering the equipment building.
2. Install an internal ground bus using no. 2 or larger solid-copper wire. (At transmission facilities, use copper strap that is at least 3-in wide.) Form a star grounding system. At larger installations, form a star-of-stars configuration. Do not allow ground loops to exist in the internal ground bus. Connect the following items to the building internal ground system:

   - Chassis racks and cabinets of all hardware

   - All auxiliary equipment

   - Battery charger

   - Switchboard

   - Conduit

   - Metal raceway and cable tray

3. Connect outside metal structures to the earth ground array (towers, metal fences, metal buildings, and guy-anchor points).
4. Connect the power-line ground to the array. Strictly follow local electrical code.
5. Connect the bulkhead to the ground array through a low-inductance, low-resistance bond.
6. Do not use soldered-only connections outside the equipment building. Crimped, brazed, and exothermic (Cadwelded) connections are preferable. For a proper bond, all metal surfaces must be cleaned, finishes removed to bare metal, and surface preparation compound applied (where necessary). Protect all connections from moisture by appropriate means (sealing compound and heat sink tubing).

5.3   AC Power Distribution and Control

All ac wiring within a facility should be performed by an experienced electrical contractor, and always fully within the local electrical code. Confirm that all wiring is sized properly for the load current. Table 5.2 lists the physical characteristics for various wire sizes. The current-carrying capability (ampacity) of single conductors in

Table 5.2 Physical Characteristics of Standard Sizes of Copper Cable (at 25° C)

Wire size (AWG/ MCM)	Area (cmil)	Number of Conductors	Diameter Each Conductor	DC resistance Ω/1,000 ft	
				Copper	Aluminum
12	6,530	1	0.0808	1.62	2.66
10	10,380	1	0.1019	1.018	1.67
8	16,510	1	0.1285	0.6404	1.05
6	26,240	7	0.0612	0.410	0.674
4	41,740	7	0.0772	0.259	0.424
3	52,620	7	0.0867	0.205	0.336
2	66,360	7	0.0974	0.162	0.266
1	83,690	19	0.0664	0.129	0.211
0	105,600	19	0.0745	0.102	0.168
00	133,100	19	0.0837	0.0811	0.133
000	167,800	19	0.0940	0.0642	0.105
0000	211,600	19	0.1055	0.0509	0.0836
250	250,000	37	0.0822	0.0431	0.0708
300	300,000	37	0.0900	0.0360	0.0590
350	350,000	37	0.0973	0.0308	0.0505
400	400,000	37	0.1040	0.0270	0.0442
500	500,000	37	0.1162	0.0216	0.0354
600	600,000	61	0.0992	0.0180	0.0295
700	700,000	61	0.1071	0.0154	0.0253
750	750,000	61	0.1109	0.0144	0.0236
800	800,000	61	0.1145	0.0135	0.0221
900	900,000	61	0.1215	0.0120	0.0197
1000	1,000,000	61	0.1280	0.0108	0.0177

free air is listed in Table 5.3. The ampacity of conductors in a raceway or cable (three or fewer conductors) is listed in Table 5.4.

Synthetic insulation for wire and cable is classified into two broad categories: (1) *thermosetting*, and (2) *thermoplastic*. A wide variety of chemical mixtures can be found within each category. Most insulation is composed of compounds made from synthetic rubber polymers (thermosetting) and from synthetic materials (thermoplastics). Various materials are combined to provide specific physical and electrical properties. Thermosetting compounds are characterized by their ability to be stretched, com-

Table 5.3 Permissible Ampacities of Single Conductors in Free Air

Wire size	Copper wire with		Aluminum wire with	
	R, T, TW Insulation	RH, RHW, TH, THW Insulation	R, T, TW Insulation	RH, RHW, TH, THW Insulation
12	25	25	20	20
10	40	40	30	30
8	55	65	45	55
6	80	95	60	75
4	105	125	80	100
3	120	145	95	115
2	140	170	110	135
1	165	195	130	155
0	195	230	150	180
00	225	265	175	210
000	260	310	200	240
0000	300	360	230	280
250	340	405	265	315
300	375	445	290	350
350	420	505	330	395
400	455	545	355	425
500	515	620	405	485
600	575	690	455	545
700	630	755	500	595
750	655	785	515	620
800	680	815	535	645
900	730	870	580	700
1000	780	935	625	750

pressed, or deformed within reasonable limits under mechanical stress, and then to return to their original shape when the stress is removed. Thermoplastic insulation materials are best known for their electrical characteristics and relatively low cost. Thermoplastics permit insulation thickness to be reduced while maintaining good electrical properties.

Many different types of insulation are used for electric conductors. The operating conditions determine the type of insulation used. Insulation types are identified by ab-

Table 5.4 Permissible Ampacities of Conductors in a Raceway or Cable (three or fewer conductors total)

Wire size	Copper wire with		Aluminum wire with	
	R, T, TW Insulation	RH, RHW, TH, THW Insulation	R, T, TW Insulation	RH, RHW, TH, THW Insulation
12	20	20	15	15
10	30	30	25	25
8	40	45	30	40
6	55	65	40	50
4	70	85	55	65
3	80	100	65	75
2	95	115	75	90
1	110	130	85	100
0	125	150	100	120
00	145	175	115	135
000	165	200	130	155
0000	195	230	155	180
250	215	255	170	205
300	240	285	190	230
350	260	310	210	250
400	280	335	225	270
500	320	380	260	310
600	355	420	285	340
700	385	460	310	375
750	400	475	320	385
800	410	490	330	395
900	435	520	355	425
1000	455	545	375	445

breviations established in the National Electrical Code (NEC). The most popular types are:

- **R**: Rubber, rated for 140° F
- **RH**: Heat-resistant rubber, rated for 167° F
- **RHH**: Heat-resistant rubber, rated for 194° F

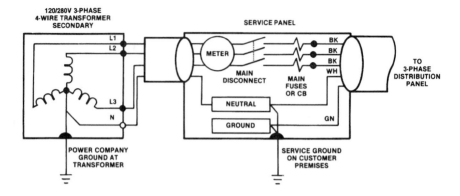

Figure 5.24 Connection arrangement for a three-phase utility company service panel.

- **RHW**: Moisture- and heat-resistant rubber, rated for 167° F

- **T**: Thermoplastic, rated for 140° F

- **THW**: Moisture- and heat-resistant thermoplastic, rated for 167° F

- **THWN**: Moisture- and heat-resistant thermoplastic with nylon, rated for 194° F

5.3.1 Utility Service Entrance

Figure 5.24 shows a typical service entrance, with the neutral line from the utility company tied to ground and to a ground rod at the meter panel. Where permitted by the local code, this should be the only point at which neutral is tied to ground in the ac distribution system.

Figure 5.25 shows a three-phase power-distribution panel. Note that the neutral and ground connections are kept separate. Most ac distribution panels give the electrical contractor the ability to lift the neutral from ground by removing a short-circuiting screw in the breaker-panel chassis. Where permitted by local code, insulate the neutral lines from the cabinet. Bond the ground wires to the cabinet for safety. Always run a separate, insulated green wire for ground. Never rely on conduit or other mechanical structures to provide an ac system ground to electric panels or equipment.

A single-phase power-distribution panel is shown in Figure 5.26. Note that neutral is insulated from ground and that the insulated green ground wires are bonded to the panel chassis.

Conduit runs often are a source of noise. Corrosion of the steel-to-steel junctions can act as an RF detector. Conduit feeding sensitive equipment may contact other conduit runs powering noisy devices, such as elevators or air conditioners. Where possible (and permitted by the local code), eliminate this problem by using PVC pipe, Romex, or jacketed cable. If metal pipe must be used, send the noise to the power ground rods by

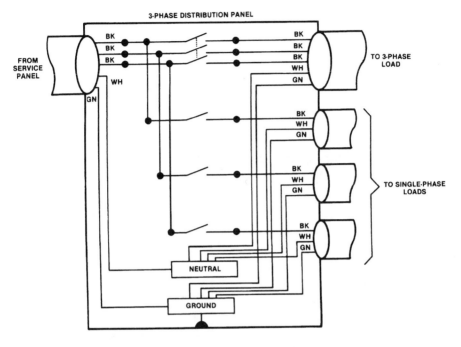

Figure 5.25 Connection arrangement of the neutral and green-wire ground system for a three-phase ac distribution panel.

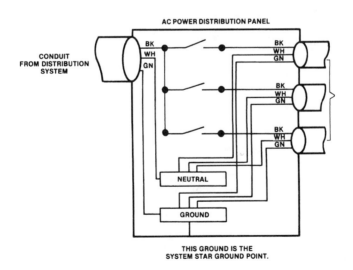

Figure 5.26 Connection arrangement of the neutral and green-wire ground system for a single-phase ac distribution panel.

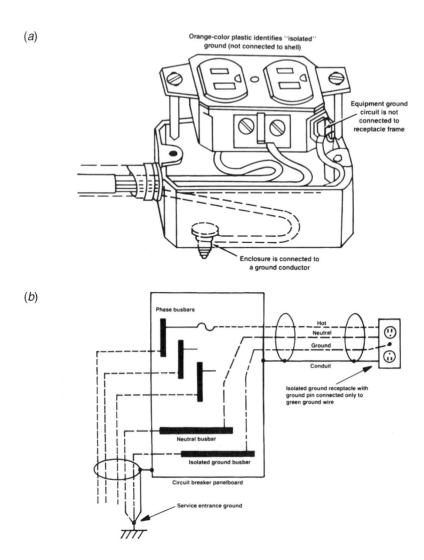

Figure 5.27 Installation requirements for an orange receptacle ac outlet: (*a*) mechanical configuration (note that at least two ground conductor paths are required—one for the receptacle ground pin and one for the receptacle enclosure); (*b*) circuit arrangement for the isolated ground outlet; (*c*, next page) circuit arrangement for a conventional ac outlet. (*After* [3].)

isolating the green ground wire from the conduit with a ground-isolating (orange) receptacle. When using an orange receptacle, a second ground wire is required to bond the enclosure to the ground system, as shown in Figure 5.27. In a new installation, isolate the conduit from building metal structures or other conduit runs. Consult the local electrical code and an experienced electrical contractor before installing or modifying

(c)

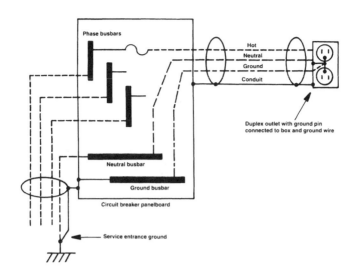

any ac power-system wiring. Make sure to also secure any necessary building permits for such work.

5.3.2 Fault Tolerance as a Design Objective

To achieve high levels of power system reliability—with the ultimate goal being 24-hour-per-day availability, 365 days per year—some form of power system redundancy is required, regardless of how reliable the individual power system components may be [4]. Redundancy, if properly implemented, also provides power distribution flexibility. By providing more than one path for power flow to the load, the key elements of a system can be shifted from one device or branch to another as required for load balancing, system renovations or alterations, or equipment failure isolation. Redundancy also provides a level of fault tolerance. Fault tolerance can be divided into three basic categories:

- Rapid recovery from failures

- Protection against "slow" power system failures, where there is enough warning of the condition to allow intervention

- Protection against "fast" power system failures, where no warning of the power failure is given

As with many corrective and preventive measures, the increasing costs must be weighed against the benefits.

For example, recent developments in large UPS system technologies have provided the capability to operate two independent UPS systems in parallel, either momentarily or continuously. The ability to momentarily connect two UPS systems allows critical loads to be transferred from one UPS system to the other without placing the UPS sys-

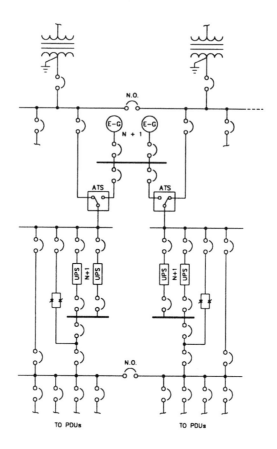

Figure 5.28 Power distribution system featuring redundancy and high reliability. Of particular interest is the ability to parallel UPS systems as required by operational conditions. (*After* [4].)

tems in bypass, thereby maintaining continuous UPS protection of the loads. Continuous paralleling of the two UPS systems, on the other hand, can be used to create a single redundant UPS system from two otherwise nonredundant systems when multiple UPS modules are out of service (because of failures or maintenance). Figure 5.28 illustrates one such implementation.

5.3.3 Critical System Bus

Many facilities do not require the operation of all equipment during a power outage. Rather than use one large standby power system, key pieces of equipment can be protected with small, dedicated, uninterruptible power systems. Small UPS units are available with built-in battery supplies for computer systems and other hardware. If cost prohibits the installation of a system-wide standby power supply (using generator

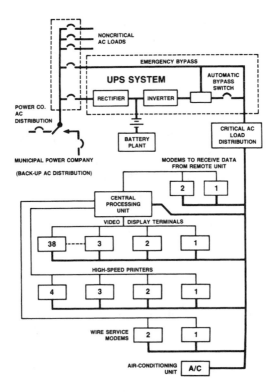

Figure 5.29 An application of the critical-load power bus concept. In the event of a power failure, all equipment necessary for continued operation is powered by the UPS equipment. Noncritical loads are dropped until commercial ac returns.

or solid-state UPS technologies), consider establishing a *critical load bus* that is connected to a UPS system or generator via an automatic transfer switch. This separate power supply is used to provide ac to critical loads, thus keeping the protected systems up and running. The concept is illustrated in Figure 5.29. Unnecessary loads are dropped in the event of a power failure.

A standby system built on the critical load principle can be a cost-effective answer to the power-failure threat. The first step in implementing a critical load bus is to accurately determine the power requirements for the most important equipment. Typical power consumption figures can be found in most equipment instruction manuals. If the data is not listed or available from the manufacturer, it can be measured using a wattmeter.

When planning a critical load bus, be certain to identify accurately which loads are critical, and which can be dropped in the event of a commercial power failure. If air conditioning is interrupted but the computer equipment at a large data processing center continues to run, temperatures will rise quickly to the point at which system components may be damaged or the hardware automatically shuts down. It may not be necessary to require cooling fans, chillers, and heat-exchange pumps to run without interruption. However, any outage should be less than 1 to 2 min in duration. Air-cooled computer systems can usually tolerate 5 to 10 min of cooling interruption.

5.3.4 Power Distribution Options

There are essentially 12 building blocks that form what can be described as an assured, reliable, clean power source for computer systems, peripherals, and other critical loads [5]. They are:

- Utility and service entry (step-down transformer, main disconnect, and panelboard, switchboard, or switchgear)
- Lightning protection
- Power bus
- Facility power distribution
- Grounding
- Power conditioning equipment
- Critical load air-conditioning
- Frequency converter (if required)
- Batteries for dc backup power
- Emergency engine-generator
- Critical load power distribution network
- Emergency readiness planning

A power system to support a critical load cannot be said to be reliable unless all these components are operating as intended, not only during normal operation, but especially during an emergency.

It is easy to become complacent during periods when everything is functioning properly, because this is the usual mode of operation. An absence of contingency plans for dealing with an emergency situation, and a lack of understanding of how the entire system works, thus, can lead to catastrophic shutdowns when an emergency situation arises. Proper training, and periodic reinforcing, is an essential component of a reliable system.

5.3.5 Plant Configuration

There are any number of hardware configurations that will provide redundancy and reliability for a critical load. Each situation is unique and requires an individual assessment of the options and—more importantly—the risks. The realities of econom-

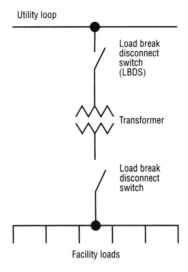

Figure 5.30 Simplified service entrance system. (*From* [5]. *Used with permission.*)

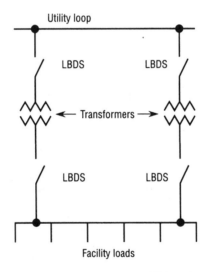

Figure 5.31 Fault-tolerant service entrance system. (*From* [5]. *Used with permission.*)

ics dictate that cost is always a factor. Through proper design, however, the expense usually can be held within an acceptable range.

Design for reliability begins at the utility service entrance [5]. The common arrangement shown in Figure 5.30 is vulnerable to interruptions from faults at the transformer and associated switching devices in the circuit. Furthermore, service entrance maintenance would require a plant shutdown. In Figure 5.31, redundancy has been pro-

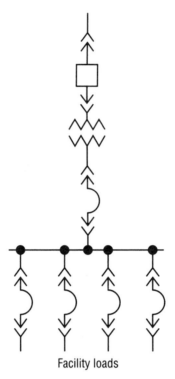

Facility loads

Figure 5.32 Secondary plant distribution using a simple radial configuration. (*From* [5]. *Used with permission.*)

vided that will prevent the loss of power should one of the devices in the line fail. Because the two transformers are located in separate physical enclosures, maintenance can be performed on one leg without dropping power to the facility.

Of equal importance is the method of distributing power within a facility to achieve maximum reliability. This task is more difficult when dealing with a campus-type facility or a process or manufacturing plant, where—instead of being concentrated in a single room or floor—the critical loads may be in a number of distant locations. Figure 5.32 illustrates power distribution through the facility using a simple radial system. An incoming line supplies the main and line feeders via a service entrance transformer. This system is suitable for a single building or a small process plant. It is simple, reliable, and lowest in cost. However, such a system must be shut down for routine maintenance, and it is vulnerable to single-point failure. Figure 5.33 illustrates a distributed and redundant power distribution system that permits transferring loads as required to patch around a fault condition. This configuration also allows portions of the system to be de-energized for maintenance or upgrades without dropping the entire facility. Note the loop arrangement and associated switches that permit optimum flexibility during normal and fault operating conditions.

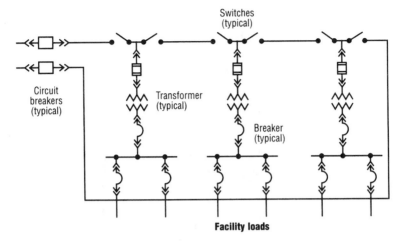

Switches (typical)

Circuit breakers (typical)

Transformer (typical)

Breaker (typical)

Facility loads

Figure 5.33 A redundant, fault-tolerant secondary plant distribution system. (*From* [5]. *Used with permission.*)

5.4 Equipment Rack Enclosures and Devices

In a professional facility, most equipment will have to be rack-mountable. To assemble the equipment in racks, the installer needs to know the exact physical location of each piece of hardware, and all information necessary to assemble and wire the equipment. This includes the placement of terminal blocks, power wiring, cooling devices, and all signal cables within the rack. Equipment locations can be shown on a rack elevation form. Other forms and drawings can specify terminal block wiring, ac power connections, patch panel assignments, and signal cable connections. An example of an equipment location drawing is shown in Figure 5.34.

Drawings showing the details of assembly, mounting hardware, and power wiring generally will not change from rack to rack. Therefore, they can be standardized for all racks to avoid having to repeat this part of the design process. Exceptions can be shown on a separate detailed drawing. This approach is illustrated in Figure 5.35.

When more than one rack is to be assembled side by side, it is normal practice to show the entire row on one drawing. The relationship of all of the equipment in adjacent racks can then be easily seen on the drawing (see Figure 5.36).

5.4.1 Industry Standard Equipment Enclosures

The modular equipment enclosure, frame, or equipment rack is one of the most convenient and commonly used methods for assembling the equipment and components that make up a technical facility. The ANSI/EIAJ RS-310-C standard for racks provides the dimensions and specifications for racks, panels, and associated hardware. Other specifications, such as the European International Electrotechnical Commis-

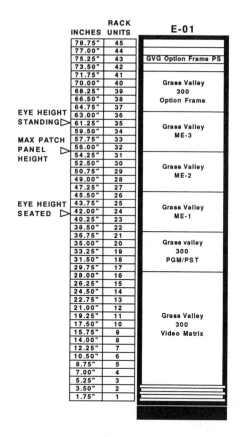

INCHES	RACK UNITS	E-01
78.75"	45	
77.00"	44	
75.25"	43	GVG Option Frame PS
73.50"	42	
71.75"	41	
70.00"	40	Grass Valley
68.25"	39	300
66.50"	38	Option Frame
64.75"	37	
63.00"	36	
61.25"	35	Grass Valley
59.50"	34	ME-3
57.75"	33	
56.00"	32	
54.25"	31	
52.50"	30	Grass Valley
50.75"	29	ME-2
49.00"	28	
47.25"	27	
45.50"	26	
43.75"	25	Grass Valley
42.00"	24	ME-1
40.25"	23	
38.50"	22	
36.75"	21	
35.00"	20	Grass valley
33.25"	19	300
31.50"	18	PGM/PST
29.75"	17	
28.00"	16	
26.25"	15	
24.50"	14	
22.75"	13	
21.00"	12	
19.25"	11	Grass Valley
17.50"	10	300
15.75"	9	Video Matrix
14.00"	8	
12.25"	7	
10.50"	6	
8.75"	5	
7.00"	4	
5.25"	3	
3.50"	2	
1.75"	1	

EYE HEIGHT STANDING ▷ (at 63.00"/36)

MAX PATCH PANEL HEIGHT ▷ (at 56.00"/32)

EYE HEIGHT SEATED ▷ (at 42.00"/24)

Figure 5.34 Equipment location drawing for a rack enclosure.

sion (IEC) Publication Number 297-1 and 297-2 and West German Industrial Standard DINJ 41494 Part 1, have matching dimensions and specifications.

Applicable standards for equipment racks include the following:

- UL-listed type 12 enclosures
- NEMA type 12 enclosures
- NEMAJ type 4 enclosures
- IEC 297-2 specifications
- IEC 297-3 specifications
- IP 55/NEMA type 12/13 enclosures
- DINJ 41494 Part 1

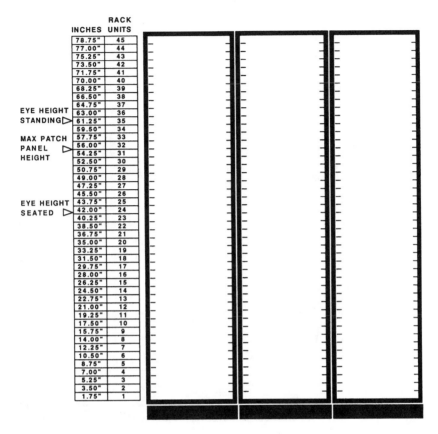

INCHES	RACK UNITS
78.75"	45
77.00"	44
75.25"	43
73.50"	42
71.75"	41
70.00"	40
68.25"	39
66.50"	38
64.75"	37
63.00"	36
61.25"	35
59.50"	34
57.75"	33
56.00"	32
54.25"	31
52.50"	30
50.75"	29
49.00"	28
47.25"	27
45.50"	26
43.75"	25
42.00"	24
40.25"	23
38.50"	22
36.75"	21
35.00"	20
33.25"	19
31.50"	18
29.75"	17
28.00"	16
26.25"	15
24.50"	14
22.75"	13
21.00"	12
19.25"	11
17.50"	10
15.75"	9
14.00"	8
12.25"	7
10.50"	6
8.75"	5
7.00"	4
5.25"	3
3.50"	2
1.75"	1

EYE HEIGHT STANDING ▷ (at 61.25")
MAX PATCH PANEL HEIGHT ▷ (at 54.25")
EYE HEIGHT SEATED ▷ (at 42.00")

Figure 5.35 Hardware location template for a series of rack enclosures being installed at a facility. Deviations from the standard template are shown as drawing details.

The chassis of most of the electronic equipment used for industrial electronics and professional audio/video have front panel dimensions that conform to the EIA specifications for mounting in standard modular equipment enclosures. Figure 5.37 shows the standard RS-310-C rack-mounting hole dimensions.

Blank panels, drawers, shelves, guides, and other accessories are designed and built to conform to the EIA standards. Rack-mounted hardware for interconnecting and supporting the wiring is also available. Figure 5.38 shows some of the hardware available for use with standard equipment enclosures.

5.4.2 Types of Rack Enclosures

There are two main types of racks. The first is the floor-mounted open frame, shown in Figure 5.39. This EIA standard equipment enclosure consists of two vertical channels (with mounting holes), separated at the top and bottom by support channels. The

INCHES	RACK UNITS	D-05	D-06	D-07	
78.75"	45				
77.00"	44				
75.25"	43				
73.50"	42	Yamaha P2075	ASACA		
71.75"	41	Amplifier	CMM20-11(U)		
70.00"	40	VU Meters X4	20" Color Picture		
68.25"	39		Monitor		
66.50"	38	Leitch SCH-730N			
64.75"	37				GVG
63.00"	36		Tek 1720		HX-UCP
61.25"	35	EYE HEIGHT STANDING ▷	WFM		
59.50"	34				
57.75"	33		Tektronix	Sony	
56.00"	32	MAX PATCH PANEL HEIGHT ▷ Tektronix 1480R	520A	BVU-800	
54.25"	31	Waveform Monitor	Vectorscope	3/4U VCR	
52.50"	30		GVG | GVG		
50.75"	29	GVG 3240-20	HX-UCP | HX-UCP		
49.00"	28	Proc Amp	Cox203 NTSC Encoder		
47.25"	27			Sony BVT-810 TBC	
45.50"	26	GVG 3240-20	Cox203 NTSC Encoder		
43.75"	25	EYE HEIGHT SEATED ▷ Proc Amp		ACR	
42.00"	24	Tropeter JSI-52	Tropeter JSI-52	Tropeter JSI-52	
40.25"	23	Video Patch Panel	Video Patch Panel	Video Patch Panel	
38.50"	22	Tropeter JSI-52	Tropeter JSI-52	Tropeter JSI-52	
36.75"	21	Video Patch Panel	Video Patch Panel	Video Patch Panel	
35.00"	20	Tropeter JSI-52	Tropeter JSI-52	Tropeter JSI-52	
33.25"	19	Video Patch Panel	Video Patch Panel	Video Patch Panel	
31.50"	18	Tropeter JSI-52	Tropeter JSI-52	Tropeter JSI-52	
29.75"	17	Video Patch Panel	Video Patch Panel	Video Patch Panel	
28.00"	16	Tropeter JSI-52	Tropeter JSI-52	ADC Audio Patch	
26.25"	15	Video Patch Panel	Video Patch Panel		
24.50"	14				
22.75"	13				
21.00"	12				
19.25"	11				
17.50"	10	GVG 100	GVG 100		
15.75"	9	Prod Switcher	Prod Switcher		
14.00"	8				
12.25"	7				
10.50"	6	GVG 8500 Video DA's	GVG 8500 Video DA's	GVG 8500 Video DA's	
8.75"	5				
7.00"	4	GVG 8500 Video DA's	GVG 8500 Video DA's	GVG 8500 Video DA's	
5.25"	3				
3.50"	2				
1.75"	1				

Figure 5.36 Equipment rack drawing for a group of enclosures, showing the overall assembly.

frame is supported in the free-standing mode by a large base, which provides front-to-back stability. The rack can also be permanently secured to the floor by bolts, eliminating the need for a base. Equipment is mounted directly to the vertical members and is accessible from the front, side, and rear.

The second type of rack is a box frame that is free-standing, with front and, optionally, rear equipment-mounting hardware. This is illustrated in Figure 5.40. This frame can be completely enclosed by installing optional side, rear, top, and bottom panels or access doors. Horizontal brackets mounted on the left and right sides of the frame increase rigidity and provide support for vertical mounting angles and other accessories.

Standard racks are available in widths of 19-, 24-, and 30-in. The preferred, and most widely used, width is 19-in (482.6 mm). The racks are designed to hold equipment and panels that have vertical heights of 1.75-in (44.45 mm) or more, in increments of 1.75-in. One *rack unit* (RU) is defined as 1.75-in (44.45 mm). The rack height is usually specified in rack units. Holes or slots along the left and right edges of the equipment

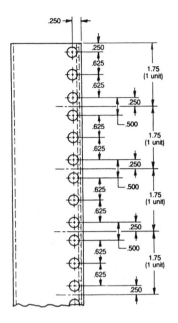

Figure 5.37 Standard equipment-mounting dimensions for RS-310-C rack enclosures.

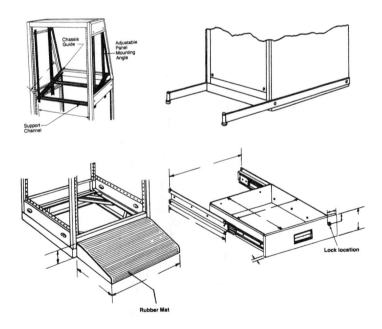

Figure 5.38 Common accessories available for use with rack enclosures. (*Courtesy of Emcor Products.*)

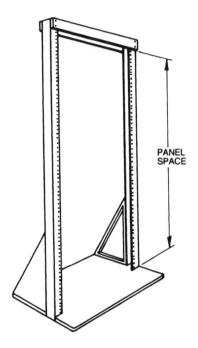

Figure 5.39 Standard open-frame equipment rack.

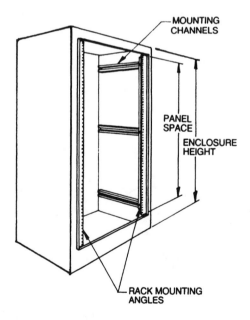

Figure 5.40 Standard enclosed equipment rack.

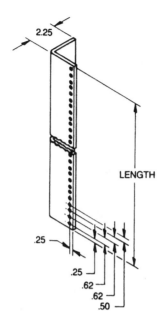

Figure 5.41 "L" equipment mounting bracket, used to support heavy instruments in a rack.

front panel or support panel are provided for screws to fasten the unit to specific mounting holes in the rack.

The mounting hole locations are defined by EIA specification so that equipment mounts vertically only at specific heights—in 1.75-in increments—within the enclosure.

The simplest vertical mounting angles are "L"-shaped and have holes uniformly distributed along their entire length through both surfaces (see Figure 5.41). The holes on one surface are used to attach the angle vertically to the horizontal side members of the rack frame. The horizontal members usually allow the vertical mounting angle to be positioned at different depths from front to rear in the rack. The holes on the other surface of the mounting angle are used for securing the equipment. Mounting holes are arranged in groups of three, centered on each 1.75-in rack-unit interval.

Rack angles of a more complex design can be used when necessary to mount accessories, such as drawers, shelves, and guides that do not use the front-mounting angles and, therefore, must be secured by an alternate means. Figure 5.42 shows one common rack angle of this type.

To secure equipment to vertical supports or other mounting hardware, 10-32 UNF-2B threaded clip nut fasteners are placed at the appropriate clearance holes in the mounting angles for each piece of equipment being mounted. Some rack enclosures provide 10-32 threaded mounting holes on the vertical support channels.

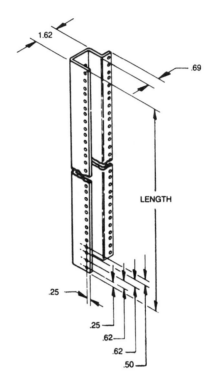

Figure 5.42 Rack angle brace used for mounting shelves, drawers, and other accessory hardware in an enclosure.

The manufacturers of equipment enclosures offer a wide variety of accessory hardware to adapt their products to varied equipment-mounting requirements. A variety of paint colors and laminate finishes are available. Vertical rack-mount support channels are available unpainted with zinc plating to provide a common ground return for the equipment chassis.

5.4.3 Rack Configuration Options

Groups of rack frames can be placed side by side in a number of custom configurations. The most common has racks arranged in a row, bolted together without panels between adjacent frames. Side panels can be mounted on each end of the row, resulting in one long enclosure. Side panels can be installed between frames in a row, if necessary, to provide electromagnetic shielding, a heat barrier, or a physical barrier. A rear door, with or without ventilation perforations or louver slots, can be installed to protect the rear wiring and provide a finished appearance. A front door can be used if access to the equipment front panel controls is not necessary. Clear or darkened Plexi-

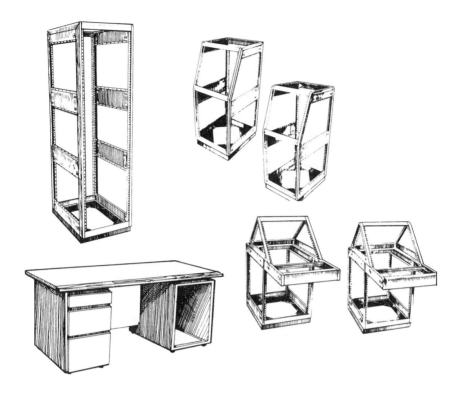

Figure 5.43 A selection of stock equipment enclosures designed for specialized installations.

glas doors can be used to allow viewing of meters or other display devices, or to showcase some aspect of the technology used in the rack.

A top panel is recommended to protect the equipment inside from falling debris and dust. Bottom panels are usually not installed unless bottom shielding is required.

Standard equipment racks provide flexibility because equipment can be mounted at any height in the rack. Many different-shaped frames are also available. These shapes conform to the same equipment-mounting and -mating dimensions, which permit assembling different frames together. Shapes are available with sloping fronts and various wedge shapes are common. Racks can be angled with respect to each other by inserting wedge-shaped frames as intermediaries between adjacent frames. Complex consoles for housing control panels and monitoring equipment can be assembled by bolting together the differently shaped frames. With these options, a complex console shape can be assembled to meet functional and human factor requirements. Figure 5.43 illustrates several of the stock configurations.

Although control consoles can be assembled from standard components that conform to standard enclosure dimensions, in many instances, custom-made consoles are desirable. These are helpful to achieve a more efficient layout for controls, or to develop a more sophisticated appearance within the control room environment.

Some equipment enclosure manufacturers offer an intermediate step between a stock rack and a custom-made console. By using off-the-shelf rack elements, the customer can specify the exact size and configuration required. After the dimensions have been provided to the manufacturer, the individual supporting rails and frames are cut to specification, and the unit is assembled.

5.4.4 Selecting an Equipment Rack

When selecting the model of rack that will be used in a facility, the physical dimensions and weight of equipment to be mounted will be needed. Specify racks with enough depth to accommodate the deepest piece of equipment that will be installed. At the same time, allow ventilating air to flow freely past and through the equipment. Allow additional clearance at the rear of the equipment chassis for connectors, and allow enough space for the minimum bend radius of the largest cable. Additional depth may also be required for cable bundles that must pass behind a deep piece of equipment.

Select a rack model that has sufficient strength to support the full array of anticipated equipment. Also, allow a margin of error for future expansion.

Select paint and laminate colors and textures for the rack assemblies and hardware. This information should be included in the specifications for the racks, which is included in the project manual.

5.4.5 Equipment Rack Layout

When specifying the location of equipment within racks and consoles, give careful consideration to the following factors:

- Physical equipment size and weight
- Power consumption
- Ventilation needs
- Mechanical noise

Human factors also must be considered. Equipment placement should be governed by the operational use of the equipment. Human factors that need consideration include:

- Accessibility to controls
- Height with respect to the operating position
- Line of sight to controls, meters, and display devices, from the operator's point of view
- Reflections on display devices from room lighting or windows
- Noise generated by the equipment

Do not completely fill a given rack with equipment. From a practical point of view, leaving blank spaces will allow for future equipment expansion and replacement.

Provide storage spaces in the racks, if required. For example, if a tape or disk pack is loaded on a machine, a location should be available for holding its container. Rack-mountable shelves and drawers are available in different sizes for this purpose.

Avoid cable clutter by providing easy access to wiring and connections. This will make installation, maintenance, and modifications easier throughout the life of the system.

Place the tops of jack fields at or below eye level. Jack field labels must be readable. The average eye level of males is 65.4-in (1660 mm) and females 61.5-in (1560 mm). Lining up the tops of jack fields that are mounted horizontally across several racks will create a neat appearance. If room is available, place blank panels between patch panels to space them vertically and to allow room for access from the front and rear.

Keep the field as confined as possible to allow the use of the shortest possible patch cords. This is especially critical when using phase-matched video patch panels. These require that patch cords be a fixed, short length.

Provide a pair of rear vertical mounting angles for supporting heavy or deep equipment. Eliminate them if they are not required. Mount heavy equipment in the lower part of the rack to facilitate easier installation and replacement. One exception might be a piece of equipment that generates excessive heat. Mounting it at the top of the rack will allow the heat to escape by convection, without heating other equipment (power supplies are a good example).

Cooling Considerations

It is a normal practice to cool the room in which technical equipment is installed. At the same time, comfort of the personnel in the room must be ensured and usually takes precedence over the comfort of the equipment. Additional steps should be taken to control heat build-up and hot spots within equipment racks and consoles. Use all possible heat-removing techniques within the racks before installing fans for that purpose. Fans cost money, consume power, take up space, are noisy, and will eventually fail. Dust drawn through the fan will collect on something. If that something is a filter, it must be cleaned or replaced periodically. If the dust collects on equipment, overheating may occur. Some steps that can be taken by the system engineer in the design phase include the following:

- Limit the density of heat-producing equipment installed in each rack.

- Leave adequate space for the free movement of air around the equipment. This will help the normal convection flow of air upward as it is heated by the equipment.

- Specify perforated or louvered blank panels above or below heat-producing equipment. A perforated or louvered rear door may also be installed to improve air flow into and out of the rack.

- When alternative equivalent products are available, select equipment that generates the least amount of heat. This will usually result from lower power consumption—a desirable feature.

- If a choice exists among equivalent units, select the one that does not require a built-in fan. Units without fans may be of a low power consumption design, which implies (but does not guarantee) good engineering practice.

- Balance heat loads by placing high heat-producing equipment in another rack to eliminate hot spots.

- Place equipment in a separate air-conditioned equipment room to reduce the heat load in occupied control rooms.

- Remove the outer cabinets of equipment or modify mounting shelves and chassis to improve air flow through the equipment. Consult the original equipment manufacturer, however, before operating a piece of hardware with the cover removed. The cover is often used to channel cooling air throughout the instrument, or to provide necessary electrical shielding.

- When specifying new equipment designs, describe the environment in which the equipment will be required to operate. Stipulate the maximum temperature that can be tolerated.

- If a forced-air design is necessary, pressurize each rack with filtered cooling air, which is brought in at the bottom of the rack and allowed to flow out only at the top of the rack.

- When forced-air cooling is used, provide a means of adjusting the air flow into each rack to balance the volume of air moving through the enclosures. This will control the amount of cooling and concentrate it in the racks where it is needed most. Adjust the air flow to the minimum required to properly cool the equipment. This will minimize the wind noise produced by air being forced through openings in the equipment.

- Install air directors, baffles, or vanes to direct the air flow within the rack. This strategy works for controlling convection and forced-air flow.

Provide a minimum of 3 ft (1 m) clearance at the rear of equipment racks. Besides enabling the enclosure door to swing fully open, this will facilitate efficient cooling and easy equipment installation and maintenance.

If required in a given installation, cooling fans and devices are available for equipment racks. Common types are shown in Figure 5.44.

5.4.6 Single-Point Ground

Equipment racks and peripheral hardware must be properly grounded for reliable operation. Single-point grounding is the basis of any properly designed technical system ground network. Fault currents and noise should have only one path to the facility

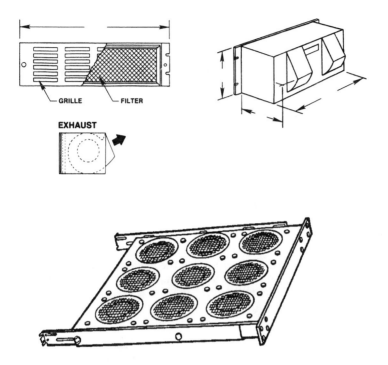

Figure 5.44 Rack accessories used for cooling equipment enclosures.

ground. Single-point grounds can be described as *star* systems, whereby radial elements circle out from a central hub. A star system is illustrated in Figure 5.45. Note that all equipment grounds are connected to a *main ground point*, which is then tied to the facility ground system. Multiple ground systems of this type can be cascaded as needed to form a *star-of-stars*. The object is to ensure that each piece of equipment has one ground reference. Fault energy and noise then are efficiently drained to the outside earth ground system.

Technical Ground System

Figure 5.46 illustrates a star grounding system as applied to an ac power-distribution transformer and circuit-breaker panel. Note that a central ground point is established for each section of the system: one in the transformer vault and one in the circuit-breaker box. The breaker ground ties to the transformer vault ground, which is connected to the building ground system. Figure 5.47 shows single-point grounding applied to a data processing center. Note how individual equipment groups are formed into a star grounding system, and how different groups are formed into a star-of-stars configuration. A similar approach can be taken for a data processing cen-

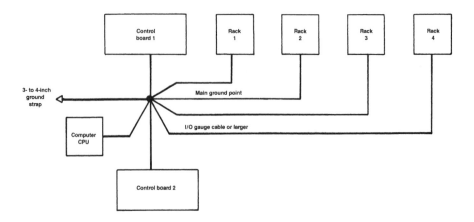

Figure 5.45 Typical facility grounding system. The *main facility ground point* is the reference from which all grounding is done at the plant. If a bulkhead entrance panel is used, it will function as the main ground point.

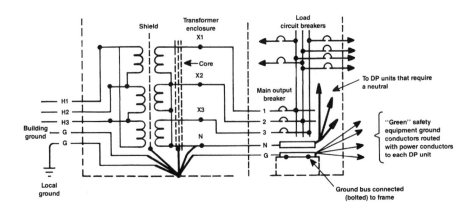

Figure 5.46 Single-point grounding applied to a power-distribution system. (*After*[6].)

ter using multiple modular power center (MPC) units. This is shown in Figure 5.48. The terminal mounting wall is the reference ground point for the facility.

Figure 5.49 shows the recommended grounding arrangement for a typical broadcast or audio/video production facility. The building ground system is constructed using heavy-gauge copper wire (no. 4 gauge or larger) if the studio is not located in an RF field, or a wide copper strap (3-in minimum) if the facility is located near an RF energy source. The copper strap is required because of the *skin effect*.

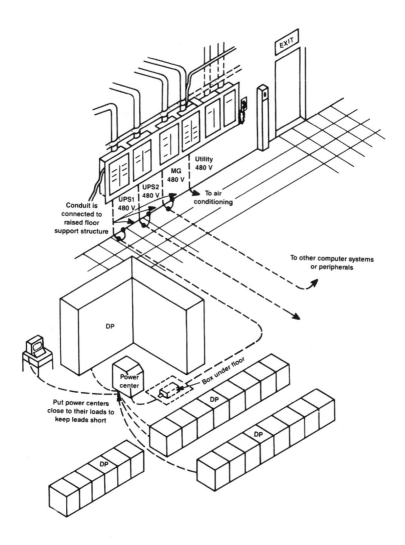

Figure 5.47 Configuration of a star-of-stars grounding system at a data processing facility. (*After* [6].)

Run the strap or cable from the perimeter ground to the main facility ground point. Branch out from the main ground point to each major piece of equipment, and to the various equipment rooms. Establish a *local ground point* in each room or group of racks. Use a separate ground cable for each piece of equipment (no. 12 gauge or larger). Figure 5.50 shows the grounding plan for a communications facility. Equipment grounding is handled by separate conductors tied to the bulkhead panel or entry plate.

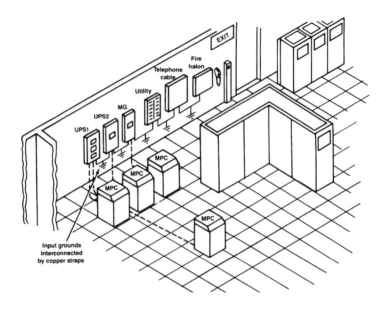

Figure 5.48 Establishing a star-based single-point ground system using multiple modular power centers. (*After* [6].)

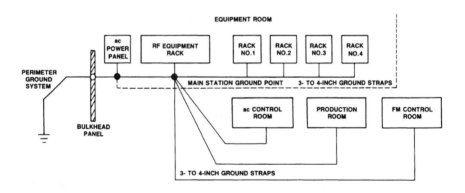

Figure 5.49 Typical grounding arrangement for individual equipment rooms at a communications facility. The ground strap from the main ground point establishes a *local ground point* in each room, to which all electronic equipment is bonded.

A *halo* ground is constructed around the perimeter of the room. Cable trays are tied into the halo. All electronic equipment is grounded to the bulkhead to prevent ground-loop paths. Figure 5.51 shows a top-down view of a bulkhead system ground.

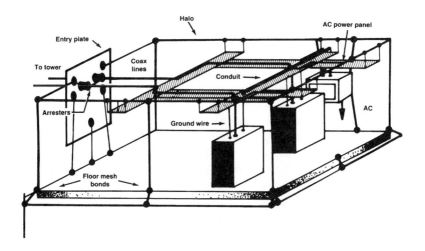

Figure 5.50 Bulkhead-based ground system, including a grounding halo.

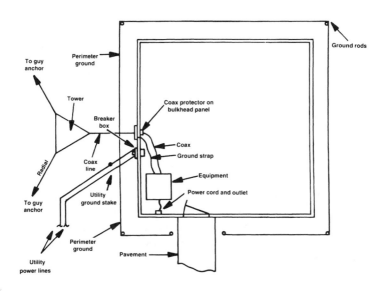

Figure 5.51 A bulkhead ground system integrating all elements of a communications facility.

The ac line ground connection for individual pieces of equipment often presents a built-in problem for the system designer. If the equipment is grounded through the chassis to the equipment room ground point, a ground loop may be created through the green-wire ground connection when the equipment is plugged in. The solution involves careful design and installation of the ac power distribution system to minimize ground-loop currents, while providing the required protection against ground faults. Some equipment manufacturers provide a convenient solution to the ground-loop problem by isolating the signal ground from the ac and chassis ground. This feature offers the user the best of both worlds: the ability to create a signal ground system and ac ground system free of interaction and ground-loops.

It should be emphasized that the design of a ground system must be considered as an integrated package. Proper procedures must be used at all points in the system. It takes only one improperly connected piece of equipment to upset an otherwise perfect ground system. The problems generated by a single grounding error can vary from trivial to significant, depending on where in the system the error exists. This consideration naturally leads to the concept of ground-system maintenance for a facility. Check the ground network from time to time to ensure that no faults or errors have occurred. Any time new equipment is installed or old equipment is removed from service, give careful attention to the possible effects that such work will have on the ground system.

Grounding Conductor Size

The NEC and local electrical codes specify the minimum wire size for grounding conductors. The size varies, depending on the rating of the current-carrying conductors. Code typically permits a smaller ground conductor than hot conductors. It is recommended, however, that the same size wire be used for ground lines and hot lines. The additional cost involved in the larger ground wire is often offset by the use of one size of cable. Further, better control over noise and fault currents is achieved with a larger ground wire.

Separate insulated ground wires should be used throughout the ac distribution system. Do not rely on conduit or raceways to carry the ground connection. A raceway interface that appears to be mechanically sound may not provide the necessary current-carrying capability in the event of a phase-to-ground fault. Significant damage can result if a fault occurs in the system. When the electrical integrity of a breaker panel, conduit run, or raceway junction is in doubt, fix it. Back up the mechanical connection with a separate ground conductor of the same size as the current-carrying conductors. Loose joints have been known to shower sparks during phase-to-ground faults, creating a fire hazard. Secure the ground cable using appropriate hardware. Clean paint and dirt from attachment points. Properly label all cables.

Structural steel, compared with copper, is a poor conductor at any frequency. At dc, steel has a resistivity 10 times that of copper. As frequency rises, the skin effect is more pronounced because of the magnetic effects involved. Furthermore, because of their bolted, piecemeal construction, steel racks and building members should not be depended upon alone for circuit returns.

Power-Center Grounding

A modular power center (MPC), commonly found in computer-room installations, provides a comprehensive solution to ac power distribution and ground-noise considerations. Such equipment is available from several manufacturers, with various options and features. A computer power distribution center generally includes an isolation transformer designed for noise suppression, distribution circuit breakers, power supply cables, and a status monitoring unit. The system concept is shown in Figure 5.52. Input power is fed to an isolation transformer with primary taps to match the ac voltage required at the facility. A bank of circuit breakers is included in the chassis, and individual pre-assembled and terminated cables supply ac power to the various loads. A status monitoring circuit signals the operator of any condition that is detected outside normal parameters.

The ground system is an important component of the MPC. A unified approach, designed to prevent noise or circulating currents, is taken to grounding for the entire facility. This results in a clean ground connection for all on-line equipment.

The use of a modular power center can eliminate the inconvenience associated with rigid conduit installations. Distribution systems also are expandable to meet future facility growth. If the plant is ever relocated, the power center can move with it. MPC units usually are expensive. However, considering the installation costs by a licensed electrician of circuit-breaker boxes, conduit, outlets, and other hardware on-site, the power center approach may be economically attractive. The use of a power center also will make it easier to design a standby power system for the facility. Many computer-based operations do not have a standby generator on site. Depending on the location of the facility, it may be difficult or even impossible to install a generator to provide standby power in the event of a utility company outage. However, by using the power center approach to ac distribution for computer and other critical-load equipment, an uninterruptible power system may be installed easily to power only the loads that are required to keep the facility operating. With a conventional power distribution system—where all ac power to the building or a floor of the building is provided by a single large circuit breaker panel—separating the critical loads from other nonessential loads (such as office equipment, lights, and air conditioning/heating equipment) can be an expensive detail.

5.4.7 Isolation Transformers

One important aspect of an MPC is the isolation transformer. The transformer serves to:

• Attenuate transient disturbances on the ac supply lines.

• Provide voltage correction through primary-side taps.

• Permit the establishment of an isolated ground system for the facility served.

Whether or not an MPC is installed at a facility, consideration should be given to the appropriate use of an isolation transformer near a sensitive load.

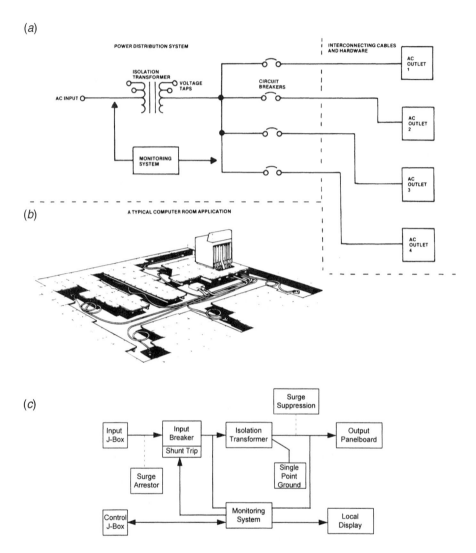

Figure 5.52 The basic concept of a computer-room modular power center:(a) basic line drawing of the system, (b) typical physical implementation, (c) functional block diagram. Both single- and multi-phase configurations are available. When ordering an MPC, the customer can specify cable lengths and terminations, making installation quick and easy.

The ac power supply for many buildings often originates from a transformer located in a basement utility room. In large buildings, the ac power for each floor may be supplied by transformers closer to the loads they serve. Most transformers are 208 Y/120 V three-phase. Many fluorescent lighting circuits operate at 277 V, supplied by a 480 Y/277 V transformer. Long feeder lines to data processing (DP) systems and other sen-

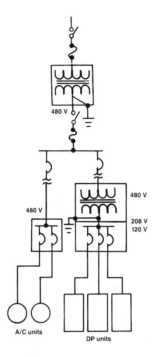

Figure 5.53 Preferred power distribution configuration for a DP site. (*After* [6].)

sitive loads raise the possibility of voltage fluctuations based on load demand and ground-loop-induced noise.

Figure 5.53 illustrates the preferred method of power distribution in a building. A separate dedicated isolation transformer is located near the DP equipment. This provides good voltage regulation and permits the establishment of an effective single-point star ground in the DP center. Note that the power distribution system voltage shown in the figure (480 V) is maintained at 480 V until it reaches the DP step-down isolation transformer. Use of this higher voltage provides more efficient transfer of electricity throughout the plant. At 480 V, the line current is about 43 percent of the current in a 208 V system for the same conducted power.

5.4.8 Grounding Equipment Racks

The installation and wiring of equipment racks must be planned carefully to avoid problems during day-to-day operations. Figure 5.54 shows the recommended approach. Bond adjacent racks together with 3/8 to 1/2in-diameter bolts. Clean the contacting surfaces by sanding down to bare metal. Use lock washers on both ends of the bolts. Bond racks together using at least six bolts per side (three bolts for each vertical rail).

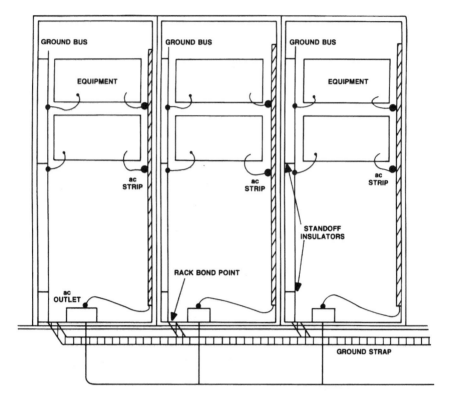

Figure 5.54 Recommended grounding method for equipment racks. To make assembly of multiple racks easier, position the ground connections and ac receptacles at the same location in all racks.

Run a ground strap from the *main facility ground point*, and bond the strap to the base of each rack. Spot-weld the strap at a convenient location on one side of the rear portion of each rack. Secure the strap at the same location for each rack used. A mechanical connection between the rack and the ground strap may be made using bolts and lock washers, if necessary. Be certain, however, to sand down to bare metal before making the ground connection. Because of the importance of the ground connection, it is recommended that each attachment be made with a combination of crimping and silver-solder.

Install a vertical ground bus in each rack (as illustrated in Figure 5.54). Use about 1-1/2-in wide, 1/4-in thick copper busbar. Size the busbar to reach from the bottom of the rack to about 1 ft short of the top. The exact size of the busbar is not critical, but it must be sufficiently wide and rigid to permit the drilling of 1/8-in holes without deforming.

Mount the ground busbar to the rack using insulated standoffs. Porcelain standoffs commonly found in high-voltage equipment are useful for this purpose. Porcelain standoffs are readily available and reasonably priced. Attach the ground busbar to the rack at the point that the facility ground strap attaches to the rack. Silver-solder the busbar to the rack and strap at the same location in each rack used.

Install an ac receptacle box at the bottom of each rack. Isolate the conduit from the rack. The easiest approach is to use an insulated bushing between the conduit and the receptacle box. With this arrangement, the ac outlet box can be mounted directly to the bottom of the rack near the point that the ground strap and ground busbar are bonded to the rack. An alternative approach is to use an orange-type receptacle. This type of outlet isolates the green-wire power ground from the receptacle box. Use insulated standoffs to mount the ac outlet box to the rack. Bring out the green-wire ground, and bond it to the rack near the point that the ground strap and ground busbar are silver-soldered to the rack. The goal of this configuration is to keep the green-wire ac and facility system grounds separate from the ac distribution conduit and metal portions of the building structure. Carefully check the local electrical code to ensure that such configurations are legal.

Although the foregoing procedure is optimum from a signal-grounding standpoint, note that under a ground fault condition, performance of the system may be unpredictable if high currents are being drawn in the current-carrying conductors supplying the load. Vibration of ac circuit elements resulting from the magnetic field effects of high-current-carrying conductors is insignificant as long as all conductors are within the confines of a given raceway or conduit. A ground fault will place return current outside of the normal path. If sufficiently high currents are being conducted, the consequences can be devastating. "Sneak" currents from ground faults have been known to destroy wiring systems that were installed exactly to code.

The fail-safe wiring method for equipment-rack ac power is to use orange-type outlets, with the receptacle green-wire ground routed back to the breaker-panel star ground system. Insulate the receptacle box from the rack to prevent conduit-based noise currents from contaminating the rack ground system. Try to route the power conduit and facility ground cable or strap via the same path, if such a compromise configuration is necessary. Remember to keep metallic conduit and building structures insulated from the facility ground line, except at the bulkhead panel (main grounding point).

Mount a vertical ac strip inside each rack to power the equipment. (See Figure 5.55.) Insulate the power strip from the rack using porcelain standoffs. Power equipment from the strip using standard three-prong grounding ac plugs. Do not defeat the safety ground connection. Equipment manufacturers use this ground to drain transient energy. Defeating the green wire ground violates building codes and is dangerous.

Mount equipment in the rack using normal metal mounting screws. If the location is in a high-RF field, clean the rack rails and equipment panel connection points to ensure a good electrical bond. This is important because, in a high-RF field, detection of RF energy can occur at the junctions between equipment chassis and the rack.

Connect a separate ground wire from each piece of equipment in the rack to the vertical ground busbar. Use no. 12 gauge stranded copper wire (insulated) or larger. Con-

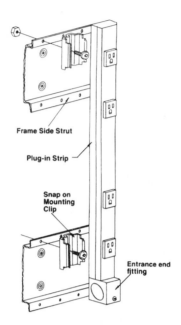

Figure 5.55 Detail of ac line power strip rack attachment.

nect the ground wire to the busbar by drilling a hole in the busbar at a convenient elevation near the equipment. Fit one end of the ground wire with an enclosed-hole
solderless terminal connector (no. 10-sized hole or larger). Attach the ground wire to
the busbar using appropriate hardware. Use an internal-tooth lock washer between the
busbar and the nut. Fit the other end of the ground wire with a terminal that will be accepted by the equipment. If the equipment has an isolated signal ground terminal, tie it
to the ground busbar.

Figure 5.56 shows each of the grounding elements that are discussed in this section
integrated into one diagram. This approach fulfills the requirements of personnel
safety and equipment performance.

Follow similar grounding rules for simple one-rack equipment installations. Figure
5.57 illustrates the grounding method for a single open-frame equipment rack. The vertical ground bus is supported by insulators, and individual jumpers are connected from
the ground rail to each chassis.

5.4.9 Computer Floors

Many large technical centers, particularly DP facilities, are built on raised "computer
floors." A computer cellular floor is a form of ground plane. Grid patterns on 2 ft centers are common. The basic open grid electrically functions as a continuous ground
plane for frequencies below approximately 20 MHz. To be effective, the floor junc-

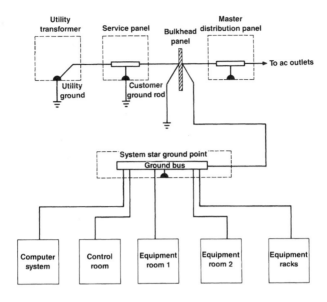

Figure 5.56 Equivalent ground circuit diagram for a medium-size commercial/industrial facility.

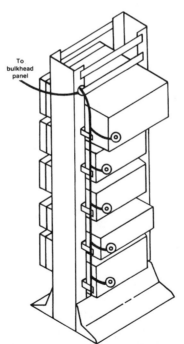

Figure 5.57 Ground bus for an open-frame equipment rack.

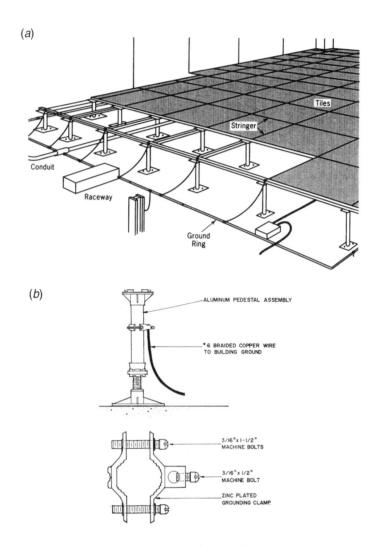

Figure 5.58 Grounding system for raised computer floor construction: (a) grounding of metal supports for raised floor; (b) ground wire clamp detail.

tions must be bonded together. The mating pieces should be plated to prevent corrosion, oxidation, and electrolytic (galvanic) action.

Floor tiles are typically backed with metal to meet fire safety requirements. Some tiles are of all-metal construction. The tiles, combined with the grounded grid structure, provide for effective *electrostatic discharge* (ESD) protection in equipment rooms.

Figure 5.58 illustrates the interconnection guidelines for a grounded raised computer floor. Note that the grid structure is connected along each side of the room to a ground ring. The ground ring, in turn, is bonded to the room's main ground conductor. Note also that all cabling enters and leaves the facility in one area, along one wall, forming a bulkhead panel for the room.

In a small facility, where one computer-floor-based center feeds various peripheral equipment, connection of the grounding bulkhead to structural steel, conduit, and raceways would not be made. Grounding would be handled by the main facility ground conductor. If the layout is sufficiently simple, and cable trays and conduit do not overlap each other, textbook single-point grounding is practical. In large facilities, however, it is impractical to isolate conduit, cable trays, and the main facility ground conductor from the structural steel of the building. In such cases, it is necessary to bond these elements together outside of individual equipment rooms. Within the rooms, however, maintain the single-point ground scheme. Bond any cable tray or conduit entering the room to the bulkhead panel.

In a large facility, when possible, use common paths—with adequate separation to prevent noise on signal-carrying lines—for cable trays, conduit, and the main facility ground conductor.

5.5 Equipment Cooling

In the commonly used model for materials, heat is a form of energy associated with the position and motion of the molecules, atoms, and ions of the material [7]. The position is analogous with the state of the material and is *potential energy*, while the motion of the molecules, atoms and ions is *kinetic energy*. Heat added to a material makes it hotter, and heat withdrawn from a material makes it cooler. Heat energy is measured in *calories* (cal), *British Thermal Units* (Btu), or *joules*. A calorie is the amount of energy required to raise the temperature of one gram of water one degree Celsius (14.5 to 15.5 °C). A Btu is the unit of energy necessary to raise the temperature of one pound of water by one degree Fahrenheit. A joule is an equivalent amount of energy equal to the work done when a force of one newton acts through a distance of one meter.

Temperature is a measure of the average kinetic energy of a substance. It can also be considered a relative measure of the difference of the heat content between bodies.

Heat capacity is defined as the amount of heat energy required to raise the temperature of one mole or atom of a material by one °C without changing the state of the material. Thus, it is the ratio of the change in heat energy of a unit mass of a substance to its change in temperature. The heat capacity, often referred to as *thermal capacity*, is a characteristic of a material and is measured in cal/gram per °C or Btu/lb per °F.

Specific heat is the ratio of the heat capacity of a material to the heat capacity of a reference material, usually water. Because the heat capacity of water is one Btu/lb and one cal/gram, the specific heat is numerically equal to the heat capacity.

Heat transfers through a material by conduction resulting when the energy of atomic and molecular vibrations is passed to atoms and molecules with lower energy. As heat is added to a substance, the kinetic energy of the lattice atoms and molecules increases. This, in turn, causes an expansion of the material that is proportional to the temperature

Table 5.5 Thermal Conductivity of Common Materials

Material	Btu/(h·ft·°F)	W/(m·°C)
Silver	242	419
Copper	228	395
Gold	172	298
Beryllia	140	242
Phosphor bronze	30	52
Glass (borosilicate)	0.67	1.67
Mylar	0.11	0.19
Air	0.015	0.026

change, over normal temperature ranges. If a material is restrained from expanding or contracting during heating and cooling, internal stress is established in the material.

5.5.1 Heat Transfer Mechanisms

The process of heat transfer from one point or medium to another is a result of temperature differences between the two. Thermal energy can be transferred by any of three basic modes:

- Conduction
- Convection
- Radiation

A related mode is the convection process associated with the change of phase of a fluid, such as condensation or boiling.

Conduction

Heat transfer by conduction in solid materials occurs whenever a hotter region with more rapidly vibrating molecules transfers a portion of its energy to a cooler region with less rapidly vibrating molecules. Conductive heat transfer is the most common form of thermal exchange in electronic equipment. Thermal conductivity for solid materials used in electronic equipment spans a wide range of values, from excellent (high conductivity) to poor (low conductivity). Generally speaking, metals are the best conductors of heat, whereas insulators are the poorest. Table 5.5 lists the thermal conductivity of materials commonly used in the construction (and environment) of power vacuum tubes. Table 5.6 compares the thermal conductivity of various substances as a percentage of the thermal conductivity of copper.

Table 5.6 Relative Thermal Conductivity of Various Materials As a Percentage of the Thermal Conductivity of Copper

Material	Relative Conductivity
Silver	105
Copper	100
Berlox high-purity BeO	62
Aluminum	55
Beryllium	39
Molybdenum	39
Steel	9.1
High-purity alumina	7.7
Steatite	0.9
Mica	0.18
Phenolics, epoxies	0.13
Fluorocarbons	0.05

Convection

Heat transfer by natural convection occurs as a result of a change in the density of a fluid (including air), which causes fluid motion. Convective heat transfer between a heated surface and the surrounding fluid is always accompanied by a mixing of fluid adjacent to the surface. Electronic devices relying on convective cooling invariably utilize forced air or water passing through a heat-transfer element [8]. This *forced convection* provides for a convenient and relatively simple cooling system. In such an arrangement, the temperature gradient is confined to a thin layer of fluid adjacent to the surface so that the heat flows through a relatively thin *boundary layer*. In the main stream outside this layer, isothermal conditions exist.

Radiation

Cooling by radiation is a function of the transfer of energy by electromagnetic wave propagation. The wavelengths between 0.1 and 100 m are referred to as *thermal radiation wavelengths*. The ability of a body to radiate thermal energy at any particular wavelength is a function of the body temperature and the characteristics of the surface of the radiating material. Figure 5.59 charts the ability to radiate energy for an ideal radiator, a *blackbody*, which, by definition, radiates the maximum amount of energy at any wavelength. Materials that act as perfect radiators are rare. Most materials radiate energy at a fraction of the maximum possible value. The ratio of the energy radiated by a given material to that emitted by a blackbody at the same temperature is termed *emissivity*. Table 5.7 lists the emissivity of various common materials.

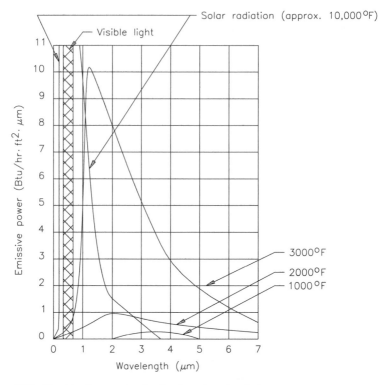

Figure 5.59 Blackbody energy distribution characteristics.

Table 5.7 Emissivity Characteristics of Common Materials at 80°F

Surface Type	Finish	Emissivity
Metal	Copper (polished)	0.018
Metal	Nickel	0.21
Metal	Silver	0.10
Metal	Gold	0.04–0.23
Glass	Smooth	0.9–0.95
Ceramic	Cermet[1]	0.58
[1] Ceramic containing sintered metal		

5.5.2 The Physics of Boiling Water

The *Nukiyama curve* shown in Figure 5.60 charts the heat-transfer capability (measured in watts per square centimeter) of a heated surface, submerged in water at vari-

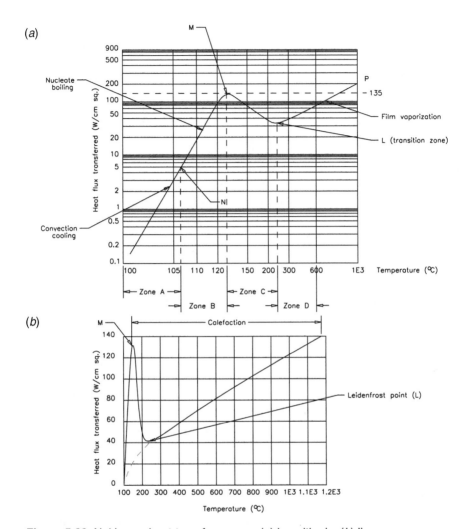

Figure 5.60 Nukiyama heat-transfer curves: (a) logarithmic, (b) linear.

ous temperatures [8]. The first portion of the curve—*zone A*—indicates that from 100 to about 108°C, heat transfer is a linear function of the temperature differential between the hot surface and the water, reaching a maximum of about 5 W/cm² at 108°C. This linear area is known as the *convection cooling zone*. Boiling takes place in the heated water at some point away from the surface.

From 108 to 125°C—*zone B*—heat transfer increases as the fourth power of ΔT until, at 125°C, it reaches 135 W/cm². This zone is characterized by *nucleate boiling*; that is, individual bubbles of vapor are formed at the hot surface, break away, and travel upward through the water to the atmosphere.

Above 125°C, an unstable portion of the Nukiyama curve is observed, where increasing the temperature of the heated surface actually reduces the unit thermal conductivity. At this area—*zone C*—the vapor partially insulates the heated surface from the water until a temperature of approximately 225°C is reached on the hot surface. At this point—called the *Leidenfrost point*—the surface becomes completely covered with a sheath of vapor, and all heat transfer is accomplished through this vapor cover. Thermal conductivity of only 30 W/cm^2 is realized at this region.

From the Leidenfrost point on through *zone D*—the *film vaporization zone*—heat transfer increases with temperature until at about 1000°C the value of 135 W/cm^2 again is reached.

The linear plot of the Nukiyama curve indicates that zones *A* and *B* are relatively narrow areas and that a heated surface with unlimited heat capacity will tend to pass from zone *A* to zone *D* in a short time. This irreversible superheating is known as *calefaction*. For a cylindrical vacuum tube anode, for example, the passing into total calefaction would not be tolerated, because any unit heat-transfer density above 135 W/cm^2 would result in temperatures above 1000°C, well above the safe limits of the tube.

5.5.3 Application of Cooling Principles

Excessive operating temperature is perhaps the single greatest cause of catastrophic failure in an electronic system. Temperature control is important because the properties of many of the materials used to build individual devices change with increasing temperature. In some applications, these changes are insignificant. In others, however, such changes can result in detrimental effects, leading to—in the worst case—catastrophic failure. Table 5.8 details the variation of electrical and thermal properties with temperature for various substances.

Forced-Air Cooling Systems

Air cooling is the simplest and most common method of removing waste heat from an electronic device or system [8]. The normal flow of cooling air is upward, making it consistent with the normal flow of convection currents. Attention must be given to airflow efficiency and turbulence in the design of a cooling system. Consider the case shown in Figure 5.61. Improper layout has resulted in inefficient movement of air because of circulating thermal currents. The cooling arrangement illustrated in Figure 5.62 provides for the uniform passage of cooling air over the device.

Long-term reliability of an electronic system requires regular attention to the operating environment. Periodic tests and preventive maintenance are important components of this effort. Optimum performance of the cooling system can be achieved only when all elements of the system are functioning properly.

Table 5.8 Variation of Electrical and Thermal Properties of Common Insulators As a Function of Temperature

Parameters		20°C	120°C	260°C	400°C	538°C
Thermal conductivity[1]	99.5% BeO	140	120	65	50	40
	99.5% Al_2O_3	20	17	12	7.5	6
	95.0% Al_2O_3	13.5				
	Glass	0.3				
Power dissipation[2]	BeO	2.4	2.1	1.1	0.9	0.7
Electrical resistivity[3]	BeO	10^{16}	10^{14}	5×10^{12}	10^{12}	10^{11}
	Al_2O_3	10^{14}	10^{14}	10^{12}	10^{12}	10^{11}
	Glass	10^{12}	10^{10}	10^{8}	10^{6}	
Dielectric constant[4]	BeO	6.57	6.64	6.75	6.90	7.05
	Al_2O_3	9.4	9.5	9.6	9.7	9.8
Loss tangent[4]	BeO	0.00044	0.00040	0.00040	0.00049	0.00080

[1] Heat transfer in Btu/ft²/hr/°F
[2] Dissipation in W/cm/°C
[3] Resistivity in Ω-cm
[4] At 8.5 GHz

Air-Handling System

The temperature of the intake air supply is a parameter that is usually under the control of the end user. The preferred cooling air temperature is typically no higher than 75°F, and no lower than the room dew point. The air temperature should not vary because of an oversized air-conditioning system or because of the operation of other pieces of equipment at the facility.

Another convenient method for checking the efficiency of the cooling system over a period of time involves documenting the back pressure that exists within the pressurized compartments of the equipment. This measurement is made with a *manometer*, a simple device that is available from most heating, ventilation, and air-conditioning (HVAC) suppliers. The connection of a simplified manometer to a transmitter output compartment is illustrated in Figure 5.63.

By charting the manometer readings, it is possible to accurately measure the performance of the cooling system over time. Changes resulting from the buildup of small dust particles (*microdust*) may be too gradual to be detected except through back-pressure charting. Deviations from the typical back-pressure value, either higher or lower, could signal a problem with the air-handling system. Decreased input or output com-

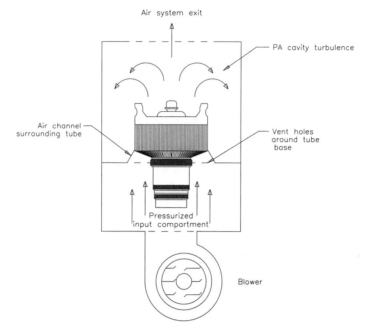

Figure 5.61 A poorly designed cooling system in which circulating air in the output compartment reduces the effectiveness of the heat-removal system.

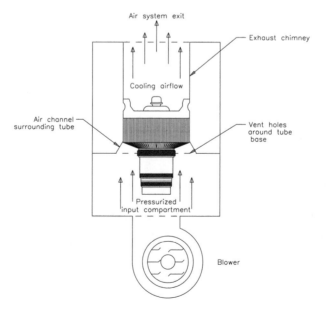

Figure 5.62 The use of a chimney to improve cooling of a power grid tube.

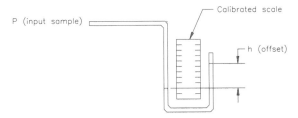

Figure 5.63 A manometer, used to measure air pressure.

partment back pressure could indicate a problem with the blower motor or an accumulation of dust and dirt on the blades of the blower assembly. Increased back pressure, on the other hand, could indicate dirty or otherwise restricted cooling fins and/or exhaust ducting. Either condition is cause for concern. Cooling problems do not improve with time; they always get worse.

Failure of a pressurized compartment air-interlock switch to close reliably may be an early indication of impending trouble in the cooling system. This situation could be caused by normal mechanical wear or vibration of the switch assembly, or it may signal that the compartment air pressure has dropped. In such a case, documentation of manometer readings will show whether the trouble is caused by a failure of the air pressure switch or a decrease in the output of the air-handling system.

Air Cooling System Design

Cooling system performance in electronic equipment is not necessarily related to airflow volume. The cooling capability of air is a function of its mass, not its volume. The designer must determine an appropriate airflow rate within the equipment and establish the resulting resistance to air movement. A specified *static pressure* that should be present within the ducting of the system can be a measure of airflow. For any given combination of ducting, filters, heat sinks, RFI honeycomb shielding, and other elements, a specified system resistance to airflow can be determined. It is important to realize that any changes in the position or number of restricting elements within the system will change the system resistance and, therefore, the effectiveness of the cooling. The altitude of operation is also a consideration in cooling system design. As altitude increases, the density (and cooling capability) of air decreases. A calculated increase in airflow is required to maintain the cooling effectiveness that the system was designed to achieve.

Figure 5.64 shows a typical high-power transmitter plant. The building is oriented so that the cooling activity of the blowers is aided by normal wind currents during the summer months. Air brought in from the outside for cooling is filtered in a hooded air-intake assembly. The building includes a heater and air conditioner.

To help illustrate the importance of proper cooling system design and the real-world problems that some facilities have experienced, consider the following examples taken from actual case histories.

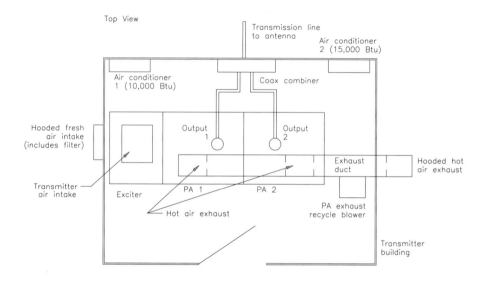

Figure 5.64 A typical heating and cooling arrangement for a high-power transmitter installation. Ducting of PA exhaust air should be arranged so that it offers minimum resistance to airflow.

Case 1

A fully automatic building ventilation system (Figure 5.65) was installed to maintain room temperature at 20°C during the fall, winter, and spring. During the summer, however, ambient room temperature would increase to as much as 60°C. A field survey showed that the only building exhaust route was through the transmitter. Therefore, air entering the room was heated by test equipment, people, solar radiation on the building, and radiation from the transmitter itself before entering the transmitter. The problem was solved through the addition of an exhaust fan (3000 cfm). The 1 hp fan lowered room temperature by 20°C.

Case 2

A simple remote installation was constructed with a heat-recirculating feature for the winter (Figure 5.66). Outside supply air was drawn by the transmitter cooling system blowers through a bank of air filters, and hot air was exhausted through the roof. A small blower and damper were installed near the roof exit point. The damper allowed hot exhaust air to blow back into the room through a tee duct during the winter months. For summer operation, the roof damper was switched open and the room damper closed. For winter operation, the arrangement was reversed. The facility, however, experienced short tube life during winter operation, even though the ambient room temperature during winter was not excessive.

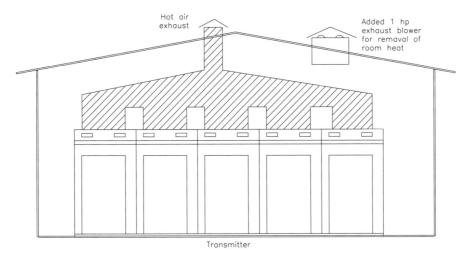

Figure 5.65 Case study in which excessive summertime heating was eliminated through the addition of a 1 hp exhaust blower to the building.

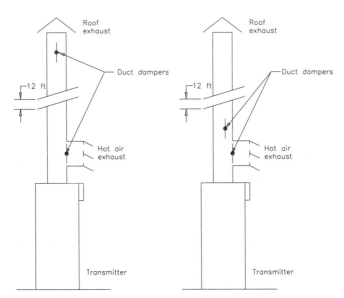

Figure 5.66 Case study in which excessive back pressure to the PA cavity was experienced during winter periods, when the rooftop damper was closed. The problem was eliminated by repositioning the damper as shown.

The solution involved moving the roof damper 12 ft down to just above the tee. This eliminated the stagnant "air cushion" above the bottom heating duct damper and signif-

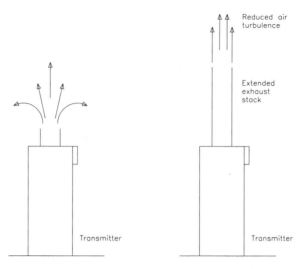

Figure 5.67 Case study in which air turbulence at the exhaust duct resulted in reduced airflow through the PA compartment. The problem was eliminated by adding a 4-ft extension to the output duct.

icantly improved airflow in the region. Cavity back pressure was, therefore, reduced. With this relatively simple modification, the problem of short tube life disappeared.

Case 3

An inconsistency regarding test data was discovered within a transmitter manufacturer's plant. Units tested in the engineering lab typically ran cooler than those at the manufacturing test facility. Figure 5.67 shows the test station difference, a 4-ft exhaust stack that was used in the engineering lab. The addition of the stack increased airflow by up to 20 percent because of reduced air turbulence at the output port, resulting in a 20°C decrease in tube temperature.

These examples point out how easily a cooling problem can be caused during HVAC system design.

5.5.4 Site Design Guidelines

There are any number of physical plant designs that will provide for reliable operation of high-power RF systems [9]. One constant, however, is the requirement for tight temperature control. Cooling designs can be divided into three broad classifications:

- Closed site design
- Open site design
- Hybrid design

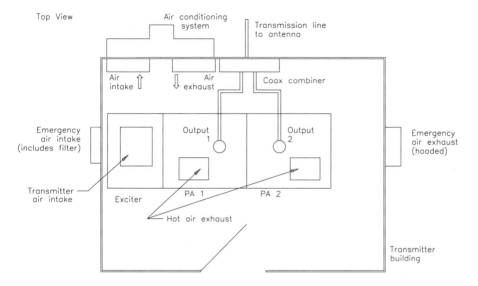

Figure 5.68 Closed site ventilation design, with a backup inlet/outlet system.

If the equipment user is to provide adequately for hot air exhaust and fresh air intake, the maximum and minimum environmental conditions in which the equipment will operate must be known. In addition, the minimum cooling requirements of the equipment must be provided by the manufacturer. The following parameters should be considered:

- Site altitude
- Maximum expected outside air temperature
- Minimum expected outside air temperature
- Total airflow through the equipment
- Air temperature rise through the equipment
- Air exhaust area

The *sensible-heat load*, then, is the sum of all additional heat loads, including:

- Solar radiation
- Heat gains from equipment and lights
- Heat gains from personnel in the area that is to be cooled

Closed Site Design

Figure 6.68 illustrates a site layout that works well in most climates, as long as the equipment package is small and the building is sealed and well insulated [9]. In fact, this closed configuration will ensure the longest possible equipment life and lowest maintenance cost. No outside air laden with moisture and contaminants circulates through the building.

At sites using this arrangement, it has been observed that periodic equipment cleaning is seldom required. In a typical closed system, the air conditioner is set to cool when the room temperature reaches 75 to 80°F. The closed system also uses a louvered emergency intake blower, which is set by its own thermostat to pull in outside air if the room temperature reaches excessive levels (above 90°F). This blower is required in a closed configuration to prevent the possibility of thermal runaway if the air conditioner fails. Without such an emergency ventilation system, the equipment would recirculate its own heated air, further heating the room. System failure probably would result.

During winter months, the closed system is self-heating (unless the climate is harsh, or the equipment generates very little heat), because the equipment exhaust is not ducted outside but simply empties into the room. Also during these months, the emergency intake blower can be used to draw cold outside air into the room instead of using the air conditioner, although this negates some of the cleanliness advantages inherent to the closed system.

An exhaust blower should not be substituted for an intake blower, because positive room pressure is desired for venting the room. This ensures that all air in the building has passed through the intake air filter. Furthermore, the louvered emergency exhaust vent(s) should be mounted high in the room, so that hot air is pushed out of the building first.

The closed system usually makes economic sense only if the transmitter exhaust heat load is relatively small.

Periodic maintenance of a closed system involves the following activities:

- Checking and changing the air conditioner filter periodically

- Cleaning the equipment air filter as needed

- Checking that the emergency vent system works properly

- Keeping the building sealed from insects and rodents

Open Site Design

Figure 5.69 depicts a site layout that is the most economical to construct and operate [9]. The main attribute of this approach is that the equipment air supply is not heated or cooled, resulting in cost savings. This is not a closed system; outdoor air is pumped into the equipment room through air filters. The equipment then exhausts the hot air. If the duct work is kept simple and the equipment has a dedicated exhaust port, such a direct exhaust system works adequately. Many types of equipment do not lend themselves to a direct exhaust connection, however, and a hood mounted over the hardware may be required to collect the hot air, as illustrated in Figure 5.70. With a hooded arrangement, it may be necessary to install a booster fan in the system, typically at the wall or roof exit, to avoid excessive back pressure.

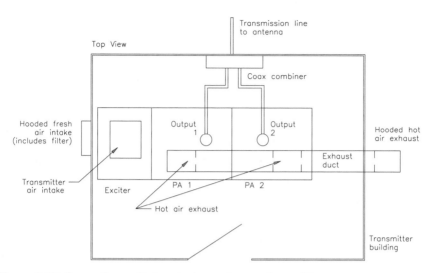

Figure 5.69 Open site ventilation design, using no air conditioner.

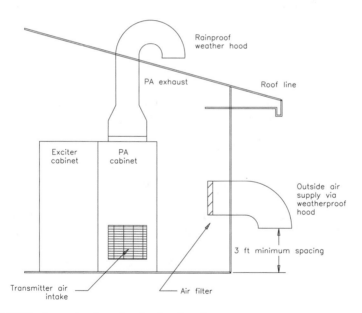

Figure 5.70 Equipment exhaust-collection hood.

When building an open air-circulation system, there are some important additional considerations. The room must be positively pressurized so that only filtered air, which has come through the intake blower and filter, is available to the equipment. With a negatively pressurized room, air will enter through every hole and crack in the building, and will not necessarily be filtered. Under negative pressure, the equipment blower also will have to work harder to exhaust air.

The intake blower should be a "squirrel-cage" type rather than a fan type. A fan is meant to move air within a pressurized environment; it cannot compress the air. A squirrel-cage blower will not only move air, but also will pressurize the area into which the air is directed. Furthermore, the intake air blower must be rated for more cubic feet per minute (cfm) airflow than the equipment will exhaust outside. A typical 20 kW FM broadcast transmitter, for example, will exhaust 500 cfm to 1000 cfm. A blower of 1200 cfm, therefore, would be an appropriate size to replenish the transmitter exhaust and positively pressurize the room.

In moderate and warm climates, the intake blower should be located on a north-facing outside wall. If the air intake is on the roof, it should be elevated so that it does not pick up air heated by the roof surface.

High-quality pleated air filters are recommended. Home-style fiberglass filters are not sufficient. Local conditions may warrant using a double filtration system, with coarse and fine filters in series.

Secure the advice of a knowledgeable HVAC shop when designing filter boxes. For a given cfm requirement, the larger the filtration area, the lower the required air velocity through the filters. This lower velocity results in better filtration than forcing more air through a small filter. In addition, the filters will last longer. Good commercial filtration blowers designed for outdoor installation are available from industrial supply houses and are readily adapted to electronics facility use.

The equipment exhaust can be ducted through a nearby wall or through the roof. Avoid ducting straight up, however. Many facilities have suffered water damage to equipment in such cases because of the inevitable deterioration of roofing materials. Normally, ducting the exhaust through an outside wall is acceptable. The duct work typically is bent downward a foot or two outside the building to keep direct wind from creating back pressure in the exhaust duct. Minimize all bends in the duct work. If a 90° bend must be made, it should be a large-radius bend with curved *helper* vanes inside the duct to minimize turbulence and back pressure. A 90° L- or T-bend is not recommended, unless oversized and equipped with internal vanes to assist the turning airflow.

For moderate and cool climates, an automatic damper can be employed in the exhaust duct to direct a certain amount of hot exhaust air back into the equipment building as needed for heating. This will reduce outside air requirements, providing clean, dry, heated air to the equipment during cold weather.

Hybrid Design

Figure 5.71 depicts a hybrid site layout that is often used for high-power transmitters or sites supporting multiple transmitters [9]. As with the layout in Figure 5.69, the room is positively pressurized with clean, filtered air from the outside. A portion of this outside air is then drawn through the air conditioner for cooling before delivery to the equipment area. Although not all of the air in the room goes through the air conditioner, enough does to make a difference in the room temperature. In humid areas,

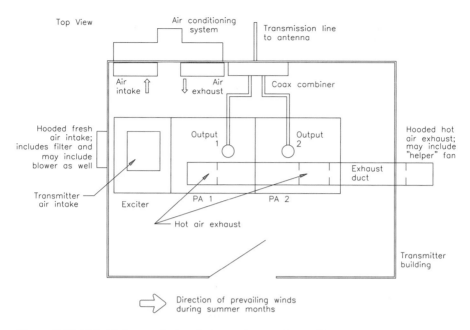

Figure 5.71 Hybrid site ventilation design, using an air conditioner in addition to filtered outside air.

much of the moisture is removed by the air conditioner. The equipment exhaust is directed outside.

Such a hybrid system is often the choice for larger equipment sites where a closed system would prove too costly, but an unconditioned system would run excessively hot during the summer months.

Some closed or hybrid systems use two parallel air conditioners. Most of the time, only one is in use. The thermostats of the units are staggered so that if one cannot keep the air below the ideal operating point, the other will turn on to assist.

5.6   References

1. Carpenter, Roy, B., "Improved Grounding Methods for Broadcasters," *Proceedings, SBE National Convention*, Society of Broadcast Engineers, Indianapolis, 1987.
2. DeDad, John A., (ed.), "Basic Facility Requirements," in *Practical Guide to Power Distribution for Information Technology Equipment*, PRIMEDIA Intertec, Overland Park, KS, pp. 24, 1997.
3. Federal Information Processing Standards Publication No. 94, *Guideline on Electrical Power for ADP Installations*, U.S. Department of Commerce, National Bureau of Standards, Washington, D.C., 1983.

4. Gruzs, Thomas M., "High Availability, Fault-Tolerant AC Power Distribution Systems for Critical Loads, *Proceedings, Power Quality Solutions/Alternative Energy*, Intertec International, Ventura, CA, pp. 20–22, September, 1996.
5. DeDad, John A., "Considerations in Designing a Reliable Power Distribution System," in *Practical Guide to Power Distribution for Information Technology Equipment*, PRIMEDIA Intertec, Overland Park, KS, pp. 4–8, 1997.
6. Federal Information Processing Standards Publication No. 94, *Guideline on Electrical Power for ADP Installations*, U.S. Department of Commerce, National Bureau of Standards, Washington, D.C., 1983.
7. Besch, David F., "Thermal Properties," in *The Electronics Handbook*, Jerry C. Whitaker (ed.), CRC Press, Boca Raton, FL, pp. 127–134, 1996.
8. Laboratory Staff, *The Care and Feeding of Power Grid Tubes*, Varian Associates, San Carlos, CA, 1984.
9. Harnack, Kirk, "Airflow and Cooling in RF Facilities," *Broadcast Engineering*, Intertec Publishing, Overland Park, KS, pp. 33–38, November 1992.

5.7   Bibliography

Benson, K. B., and J. Whitaker: *Television and Audio Handbook for Engineers and Technicians*, McGraw-Hill, New York, 1989.

Block, Roger, "How to Ground Guy Anchors and Install Bulkhead Panels," *Mobile Radio Technology*, Intertec Publishing, Overland Park, KS, February 1986.

Block, Roger: "The Grounds for Lightning and EMP Protection," PolyPhaser Corporation, Gardnerville, NV., 1987.

Defense Civil Preparedness Agency, "EMP Protection for AM Radio Stations," Washington, D.C., TR-61-C, May 1972.

Fardo, S., and D. Patrick: *Electrical Power Systems Technology*, Prentice-Hall, Englewood Cliffs, NJ, 1985.

Hill, Mark, "Computer Power Protection," *Broadcast Engineering*, Intertec Publishing, Overland Park, KS, April 1987.

Lanphere, John: "Establishing a Clean Ground," *Sound & Video Contractor*, Intertec Publishing, Overland Park, KS, August 1987.

Lawrie, Robert: *Electrical Systems for Computer Installations*, McGraw-Hill, New York, N.Y., 1988.

Little, Richard: "Surge Tolerance: How Does Your Site Rate?," *Mobile Radio Technology*, Intertec Publishing, Overland Park, KS, June 1988.

Morrison, Ralph, and Warren Lewis: *Grounding and Shielding in Facilities*, John Wiley & Sons, New York, 1990.

Mullinack, Howard G.: "Grounding for Safety and Performance," *Broadcast Engineering*, Intertec Publishing, Overland Park, KS, October 1986.

Schneider, John: "Surge Protection and Grounding Methods for AM Broadcast Transmitter Sites," *Proceedings of the SBE National Convention*, Society of Broadcast Engineers, Indianapolis, 1987.

Technical Reports LEA-9-1, LEA-0-10 and LEA-1-8, Lightning Elimination Associates, Santa Fe Springs, CA.

Whitaker, Jerry C., *AC Power Systems*, 2nd ed., CRC Press, Boca Raton, FL, 1998.

Whitaker, Jerry C., *Maintaining Electronic Systems*, CRC Press, Boca Raton, FL, 1992.

6

Wiring Practices

6.1　Introduction

All signal-transmission media impair—to some extent—an input electrical signal as it is transmitted, whether analog or digital. Foremost among the impairments is attenuation. The distance over which transmission is possible is determined technically by the threshold sensitivity of the signal receiver. Subjectively, the maximum distance is determined by user-established specifications for tolerable signal bandwidth reduction and S/N (signal-to-noise ratio) increase. Noise in this analysis is a generic term that includes Gaussian noise present in all active components in the transmission system, unwanted signals (*crosstalk*) coupled from parallel signal-transmission circuits, EMI (*electromagnetic interference*), and RFI (*radio frequency interference*) from the total environment through which the signal passes.

All other transmission impairments can be grouped within a generic term of *non-linearities*. These include passband frequency-response flatness deviations, harmonic distortion, and aberrations detected as frequency-specific differences in signal gain and phase. The methods used to interconnect various pieces of equipment, and the hardware used to make the interconnection, determine largely how the overall system will operate. Proper cable installation and termination requires skill and experience. To ensure that the installation will be of high quality and have a neat, organized appearance, the system engineer should specify the practices to be followed by installers.

Installation specifications should be included in the project manual to guide the installers and to ensure good workmanship and adherence to industry standards. Specify how the wiring is to be bundled, supported, and routed within the racks. Group cables into bundles that are held together by cable ties or another method of harnessing. Crosstalk between cables carrying different types and levels of signals can be minimized by isolating the cables into separate groups for video, pulse, audio, control, data, and power. Audio cable should be further subdivided into the following categories:

- Low level (below −20 dBm)

- Medium level (−20 to +20 dBm)

- High level (above +20 dBm)

Table 6.1 Resistivity of Common Materials

Material	Resistivity (mΩ-cm)
Silver	1.468
Copper	1.724
Aluminum	2.828
Steel	5.88
Brass	7.5
Iron	9.8

Control cables and cables carrying dc can be bundled together.

Specify wire and cable types and colors, and identify each on drawings and cable schedules with a unique identifying number or code.

6.2 Electrical Properties of Conductors

At the heart of any facility is the cable used to tie distant parts of the system together. Conductors are rated by the American Wire Gauge (AWG) scale. The smallest is no. 36; the largest is no. 0000. There are 40 sizes in between. Sizes larger than no. 0000 AWG are specified in *thousand circular mil* units, referred to as "MCM" units (M is the roman numeral expression for 1,000). The cross-sectional area of a conductor doubles with each increase of three AWG sizes. The diameter doubles with every six AWG sizes.

Most conductors used for signal and power distribution are made of copper. Stranded conductors are used where flexibility is required. Stranded cables usually are more durable than solid conductor cables of the same AWG size.

Resistance and inductance are the basic electrical parameters of concern in the selection of wire for electronic systems. Resistivity is commonly measured in ohm-centimeters (Ω-cm). Table 6.1 lists the resistivity of several common materials.

Ampacity is the measure of the ability of a conductor to carry electrical current. Although all metals will conduct current to some extent, certain metals are more efficient than others. The three most common high-conductivity conductors are:

- Silver, with a resistivity of 9.8 Ω/circular mil-foot

- Copper, with a resistivity of 10.4 Ω/cmil-ft

- Aluminum, with a resistivity of 17.0 Ω/cmil-ft

The ampacity of a conductor is determined by the type of material used, the cross-sectional area, and the heat-dissipation effects of the operating environment. Conductors operating in free air will dissipate heat more readily than conductors placed in a larger cable or in a raceway with other conductors.

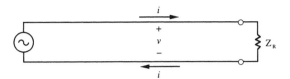

Figure 6.1 Basic transmission line circuit.

Effects of Inductance

Current through a wire results in a magnetic field. All magnetic fields store energy, and this energy cannot be changed in zero time. Any change in the field takes a finite length of time to occur. Inductance is the property of opposition to changes in energy level. The inductance of equipment interconnection cables is usually a distributed parameter.

Voltage drop in a conductor is a function of resistance and inductance. The skin effect and circuit geometry affect both parameters. For example, the inductance of a #10 conductor is approximately 3.5 µH/100-in. At 1 MHz, this translates to a resistance of 22 Ω/100-in.

6.3   Coaxial Cable

Of all of the metallic cable types used to interconnect a given facility, coaxial cable usually represents the greatest challenge. The unique properties of coax permit use over a broad range of frequencies, offering distinct advantages for the system engineer.

The motion of electrical energy requires the presence of an electric field and a magnetic field. Any two conductors can direct the flow of energy. The basic geometry for energy transport is two parallel conductors, as illustrated in Figure 6.1. The transmission line exhibits distributed capacitance C and distributed inductance L along its length. When the switch in the diagram is closed, current begins to flow, charging the capacitance. This current also establishes a magnetic field around both conductors. The energy in these two fields is supplied at a fixed rate. The voltage wave propagates down the line at a fixed velocity, given by the following equation:

$$V = \sqrt{L \times C} \tag{6.1}$$

The velocity in the conductors is typically about one-half the speed of light.

Energy is stored on the line, and—as energy is added—it must be transported past any existing storage. This requires an electrical field and a magnetic field behind the wavefront. The current I that flows in the line is given by the equation:

$$I = \frac{V}{\sqrt{L/C}} \text{ and } I = \frac{V}{Z} \tag{6.2}$$

where Z = the characteristic impedance of the line in Ω

If the transmission line were cut at some point and terminated in an impedance Z, energy would continue to flow on the line as if it had infinite length. When the wavefront reaches the termination, energy is dissipated per unit time rather than being stored per unit time.

The transmission line principles presented here represent an ideal circuit. In a practical transmission line, many factors contribute to losses and some radiation, including the following:

- Skin effect

- Dielectric and conductive losses

- Irregularities in geometry

These factors change with the frequency of the transported wave.

When a transmission line is not terminated in its characteristic impedance, reflections of the transported wave will occur. When a signal reaches an open circuit on the line, the total current flow at the open point must be zero. A reflected wave, therefore, is generated that cancels this current. If, on the other hand, a signal reaches a short circuit on the line, a reflected wave is generated that cancels the voltage. Reflections of these types return energy to the source. The signal at any point along the line is a composite of the initial signal and any reflections.

Sinusoidal signals are assumed when the input impedance of a transmission line is discussed. The input impedance is determined by the following:

- Characteristic impedance of the line

- Terminating impedance

- Applied frequency

- Length of the line

Reflected energy reaching the source modifies the voltage-current relationship. On short unterminated lines, the input impedance can vary significantly. If the reflected wave returns in phase with the input signal, no current will flow; the input impedance is infinite. If the input signal returns 90° out of phase, the line will appear as a pure reactive load to the source.

6.3.1 Operating Principles

A coaxial transmission line consists of concentric center and outer conductors that are separated by a dielectric material. When current flows along the center conductor, it establishes an electric field. The *electric flux density* and the *electric field intensity* are determined by the *dielectric constant* of the dielectric material. The dielectric material becomes polarized with positive charges on one side and negative charges on the opposite side. The dielectric, therefore, acts as a capacitor with a given capacitance per unit length of line. Properties of the field also establish a given inductance per unit length, and a given series resistance per unit length. If the transmission line resistance is negligible and the line is terminated properly, the following formula describes the characteristic impedance (Z_0) of the cable:

$$Z_0 = \sqrt{\frac{L}{C}} \qquad (6.3)$$

Where:
L = inductance in H/ft
C = capacitance in F/ft

Coaxial cables typically are manufactured with 50 Ω or 75 Ω characteristic impedances. Other characteristic impedances are possible by changing the diameter of the center and outer conductors. Figure 6.2 illustrates the relationship between characteristic impedance and the physical dimensions of the cable.

6.3.2 Selecting Coaxial Cable

The function of a coaxial cable is to carry signals from the source to the destination with a minimum of degradation. System requirements dictate the choice of cable type to be used for each application. The materials and construction of a cable determine its effect on a signal. Proper handling and installation practices can prevent damage and guarantee good performance and long life.

Cable Characteristics

Video signals are typically transported by coaxial cable with a characteristic impedance of 75 Ω. Matching the 75 Ω input and output impedance of the equipment ensures a good-quality signal transmission with no reflections. For best results, precision video cable should be used. Precision video cable is sweep-tested by the manufacturer to assure that the cable meets published specifications.

The center conductor of coaxial cable used for video signals should be solid copper in order to assure a low dc resistance. Copper- or silver-covered steel wire, which is designed for RF applications, should not be used. If either the conductor or the shield has a high dc resistance, the frequency response will roll off at low frequencies and cause distortion of the video signal. Direct current flows through the entire cross section of a conductor, but at higher frequencies the skin effect phenomenon causes current to concentrate at or near the surface of the conductor. Because of the higher resistance of the steel core compared with copper, the copper-covered steel cable will exhibit a higher resistance at low frequencies down to dc. Figure 6.3 shows the attenuation characteristics of two types of RG-59 cable.

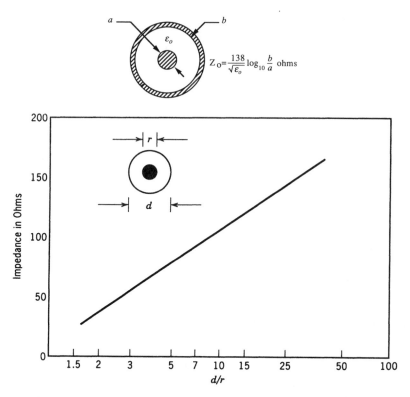

Figure 6.2 The interdependence of coaxial cable physical dimensions and characteristic impedance.

When a cable is to be used in an application where repeated flexing is anticipated, specify a stranded-center conductor, which is more flexible. This will allow the cable to bend more easily, resulting in longer life.

Shield

The shield braid should provide physical coverage of 90 percent or more to protect against electromagnetic interference. Various types of braided coaxial cable shield are illustrated in Figure 6.4. Some cables designed for cable television (CATV) applications use aluminum-foil tape shielding, which provides very good shield coverage. However, because of its severe low-frequency roll-off below 1 MHz, this should not be used for baseband video signal applications. Only use cable with solid-copper braid shielding.

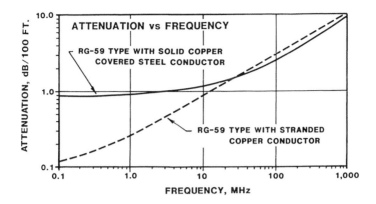

Figure 6.3 Attenuation characteristics of two types of RG-59 cable.

(a)

(b)

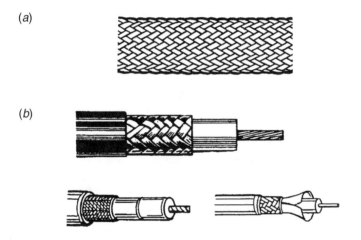

Figure 6.4 Coaxial cable braided shield: (a) basic braid weave, (b) three common applications of a braided shield.

Signal Loss

Coaxial cable attenuation varies with length and frequency. The signal loss must be considered in the facility design process. Coaxial cable attenuation is specified by the manufacturer in dB/100 ft or dB/100 m at one or more frequencies. The attenuation for a given length of cable can be calculated using the published data. If the attenuation is too high for the application, a larger (lower-loss) coaxial cable should be selected. Cable loss can also be compensated for through the addition of an equalizing DA.

Frequency response is affected by the distributed capacitance of the cable, which causes signal energy to roll off at high frequencies. Low-capacitance cable will exhibit less roll-off at high frequencies. An equalizing DA can be used to compensate for this high-frequency roll-off.

In some applications, as in the case of closed-circuit television, power may be supplied to a camera or other system through the coaxial cable. A low-resistance conductor and shield are required to keep the voltage drop within acceptable limits. It may be necessary to increase the power supply voltage to compensate for this voltage drop in long cable runs.

Cable Jacket

Video cables are available with several types of protective jacket materials. Each has particular properties that make it appropriate for specific environments. PVC-jacketed cable costs less than other materials and is used for indoor wiring. Because of its lower replacement cost, however, PVC-jacketed cables may be specified for outdoor use for temporary installations where the cable is expected to wear out from repeated mechanical abuse before sunlight and water can cause noticeable performance deterioration.

Polyethylene or high-density polyethylene has good water-resistance properties. It is the best choice for instances where the cable may be exposed to moisture or immersed in water, such as in underground conduits, or where it may be buried directly or used outdoors. Polyethylene-jacketed cable is available in many colors. This makes it easy to identify the signal type.

Most polyethylene and PVC coaxial cables have a maximum temperature rating of $60° - 80°C$. In some cases, this rating may not be high enough. Teflon cable may be required in areas where a high-ambient temperature is anticipated. (Teflon is a registered trademark of Dupont.) Teflon cable is available with temperature ratings up to 260°C and can be used in or around steam piping or heating ducts. NEC-approved Teflon plenum cables are required for installation in plenums, air ducts, or air returns, where a fire may cause the cable to burn and give off hazardous toxic fumes that could circulate through the air conditioning ducts into other rooms. Teflon cable can be laid directly in false-ceiling air-plenum areas without the need for expensive metal conduit. This cable is also resistant to chemicals but is not suitable where it may be exposed to radiation.

6.3.3 Cable-Rating Standards

The National Electrical Code requires that signal-carrying cables conform to certain rules enacted to prevent fire hazards and electric shock to humans. The code has adopted strict requirements regarding smoke emission and flame propagation, resulting from a number of tragic high-rise fires and deaths from toxic smoke. These revisions have affected the design and construction of electronic cable. Some of the tests required for video cables include:

- Conductor dc resistance

- Insulation resistance

- Heat aging properties

- Cold bend properties

- Smoke emission and flame propagation

The predominant NEC code articles applicable to video cables include the following:

- Article 725—remote control, signaling, and power-limited circuits

- Article 760—fire-protective signaling systems

- Article 770—optical-fiber cables

- Article 800—communications circuits

- Article 820—CATV systems

The four major UL classifications are:

- General purpose (no suffix)

- Plenum (P)—for use in return air plenums, ducts, and environmental air areas

6.3.4 Installing Coaxial Cable

In order to meet its performance specifications, coaxial cable must be installed properly to avoid mechanical stress and damage, which can alter its characteristic impedance and, therefore, the signal it carries. At high frequencies, the change in characteristic impedance resulting from damage or compression of the cable will cause high-frequency components to be reflected to the source. The reflected signal will be added to the instantaneous amplitude of the transmitted video or data signal and cause it to be distorted. A typical data sheet for common types of coaxial cable is shown in Table 6.2.

To avoid possible damage to a cable, its minimum bend radius must not be exceeded. Design conduit and cable trays with the minimum bend requirements in mind. The recommended bend radius of coaxial cable for a single permanent bend is 10 times the cable diameter. In installations where the cable will be repeatedly flexed, the minimum bend radius is 15 times the cable diameter. Provide a loop of slack cable to prevent sharp transitions at the point of bending. Various methods of strain relief are available, which can be used to limit the bend radius of the cable at the point where it flexes.

When pulling cables through conduit, the mechanical stress must be distributed evenly over each cable. Do not exceed the maximum allowable pulling tension for the weakest cable in the conduit, or the conductors may be stretched or broken. For cables with copper conductors, the allowable tension is 40 percent of the breaking strength. This point is the maximum pulling tension that may be applied without stretching the copper center conductor. The maximum pulling tension specification for a given cable is available from the cable manufacturer.

Table 6.2 Basic Specifications for Common Types of Coaxial Cable

Parameter	Type 8279	Type 8281	Type 9231
Standard available lengths	100, 500, 1000	500, 1000	500, 1000
AWG (stranded)	23 (7×32)	20 (solid)	20 (solid)
Insulation type	Polyethylene	Polyethylene	Polyethylene
Nominal OD (in)	0.220	0.304	0.304
Number of shields	Single	Double	Double
Shield type	Tinned braid, copper	Tinned braid, copper	Tinned braid, copper
Nominal impedance (ohms)	75	75	75
Nominal velocity of propagation (%)	66	66	66
Nominal capacitance (pF/ft)	21	21	21
Attenuation at 10 MHz (dB), 100 ft	1.3	0.78	0.78

If necessary, use a lubricant to reduce friction in conduit. Dry compounds, such as talc and powdered soapstone, are available. Liquids and pastes may also be used. Any lubricating compound must be compatible with the cable jacket material.

Pull coaxial cable by the braid. Pull the cables with a steady tension to avoid jerking the conductors. Grips or clamps, such as *Kellum grips*[1], should be used to pull cables. These grips use the Chinese finger puzzle principle to grip the cable, distributing the pulling tension evenly throughout the cable. They are reusable and easy to install. Other pulling devices are also available. Spring scales and similar tension-measuring devices can be used to ensure that the tension limit of the cable is not exceeded.

When cable is pulled over the flange of a stationary reel, the cable will be twisted 360° for every revolution around the spool, causing kinks in the cable. This twisting can damage the conductors and make the cable difficult to pull. The reel must be mounted on an arbor so that the cable can be pulled from a revolving reel (see Figure 6.5*a*).

Twisting also can be avoided by specifying that the cable be supplied in carton put-ups. The cable is laid into these cartons, not wound on reels. That way, the cable, when drawn from an opening in the carton, will not be twisted. The cartons can be stacked on each other and need less space and set-up time than arbor let-offs. No other let-off equipment is required. Inertia spills, where the reel spins, dumping cable onto

---

1 Manufactured by the Kellum Division of Hubbell, Stonington, CT.

(a)   (b)

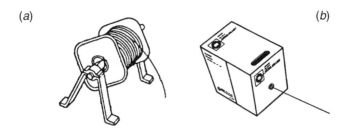

Figure 6.5 Cable packaging/pulling methods: (a) reel, (b) boxed put-up.

the floor, are eliminated when using this type of cable packaging. Figure 6.5b shows one such carton put-up.

In a permanent system installation, use continuous unbroken lengths of cable between devices. If a coaxial cable must be spliced, use coaxial cable connectors that are designed for that specific cable. Male and female cable end connectors or two male connectors with a dual female adapter are available. The proper connectors will maintain the coaxial configuration, impedance, and shielding of the cable with minimum discontinuity. If the splice will be exposed to high humidity or immersed in water, encapsulate the splice in a sealant/encapsulant, such as Scotchcast or RTV (*room temperature vulcanized*) silicone rubber, to prevent infiltration of moisture. (Scotchcast is a registered trademark of 3M.) If the cable has a polyethylene jacket, use fine sandpaper to roughen its surface before applying the sealant to provide good adhesion.

6.3.5 Installation Considerations

Do not run coaxial cable in the same wire tray with power cables. Electromagnetic coupling of the 60 Hz current in power lines can induce hum in the signal. The mixing of signal and power cables in the same cable tray may also violate local and national electrical codes.

Lightning protection is required where cables enter a building. Use coaxial-type lightning arrestors for this purpose. They provide a method for safely connecting the shield of a cable to ground during a lightning discharge.

In locations where cables must be strung between two poles or buildings, determine whether the cable can support its own weight across the span. The sag-vs.-span specification for a cable is usually available from the manufacturer. Use a steel *messenger cable* to support the line if the span is longer than the cable can support. Special hardware is available to secure signal cables to the messenger line, which will support the load.

When cable is to be stored outside, seal the ends of the cable to prevent moisture from entering and damaging it. Take care when installing cable in areas where water is present or can accumulate. If the line is going to be pulled into a conduit or tray that may be filled with water, seal the cable end first. Water can enter the cable through a tear in the jacket. The jacket must be protected during the pulling process.

Cold temperatures will cause the materials used to make most coaxial cables stiffen. At very cold temperatures, the jacket may become brittle and crack when the cable is flexed. If the cable has just been brought in from a cold area or is being installed in an unheated building, store it in a heated area before it is installed. The heat will make the cable more flexible and easier to pull. A portable heater can be used to warm the cable at the pull site. Keep the heat from being applied directly to the coax by using baffles or diffusers.

The National Electrical Code requires that signal-carrying cables conform to certain rules enacted to prevent electrical shock to humans and fire hazards. Research the applicable rules, and follow them.

6.4 Equipment Interconnection Issues

Common-mode rejection ratio (CMRR) is the measure of how well an input circuit rejects ground noise. The concept is illustrated in Figure 6.6. The input signal to a differential amplifier is applied between the plus and minus amplifier inputs. The stage will have a certain gain for this signal condition, called the *differential gain*. Because the ground-noise voltage appears on the plus and minus inputs simultaneously, it is common to both inputs.

The amplifier subtracts the two inputs, yielding only the difference between the voltages at the input terminals at the output of the stage. The gain under this condition should be zero, but in practice, it is not. CMRR is the ratio of these two gains (the differential gain and the common-mode gain) in decibels. The larger the number, the better. For example, a 60 dB CMRR means that a ground signal common to the two inputs will have 60 dB less gain than the desired differential signal. If the ground noise is already 40 dB below the desired signal level, the output noise will be 100 dB below the desired signal level. If, however, the noise is already part of the differential signal, the CMRR will do nothing to improve it.

6.4.1 Active-Balanced Input Circuit

Active-balanced I/O circuits are the basis for nearly all professional audio interconnections (except for speaker connections) and many—if not most—data connections. A wide variety of circuit designs have been devised for active-balanced inputs. All have the common goal of providing high CMRR and adequate gain for subsequent stages. All also are built around a few basic principles.

Figure 6.7 shows the simplest and least expensive approach, using a single operational amplifier (op-amp). For a unity gain stage, all of the resistors are the same value. This circuit presents an input impedance to the line that is different for the two input sides. The positive input impedance will be twice that of the negative input. The CMRR is dependent on the matching of the four resistors and the balance of the source impedance. The noise performance of this circuit, which usually is limited by the resistors, is a tradeoff between low loading of the line and low noise.

Another approach, shown in Figure 6.8, uses a buffering op-amp stage for the positive input. The positive signal is inverted by the op-amp, then added to the negative in-

(a)

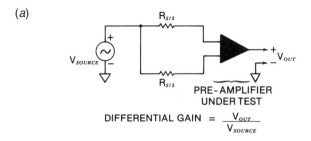

DIFFERENTIAL GAIN $= \dfrac{V_{OUT}}{V_{SOURCE}}$

(b)

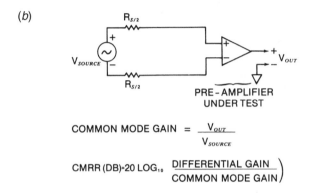

COMMON MODE GAIN $= \dfrac{V_{OUT}}{V_{SOURCE}}$

CMRR (DB)=20 LOG$_{10}$ $\left(\dfrac{\text{DIFFERENTIAL GAIN}}{\text{COMMON MODE GAIN}} \right)$

Figure 6.6 The concept of common-mode rejection ratio (CMRR) for an active-balanced input circuit: (a) differential gain measurement, (b) calculating CMRR.

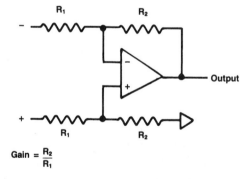

Gain $= \dfrac{R_2}{R_1}$

Figure 6.7 The simplest and least expensive active-balanced input op-amp circuit. Performance depends on resistor matching and the balance of the source impedance.

put of the second inverting amplifier stage. Any common-mode signal on the positive input (which has been inverted) will cancel when it is added to the negative input signal. Both inputs have the same impedance. Practical resistor matching limits the CMRR to

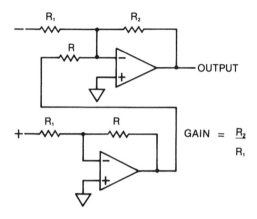

Figure 6.8 An active-balanced input circuit using two op-amps, one to invert the positive input terminal and the other to buffer the difference signal. Without adjustments, this circuit will provide about 50 dB CMRR.

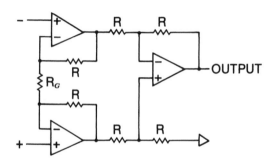

Figure 6.9 An active-balanced input using three op-amps to form an instrumentation-grade circuit. The input signals are buffered and then applied to a differential amplifier.

about 50 dB. With the addition of an adjustment potentiometer, it is possible to achieve 80 dB CMRR, but component aging will degrade this over time.

Adding a pair of buffer amplifiers before the summing stage results in an instrumentation-grade circuit, as shown in Figure 6.9. The input impedance is increased substantially, and any source impedance effects are eliminated. More noise is introduced by the added op-amp, but the resistor noise usually can be decreased by reducing impedances, causing a net improvement (reduction) in system noise.

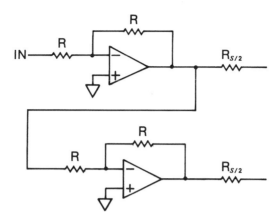

Figure 6.10 A basic active-balanced output circuit. This configuration works well when driving a single balanced load.

6.4.2 Active-Balanced Output Circuit

Early active-balanced output circuits used the approach shown in Figure 6.10. The signal is buffered to provide one phase of the balanced output. This signal then is inverted with another op-amp to provide the other phase of the output signal. The outputs are taken through two resistors, each of which represents half of the desired source impedance. Because the load is driven from the two outputs, the maximum output voltage is double that of an unbalanced stage.

The circuit shown in Figure 6.10 works reasonably well if the load is always balanced, but it suffers from two problems when the load is not balanced. If the negative output is shorted to ground by an unbalanced load connection, the first op-amp is likely to distort. This produces a distorted signal at the input to the other op-amp. Even if the circuit is arranged so that the second op-amp is grounded by an unbalanced load, the distorted output current will probably show up in the output from coupling through grounds or circuit-board traces. Equipment that uses this type of balanced stage often provides a second set of output jacks that are wired to only one amplifier for unbalanced applications.

The second problem with the circuit in Figure 6.10 is that the output does not float. If any voltage difference, such as power-line hum, exists between the local ground and the ground at the device receiving the signal, it will appear as an addition to the signal. The only ground-noise rejection will be from the CMRR of the input stage at the receive end.

The preferred output stage is the electronically balanced and floating design, shown in Figure 6.11. The circuit consists of two op-amps that are cross-coupled with positive and negative feedback. The output of each amplifier is dependent on the input signal and the signal present at the output of the other amplifier. This type of design may have gain or loss, depending on the selection of resistor values. The output impedance is set by appropriate selection of resistor values. Some resistance is needed from the output

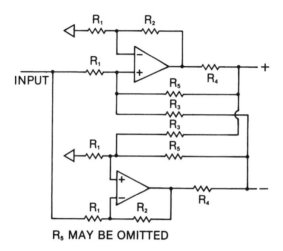

R₅ MAY BE OMITTED

Figure 6.11 An electronically balanced and floating output circuit. A stage such as this will perform well even when driving unbalanced loads.

terminal to ground to keep the output voltage from floating to one of the power-supply rails. Care must be taken to properly compensate the devices. Otherwise, stability problems may result.

6.4.3 Analyzing Noise Currents

Figure 6.12 shows a basic source and load connection. No grounds are present, and both the source and the load float. This is the optimum condition for equipment interconnection. Either the source or the load may be tied to ground with no problems, provided only one ground connection exists. Unbalanced systems are created when each piece of equipment has one of its connections tied to ground, as shown in Figure 6.13. This condition occurs if the source and load equipment have unbalanced (single-ended) inputs and outputs. This type of equipment uses chassis, or common, ground for one of the conductors. Problems are compounded when the equipment is separated by a significant distance.

As shown in the figure, a difference in ground potential causes current flow in the ground wire. This current develops a voltage across the wire resistance. The ground-noise voltage adds directly to the signal. Because the ground current is usually the result of leakage in power transformers and line filters, the 60 Hz signal gives rise to hum of one form or another. Reducing the wire resistance through a heavier ground conductor helps the situation, but it cannot eliminate the problem.

By amplifying the high side and the ground side of the source and subtracting the two to obtain a *difference signal*, it is possible to cancel the ground-loop noise. This is the basis of the *differential input* circuit, illustrated in Figure 6.14. Unfortunately, prob-

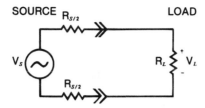

Figure 6.12 A basic source and load connection. No grounds are indicated, and both the source and the load float.

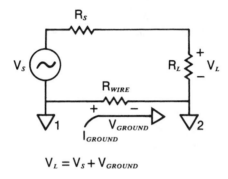

$$V_L = V_S + V_{GROUND}$$

Figure 6.13 An unbalanced system in which each piece of equipment has one of its connections tied to ground.

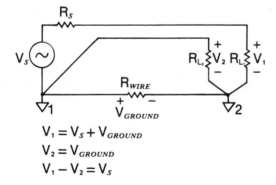

$$V_1 = V_S + V_{GROUND}$$
$$V_2 = V_{GROUND}$$
$$V_1 - V_2 = V_S$$

Figure 6.14 Cancellation of ground-loop noise by amplifying both the high and ground side of the source and subtracting the two signals.

lems still may exist with the unbalanced-source-to-balanced-load system. The reason centers on the impedance of the unbalanced source. One side of the line will have a

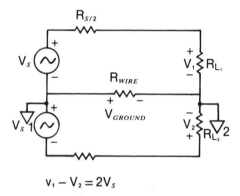

$$V_1 - V_2 = 2V_S$$

Figure 6.15 A balanced source configuration, where the inherent amplitude error of the system shown in Figure 6.14 is eliminated.

slightly lower amplitude because of impedance differences in the output lines. By creating an output signal that is out of phase with the original, a balanced source can be created to eliminate this error (see Figure 6.15). As an added benefit, for a given maximum output voltage from the source, the signal voltage is doubled over the unbalanced case.

Grounding Signal-Carrying Cables

Proper ground system installation is the key to minimizing noise currents on signal-carrying cables. Audio, video, and data lines are often subject to ac power noise currents and RFI. The longer the cable run, the more susceptible it is to disturbances. Unless care is taken in the layout and installation of such cables, unacceptable performance of the overall system may result.

Types of Noise

Open (non-coaxial) wiring can couple energy from external fields. These fields result from power lines, signal processes, and RF sources. The extent of coupling is determined by the following:

- Loop area between conductors
- Cable length
- Cable proximity
- Frequency
- Field strength

Two basic types of noise can appear on ac power, audio, video, and computer data lines within a facility: normal mode and common mode. Each type has a particular effect on sensitive load equipment. The normal-mode voltage is the potential difference

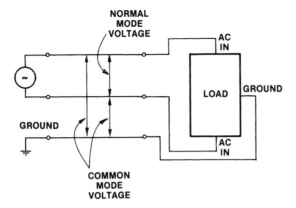

Figure 6.16 The principles of normal-mode and common-mode noise voltages as they apply to ac power circuits.

that exists between pairs of power (or signal) conductors. This voltage also is referred to as the *transverse-mode* voltage. The common-mode voltage is a potential difference (usually noise) that appears between power or signal conductors and the local ground reference. The differences between normal-mode and common-mode noise are illustrated in Figure 6.16.

The common-mode noise voltage will change depending on what is used as the ground reference point. Often, it is possible to select a ground reference that has a minimum common-mode voltage with respect to the circuit of interest, particularly if the reference point and the load equipment are connected by a short conductor. Common-mode noise can be caused by electrostatic or electromagnetic induction.

A single common-mode or normal-mode noise voltage is rarely found. More often than not, load equipment will see both types of noise signals. In fact, unless the facility wiring system is unusually well balanced, the noise signal of one mode will convert some of its energy to the other.

Common-mode and normal-mode noise disturbances typically are caused by momentary impulse voltage differences among parts of a distribution system that have differing ground potential references. If the sections of a system are interconnected by a signal path in which one or more of the conductors is grounded at each end, the ground offset voltage can create a current in the grounded signal conductor. If noise voltages of sufficient potential occur on signal-carrying lines, normal equipment operation can be disrupted (see Figure 6.17).

Electrostatic Noise

Electrostatic noise can be generated by a number of sources:

- Sparks at the armatures of motors or generators

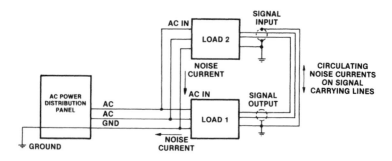

Figure 6.17 An illustration of how noise currents can circulate within a system because of the interconnection of various pieces of hardware.

- Gas-discharge lighting (neon and fluorescent) fixtures
- Portable electronic devices and appliances

Electrostatic noise can invade a signal-carrying cable by means of capacitive coupling. Electrostatic shielding, such as a metallic braided jacket, a swerved (spiral-wrapped) jacket, or a foil tape, can reduce electrostatic noise, provided the shield offers a low-resistance path to ground.

A circuit generating or carrying electrostatic noise acts as one plate of a capacitor; the signal-carrying cable can act as the other plate of the capacitor. A portion of the noise source voltage will, therefore, be electrostatically (capacitively) coupled into the cable. The nature of capacitive reactance is such that higher frequencies are more readily admitted into the cable. Moreover, the higher the impedance of the circuit, the greater the inducted noise voltage.

Wrapping the signal-carrying conductors of the cable with a grounded, electrically conductive screen (shield) offers a low-resistance path to ground. This electrostatic shielding provides protection against the noise that would otherwise be induced by electrostatic coupling. The effectiveness of the shield depends on the *percentage of coverage*. The percentage of coverage is a measure of how much space there is within the shield structure for electrostatic and electromagnetic noise to leak into the signal-carrying conductors.

Electromagnetic Noise

Electromagnetic noise can be generated by a number of electronic devices, including the following:

- Electric motors
- Fluorescent lighting ballasts
- Silicon-controlled rectifier dimmers

Electromagnetic noise can invade a signal-carrying cable by means of inductive coupling. Conventional electrostatic shielding offers no protection. Instead, solid conduit (iron or steel) or simply physical distance is required to minimize electromagnetic noise.

The magnetic fields generated by various sources cut across the conductors of a cable. Because these fields alternately build and collapse, they induce a corresponding alternating noise voltage in the cable. The induced voltage is affected by the following:

- Power line frequency—the higher the frequency, the greater the problem.

- Current flowing in the source—the greater the current, the greater the induced noise.

- Proximity of the interfering source to the cable.

- The length of cable exposed to the noise source.

The ac power line waveform in most parts of the world is 50 or 60 Hz, but this can become contaminated by a rich harmonic spectrum. The harmonics are generated by various sources, most notably by the clipped waveforms emitted by SCR dimmers. SCR dimmers are a major source of noise problems because they generate high harmonics at some settings, and because these higher frequencies more readily couple into circuits.

Although SCR dimmers are a major factor contributing to higher-order power line harmonics, they are not the only problem source of electromagnetic noise. Frequent offenders include saturated power transformer cores and reactive fluorescent lamp ballasts. The noise caused by these sources includes 60 Hz hum and also considerable energy at 120, 240, and 480 Hz. If the power utility service is three-phase, it is also possible to obtain harmonics at 180, 300, 360 Hz, and so forth. Still, it is predominantly low-frequency energy that is heard as "hum," rather than the higher order harmonic energy (as from SCRs), heard as "buzz" (in an audio circuit). The sharp turn-on point of an SCR creates a wide spectrum of noise unless the switching device is properly filtered and shielded.

Figure 16.18 shows some of the more common types of cable shielding.

6.4.4 Skin Effect

Low-level signal cables are particularly susceptible to high-frequency noise energy because of the skin effect of current-carrying conductors. When a conductor carries an alternating current, a magnetic field is produced, which surrounds the wire. This field is expanding and contracting continually as the ac current wave increases from zero to its maximum positive value and back to zero, then through its negative half-cycle. The changing magnetic lines of force cutting the conductor induce a voltage in the conductor in a direction that tends to retard the normal flow of current in the wire. This effect is more pronounced at the center of the conductor. Thus, current within the conductor tends to flow more easily toward the surface of the wire. The

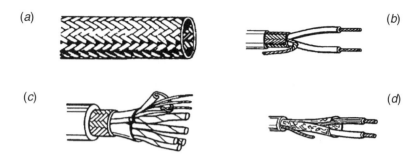

Figure 6.18 Common types of cable shielding: (*a*) basic braid material, (*b*) 2-conductor braided cable with a drain ground wire, (*c*) multi-pair cable using foil wrapped pairs enclosed in a braid, (*d*) 2-conductor cable using foil shield and a drain wire.

higher the frequency, the greater the tendency for current to flow at the surface. The depth of current flow is a function of frequency and is determined from the equation:

$$d = \frac{2.6}{\sqrt{\mu \times f}} \qquad (6.4)$$

Where:
d = depth of current in mils
μ = permeability (copper = 1, steel = 300)
f = frequency of signal in MHz

It can be calculated that at a frequency of 100 kHz, current flow penetrates a conductor by 8 mils. At 1 MHz, the skin effect causes current to travel in only the top 2.6 mils in copper, and even less in almost all other conductors. Therefore, the series impedance of conductors at high frequencies is significantly higher than at ac power line frequencies. This makes low-level signal-carrying cables particularly susceptible to disturbances resulting from RFI.

Both skin effect and self-inductance combine to reduce current flow in a conductor as the frequency is increased. If the *loop area* of the circuit is large, the self-inductance will also be large. In facilities exhibiting uncontrolled geometries, where the return path for current is undefined, the effects of self-inductance will dominate over the skin effect.

Because current penetration is a function of permeability, steel exhibits a greater skin effect than copper. This difference, however, disappears at high frequencies, because permeability rapidly falls off as frequency is increased. At frequencies greater than 250 kHz, the impedance of steel and copper will be about the same, and the inductance effects will dominate.

It follows that in a facility constructed with steel beams, the steel provides a better conductive path than a copper ground strap because of the large surface areas afforded

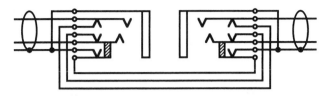

Figure 6.19 Patch-panel wiring for seven-terminal normaling jack fields. Use patch cords that connect ground (sleeve) at both ends.

by the structural steel. This large surface area reduces inductance, which is an important factor in controlling high-frequency noise. A grid formed by structural steel usually provides a better ground system than can be achieved by installing copper conductors, provided the steel elements are welded together. Bolted construction can lead to unpredictable performance, which is likely to deteriorate with time.

6.4.5 Patch-Bay Grounding

Patch panels for audio, video, and data circuits require careful attention to planning to avoid built-in grounding problems. Because patch panels are designed to tie together separate pieces of equipment, often from remote areas of a facility, the opportunity exists for ground loops. The first rule of patch-bay design is to never use a patchbay to switch low-level (microphone) signals. If mic sources must be patched from one location to another, install a bank of mic-to-line amplifiers to raise the signal levels to 0 dBm before connecting to the patchbay. Most video output levels are 1 V P-P, giving them a measure of noise immunity. Data levels are typically 5 V. Although these line-level signals are significantly above the noise floor, capacitive loading and series resistance in long cables can reduce voltage levels to a point that noise becomes a problem.

Newer-design patch panels permit switching of ground connections along with signal lines. Figure 6.19 illustrates the preferred method of connecting a patch panel into a system. Note that the source and destination jacks are *normalled* to establish ground signal continuity. When another signal is plugged into the destination jack, the ground from the new source is carried to the line input of the destination jack. With such an approach, jack cords that provide continuity between sleeve (ground) points are required.

If only older-style conventional jacks are available, use the approach shown in Figure 6.20. This configuration will prevent ground loops, but, because destination shields are not carried back to the source when normaling, noise will be higher. Bus all destination jack sleeves together, and connect them to the local (rack) ground. The wiring methods shown in Figures 6.19 and 6.20 assume balanced input and output lines with all shields terminated at the load (input) end of the equipment.

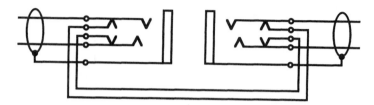

Figure 6.20 Patch-panel wiring for conventional normaling jack fields. Use patch cords that connect ground (sleeve) at both ends.

Video Patch Panel

The jacks commonly used in patch panels in the U.S. conform to Western Electric standard dimensions. The number of insertion cycles a jack can endure should be rated in the tens of thousands. The factors affecting the life and reliability of a jack include contact wear and failure of the termination switch. Desirable features include the following:

- Contacts fully isolated from the panel.

- Sealed metal housing to keep out contaminants and provide EMI protection.

- Easy replacement from the front of the panel.

- Low VSWR (below 600 MHz).

- High signal isolation (40 dB).

- 75 Ω characteristic impedance.

- Wide designation strips, making it easier to label the field and to allow more flexibility in selecting names that will fit on the labels.

If a patch cable is inserted in the signal path of a timed video system, it will delay the signal by an amount determined by its length and physical properties. The patch thereby alters the timing of the signal path. This can be avoided by using phase-matched normal-through patch panels. The design of these patch panels anticipates the delay caused by a fixed length of patch cable by including that length in the loop-through circuit.

With phase-matched panels, the normaling connection in each connector module includes a length of cable that provides a fixed delay through the panel, usually 3 ft (0.914 m). If a patch cord of the same length as the internal cable is used to make connections between patch points, the delay will be the same as that of the normal-through delay; there will be no change in the timing of the signals passing through the patch panel. When a patch cord is plugged in, it is substituted for the loop cable through the switching mechanism normally used in normalled patch connectors. Thus, critical timing relationships can be maintained. Figure 6.21 shows a phase-matched patch panel.

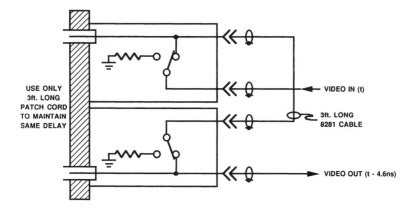

Figure 6.21 A phase-matched video patch panel.

In a normal uncompensated patch panel, when a cable is used to patch between two points on the panel, the length of the patch cord is added to that of the cables connected to the patch. The additional cable length delays the signal by approximately 1.52 ns/ft (5 ns/m). To avoid the delay problems associated with conventional patch panels, phase-matched normal-through video patch panels should be used.

If phase-matched patch panels are used, all of the patch cords must be the same length as the delay built into the patch panel. Obviously, if all of the patch cords must be the same short length for the phase-matched panel, it would not be possible to patch between panels that are separated by a longer distance than the cord can reach. This limitation should be considered when laying out patch panels in a rack.

Color-coded cables can be specified. When different-length patch cords are specified, different colors can be used to distinguish one length from another.

6.5 Computer Networks

The *open system interconnections* (OSI) model is the most broadly accepted explanation of LAN transmissions in an open system. The reference model was developed by the International Organization for Standardization (ISO) to define a framework for computer communication. The OSI model divides the process of data transmission into the following steps:

- Physical layer
- Data-link layer
- Network layer
- Transport layer
- Session layer

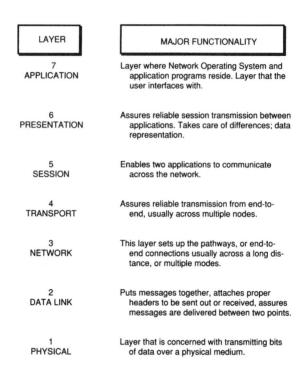

LAYER	MAJOR FUNCTIONALITY
7 APPLICATION	Layer where Network Operating System and application programs reside. Layer that the user interfaces with.
6 PRESENTATION	Assures reliable session transmission between applications. Takes care of differences; data representation.
5 SESSION	Enables two applications to communicate across the network.
4 TRANSPORT	Assures reliable transmission from end-to-end, usually across multiple nodes.
3 NETWORK	This layer sets up the pathways, or end-to-end connections usually across a long distance, or multiple modes.
2 DATA LINK	Puts messages together, attaches proper headers to be sent out or received, assures messages are delivered between two points.
1 PHYSICAL	Layer that is concerned with transmitting bits of data over a physical medium.

Figure 6.22 The OSI reference model.

- Presentation layer
- Application layer

An overview of the OSI model is illustrated in Figure 6.22.

6.5.1 Physical Layer

Layer 1 of the OSI model is responsible for carrying an electrical current through the computer hardware to perform an exchange of information. The physical layer is defined by the following parameters:

- Bit transmission rate.
- Type of transmission medium (twisted-pair, coaxial cable, or fiber-optic cable).
- Electrical specifications, including voltage- or current-based, and balanced or unbalanced.
- Type of connectors used (normally RJ-45 or DB-9).
- Many different implementations exist at the physical layer.

Installation Considerations

Layer 1 can exhibit error messages as a result of overusage. For example, if a file server is being burdened with requests from workstations, the results may show up in error statistics that reflect the server's inability to handle all incoming requests. An overabundance of response timeouts may also be noted in this situation. A response timeout (in this context) is a message sent back to the workstation stating that the waiting period allotted for a response from the file server has passed without action from the server.

Error messages of this sort, which can be gathered by any number of commercially available software diagnostic utilities, can indicate an overburdened file server or a hardware flaw within the system. Intermittent response timeout errors can be caused by a corrupted network interface card (NIC) in the server. A steady flow of timeout errors throughout all nodes on the network may indicate the need for another server or bridge. Hardware problems are among the easiest to locate. In simple configurations, where something has suddenly gone wrong, the physical and data-link layers are usually the first suspects.

6.5.2 Data Link Layer

Layer 2 of the OSI model, the data-link layer, describes hardware that enables data transmission (NICs and cabling systems). This layer integrates data packets into messages for transmission and checks them for integrity. Sometimes layer 2 will also send an "arrived safely" or "did not arrive correctly" message back to the transport layer (layer 4), which monitors this communications layer. The data-link layer must define the frame (or package) of bits that is transmitted down the network cable. Incorporated within the frame are several important fields:

- Addresses of source and destination workstations.

- Data to be transmitted between workstations.

- Error control information, such as a *cyclic redundancy check* (CRC), which assures the integrity of the data.

The data-link layer must also define the method by which the network cable is accessed, because only one workstation may transmit at a time on a baseband LAN. The two predominant schemes are:

- *Token passing*, used with the ARCnet and token-ring networks.

- *Carrier sense multiple access with collision detection* (CSMA/CD), used with Ethernet and starLAN networks.

At the data-link layer, the true identity of the LAN begins to emerge.

Installation Considerations

Because most functions of the data-link layer (in a PC-based system) take place in integrated circuits on NICs, software analysis is generally not required in the event of an installation problem. As mentioned previously, when something happens on the network, the data-link layer is among the first to be suspect. Because of the complexities of linking multiple topologies, cabling systems, and operating systems, the following failure modes may be experienced:

- RF disturbance. Transmitters, ac power controllers, and other computers can all generate energy that may interfere with data transmitted on the cable. RF interference (RFI) is usually the single biggest problem in a broadband network. This problem can manifest itself through excessive checksum errors and/or garbled data.

- Excessive cable run. Problems related to the data-link layer may result from long cable runs. Ethernet runs can stretch to 1,000 ft, depending on the cable. A typical token-ring system can stretch 600 ft, with the same qualification. The need for additional distance can be accommodated by placing a bridge, gateway, active hub, equalizer, or amplifier on the line.

The data-link layer usually includes some type of routing hardware with one or more of the following:

- Active hub

- Passive hub

- Multiple access units (for token-ring, starLAN, and ARCnet networks)

6.5.3 Network Layer

Layer 3 of the OSI model guarantees the delivery of transmissions as requested by the upper layers of the OSI. The network layer establishes the physical path between the two communicating endpoints through the *communications subnet*, the common name for the physical, data-link, and network layers taken collectively. As such, layer 3 functions (routing, switching, and network congestion control) are critical. From the viewpoint of a single LAN, the network layer is not required. Only one route—the cable—exists. Internetwork connections are a different story, however, because multiple routes are possible. The internet protocol (IP) and internet packet exchange (IPX) are two examples of layer 3 protocols.

Installation Considerations

The network layer confirms that signals get to their designated targets, and then translates logical addresses into physical addresses. The physical address determines where the incoming transmission is stored. Lost data errors can usually be traced back to the network layer, in most cases incriminating the network operating system. The network layer is also responsible for statistical tracking and communications with other environments, including gateways. Layer 3 decides which route is the best to take, given the needs of the transmission. If router tables are being corrupted or excessive time is required to route from one network to another, an operating system error on the network layer may be involved.

6.5.4 Transport Layer

Layer 4, the transport layer, acts as an interface between the bottom three and the upper three layers, ensuring that the proper connections are maintained. It does the same work as the network layer, only on a local level. The network operating system driver performs transport layer tasks.

Installation Considerations

Connection flaws between computers on a network can sometimes be attributed to the *shell driver*. The transport layer may be able to save transmissions that were en route in the case of a system crash, or reroute a transmission to its destination in case of a primary route failure. The transport layer also monitors transmissions, checking to make sure that packets arriving at the destination node are consistent with the build specifications given to the sending node in layer 2. The data-link layer in the sending node builds a series of packets according to specifications sent down from higher levels, then transmits the packets to a destination node. The transport layer monitors these packets to ensure that they arrive according to specifications indicated in the original build order. If they do not, the transport layer calls for a retransmission. Some operating systems refer to this technique as a *sequenced packet exchange* (SPX) transmission, meaning that the operating system guarantees delivery of the packet.

6.5.5 Session Layer

Layer 5 is responsible for turning communications on and off between communicating parties. Unlike other levels, the session layer can receive instructions from the application layer through the network basic input/output operation system (netBIOS), skipping the layer directly above it. The netBIOS protocol allows applications to "talk" across the network. The session layer establishes the session, or logical connection, between communicating host processors. Name-to-address translation is another important function; most communicating processors are known by a common name, rather than a numerical address.

Installation Considerations

Multi-vendor problems can often arise in the session layer. Failures relating to gateway access usually fall into layer 5 for the OSI model, and are often related to compatibility issues.

6.5.6 Presentation Layer

Layer 6 translates application layer commands into syntax that is understood throughout the network. It also translates incoming transmissions for layer 7. The presentation layer masks other devices and software functions. Reverse video, blinking cursors, and graphics also fall into the domain of the presentation layer. Layer 6 software controls printers and plotters, and may handle encryption and special file formatting. Data compression, encryption, and ASCII translations are examples of presentation layer functions.

Installation Considerations

Failures in the presentation layer are often the result of products that are not compatible with the operating system, an interface card, a resident protocol, or another application.

6.5.7 Application Layer

At the top of the seven-layer stack is the application layer. It is responsible for providing protocols that facilitate user applications. Print spooling, file sharing, and e-mail are components of the application layer, which translates local application requests into network application requests. Layer 7 provides the first layer of communications into other open systems on the network.

Installation Considerations

Failures at the application layer usually center on software quality and compatibility issues. The program for a complex network may include latent faults that will manifest only when a specific set of conditions are present. The compatibility of the network software with other programs is another source of potential complications.

6.5.8 Transmission System Options

A variety of options beyond the traditional local serial interface are available for linking intelligent devices. The evolution of wide area network (WAN) technology has permitted efficient two-way transmission of data between distant computer systems. High-speed facilities are cost-effective and widely available from the telephone company (telco) central office to the customer premises. Private communications companies also provide interconnection services.

LANs have proliferated and integrated with WANs through *bridges* and *gateways*. Interconnections via fiber-optic cable are common. Further extensions of the basic LAN include the following:

- *Campus area network* (CAN)—designed for communications within an industrial or educational campus.

- *Metropolitan area network* (MAN)—designed for communications among different facilities within a certain metropolitan area. MANs generally operate over common-carrier-owned switched networks installed in and over public rights of way.

- *Regional area network* (RAN)—interconnecting MANs within a unified geographical area, generally installed and owned by interexchange carriers (IECs).

- *Wide area network* (WAN)—communications systems operating over large geographic areas. Common carrier networks interconnect MANs and RANs within a contiguous land mass, generally within a country's political boundaries.

- *Global area network* (GAN)—networks interconnecting WANs, both across national borders and ocean floors, including between continents.

These network systems can carry a wide variety of multiplexed analog and/or digital signal transmissions on a single piece of coax or fiber.

System Design Alternatives

The signal form at the input and/or output interface of a large cable or fiber system may be either analog or digital, and the number of independent electrical signals transmitted may be one or many. Independent electrical signals may be combined into one signal for optical transmission by virtually unlimited combinations of electrical analog frequency division multiplexing (using analog AM and/or FM carriers) and digital bit stream multiplexing. Frequency division multiplexing involves the integration of two or more discrete signals into one complex electrical signal.

With the current availability of fiber-optic transmission lines, fiber interconnection of data networks is the preferred route for new systems. Three primary multiplexing schemes are used for fiber transmission:

- *Frequency division multiplexing* (FDM)

- *Time division multiplexing* (TDM)

- *Wave(length) division multiplexing* (WDM)

Frequency Division Multiplexing

The FDM technique of summing multiple AM or FM carriers is widely used in coaxial cable distribution. Unfortunately, nonlinearity of optical devices operated in the intensity-modulation mode can result in substantial—and often unacceptable—noise and intermodulation distortion in the delivered signal channels. Wide and selective spacing of carriers ameliorates this problem to some degree.

Time Division Multiplexing

TDM involves sampling the input signals at a high rate, converting the samples to high-speed digital codes, and interleaving the codes into pre-determined time slots.

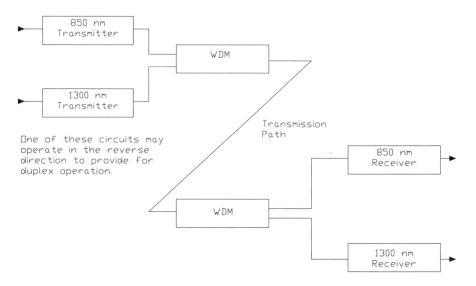

Figure 6.23 Basic operation of a wave division multiplexer. This type of passive assembly is created by fusing optical-fiber pigtails.

The principles of digital TDM are straightforward. Specific-length bit groups in a high-speed digital bit stream are repetitively allocated to carry the digital representations of individual analog signals and/or the outputs of separate digital devices.

Wave(length) Division Multiplexing

This multiplexing technique, illustrated in Figure 6.23, reduces the number of optical fibers required to meet a specific transmission requirement. Two or more complete and independent fiber-transmission systems operating at different optical wavelengths can be transported over a single fiber by combining them in a passive optical multiplexer. This device is an assembly in which pigtails from multiple optical transmitters are fused together and spliced into the transporting fiber. Demultiplexing the optical signals at the receiver end of the circuit is accomplished in an opposite-oriented passive optical multiplexer. The pigtails are coupled into photodetectors through wavelength-selective optical filters.

6.5.9 Selecting Cable for Digital Signals

Cable for the transmission of digital signals is selected on the basis of its electrical performance: the ability to transmit the required number of pulses at a specified bit rate over a specified distance, and its conformance to appropriate industry or government standards. A wide variety of data cables are available from manufacturers. Figure 6.24 illustrates some of the more common types. The type of cable chosen for an application is determined by the following:

(a)

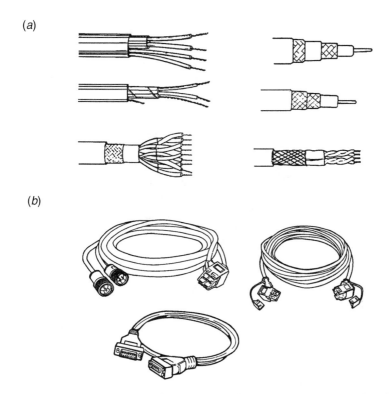

(b)

Figure 6.24 Common types of data cable: (a) shielded pair, multi-pair shielded, and coax, (b) data cable with various terminations.

Type of network involved. Different network designs require different types of cable.

- Distance to be traveled. Long runs require low-loss cable.

- Physical environment. Local and national safety codes require specific types of cable for certain indoor applications. Outdoor applications require a cable suitable for burial or exposure to the elements.

- Termination required. The choice of cable type may be limited by the required connector termination on one or both ends.

6.5.10 Data Patch Panel

The growth of LANs has led to the development of a variety of interconnection racks and patch panels. Figure 6.25 shows two common types. Select data patch panels that offer many cycles of repeated insertion and removal. Use components specifically designed for network interconnection. Such components include the following:

- Twisted-pair network patch panels

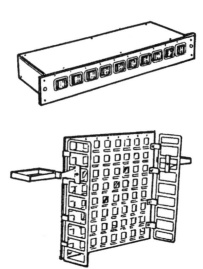

Figure 6.25 Data network patch-panel hardware.

- Coax-based network patch panels
- Fiber-based network patch panels
- Modular feed-through (normalled) patch panels
- Pre-assembled patch cables of various lengths
- Pre-assembled "Y" patch cables
- Patch cables offering different connectors on each end
- Media filter cables
- Balanced-to-unbalanced (balun) cable assemblies

Connector termination options for patch hardware include the following:

- Insulation displacement (punch-block) for twisted-pair cable
- Screw terminal (for twisted-pair)
- BNC connectors for coax
- Fiber-termination hardware

Although the cost of pre-assembled network patch panels and patch cables is higher than purchasing the individual components and then assembling them, most system engineers should specify factory-assembled hardware. Reliability is greater with pre-assembled elements, and installation is considerably faster.

6.6 Optical Cable

Fiber-optic (FO) technology offers the end-user a number of benefits over metallic cable, including:

- **Signal-carrying ability.** The bandwidth information-carrying capacity of a communications link is directly related to the operating frequency. Light carrier frequencies are several orders of magnitude higher than the highest radio frequencies. Fiber-optic systems easily surpass the information-carrying capacity of microwave radio and coaxial cable alternatives; and fiber's future carrying capacity has only begun to be used. Fiber provides bandwidths in excess of several gigahertz per kilometer, which allows high-speed transfer of all types of information. Multiplexing techniques allow many signals to be sent over a single fiber.

- **Low loss.** A fiber circuit provides substantially lower attenuation than copper cables and twisted pairs. It also requires no equalization. Attenuation below 0.5 dB/km is available for certain wavelengths.

- **Electrical isolation.** The fiber and its coating are dielectric material, and the transmitter and receiver in each circuit are electrically isolated from each other. Isolation of separated installations from respective electrical grounds is assured if the strength material (messenger) in the cable is also a dielectric. Lightwave transmission is free of spark hazards and creates no EMI. All-dielectric fiber cable may also be installed in hazardous or toxic environments.

- **Size and weight.** An optical waveguide is less than the diameter of a human hair. A copper cable is many times larger, stiffer, and heavier than a fiber that carries the same quantity of signals. Installation, duct, and handling costs are much lower for a fiber installation than for a similar coaxial system. Fiber cable is the only alternative for circuit capacity expansion when ducts are full of copper.

6.6.1 Types of Fibers

Of the many ways to classify fibers, the most informative is by *refractive index profile* and number of modes supported. The two main types of index profiles are *step* and *graded*. In a *step index* fiber, the core has a uniform index with a sharp change at the boundary of the cladding. In a *graded index* fiber, the core index is not uniform; it is highest at the center and decreases until it matches the cladding.

Step Index Multi-mode Fiber

A multi-mode step index fiber typically has a core diameter in the 50 to 1,000 micron range. The large core permits many modes of propagation. Because light will reflect differently for different modes, the path of each ray is a different length. The lowest-order mode travels down the center; higher-order modes strike the core-cladding interface at angles near the critical angle. As a result, a narrow pulse of light spreads out as it travels through this type of fiber. This spreading is called *modal dispersion* (Figure 6.26).

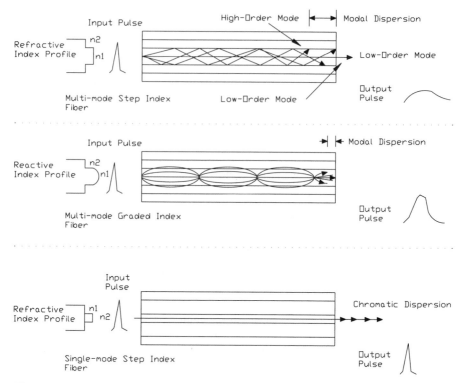

Figure 6.26 Modal dispersion in an FO cable. The core diameter and its refractive index characteristics determine the light propagation path(s) within the fiber core.

Step Index Single (Mono) -Mode Fiber

Modal dispersion can be reduced by making the fiber core small, typically 5 to 10 microns (1/6 the diameter of a human hair). At this diameter, only one mode propagates efficiently. The small size of the core makes it difficult to splice. Single mode of propagation permits high-speed, long-distance transmission.

Graded Index Multi-mode Fiber

Like the step index single-mode fiber, a graded index fiber also limits modal dispersion. The core is essentially a series of concentric rings, each with a lower refractive index. Because light travels faster in a lower-index medium, light further from the axis travels faster. Because high-order modes have a faster average velocity than low-order modes, all modes tend to arrive at a given point at nearly the same time. Rays of light are not sharply reflected by the core-cladding interface; they are refracted successively by differing layers in the core.

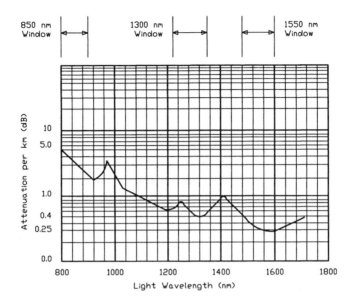

Figure 6.27 Fiber attenuation vs. light wavelength characteristics. Attenuation has been reduced steadily in the last two decades through improved fiber drawing techniques and a reduction in impurities. It has now approached the theoretical limits of silica-based glass at the 1,300 and 1,550 nm wavelengths.

6.6.2 Characteristics of Attenuation

Attenuation represents a loss of power. During transit, some of the light in a fiber-optic system is absorbed into the fiber or scattered by impurities. Attenuation for a fiber cable is usually specified in decibels per kilometer (dB/km). For commonly available fibers, attenuation ranges from approximately 0.5 dB/km for premium single-mode fibers to 1,000 dB/km for large-core plastic fibers. Because emitted light represents power, 3 dB represents a doubling or halving of any reference power level.

Attenuation and light wavelength are uniquely related in fiber-transmission systems. This is illustrated in Figure 6.27. Most fibers have a medium loss region in the 800–900 nm wavelength range (3–5 dB/km), a low loss region in the 1,150–1,350 nm range (0.6–1.5 dB/km), and a very low loss region (less than 0.5 dB/km) in the 1,550 nm range. As a result, optimum performance is achieved by careful balancing of fiber, light source wavelength, and distance requirements.

Light intensity attenuation has no direct effect on the bandwidth of the electrical signals being transported. There is a direct correlation, however, between the S/N of the fiber receiver electronic circuits and the usable recovered optical signal.

6.6.3 Types of Cable

The first step in packaging an optical fiber into a cable is the extrusion of a layer of plastic around the fiber. This layer of plastic is called a *buffer tube*; it should not be confused with the buffer coating. The buffer coating is placed on the fiber by the fiber manufacturer. The buffer tube is placed on the fiber by the cable manufacturer. This extrusion process can produce two different cable designs:

- *Tight tube design*—The inner diameter of the plastic (buffer tube) is the same size as the outer diameter of the fiber, and is in contact with the fiber around its circumference.

- *Loose tube design*—The layer of plastic is significantly larger than the fiber, and, therefore, the plastic is not in contact with the fiber around the circumference of the fiber.

The two types of fiber cable are illustrated in Figure 6.28. Note that the loose tube design is available configured either as a *single-fiber-per-tube* (SFPT) or *multiple-fibers-per-tube* (MFPT) design. The six-fibers-per-tube MFPT design is often used for data communications.

After a fiber (or group of fibers) has been surrounded by a buffer tube, it is called an *element*. The cable manufacturer uses elements to build up the desired type of cable. In building the cable from elements, the manufacturer can create six distinct designs:

- Breakout design

- MFPT, central loose tube design

- MFPT, stranded loose tube design

- SFPT, stranded loose tube design

- *Star*, or *slotted core*, design

- *Tight tube*, or *stuffed*, design

Breakout Design

In the breakout design, shown in Figure 6.29, the element or buffered fiber is surrounded with a flexible-strength member, often Kevlar. The strength member is surrounded by an inner jacket to form a subcable, as shown. Multiple subcables are stranded around a central strength member or filler to form a cable core. This cable core is held together by a binder thread or Mylar wrapping tape. The core is surrounded by an extruded jacket to form the final cable.

Optional steps for this design include additional strength members, jackets, or armor. The additional jackets may be extruded directly on top of one another or separated by additional external strength members.

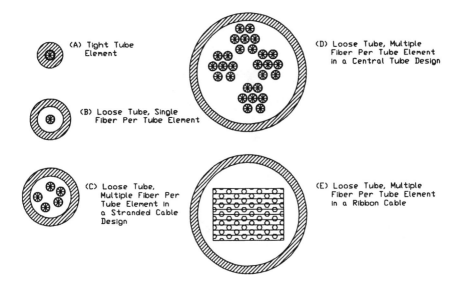

Figure 6.28 Loose-tube cables are available in either single-fiber-per-tube (SFPT) or multiple-fibers-per-tube (MFPT) designs. In both cases, the diameter of the plastic tube surrounding the core is larger than the outside diameter of the core. In a tight tube cable; the inner diameter tube is the same as the outer diameter of the fiber.

MFPT, Central Loose Tube Design

Fibers are placed together to form groups. Sometimes, the fibers are laid along a ribbon in groups of 12. These ribbons are then stacked up to 12 high and twisted. This version of the central loose tube design is referred to as a *ribbon design*, and was developed by AT&T. The space between the fibers and the tube can be filled with a water-blocking compound.

MFPT, Stranded Loose Tube Design

Multiple buffer tubes are stranded around a central strength member or filler to form a core, as illustrated in Figure 6.30. This cable core is held together by a binder thread or Mylar wrapping tape. The core is surrounded by an extruded jacket to form a finished cable. Optional jacketing, strength members, or armor can be added.

SFPT, Stranded Loose Tube Design

This type of cable is manufactured similarly to MFPT cable. The primary difference is that the cable has one fiber per tube and smaller-diameter buffer tubes.

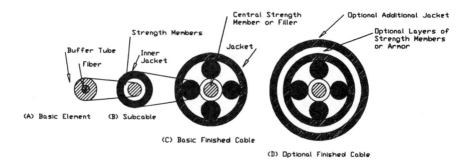

Figure 6.29 In the breakout type of cable, each element is surrounded by a flexible strength member, which is then surrounded by an inner jacket. This forms a subcable, which is incorporated into a larger cable. Optional additional jackets or armor can be applied.

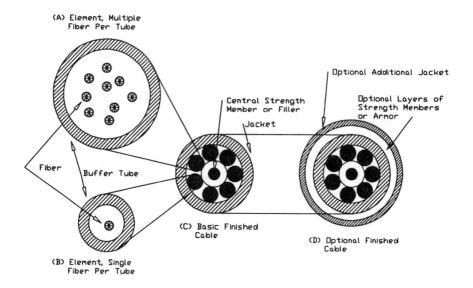

Figure 6.30 The MFPT stranded loose tube design relies on a center strength member to form the cable core. Multiple elements are then added to build up the desired cable capacity.

Star, or Slotted Core, Design

This design is seldom used in the United States. In this scheme, the buffer tubes (usually MFPT) are laid in helical grooves, which are formed in the filler in the center of

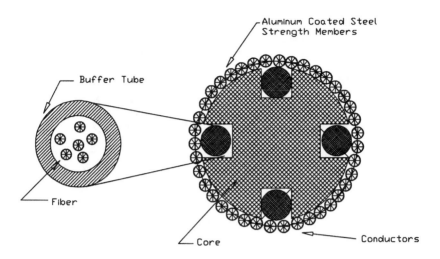

Figure 6.31 Utility companies sometimes use an optical power ground wire type of cable because it incorporates a metallic power ground wire within the design. The cable is based on a slotted-core configuration, but with the addition of helically wrapped wires around the outside for strength and conductivity.

the cable. The core is then surrounded by an extruded jacket to form a finished cable. One variation of this design, shown in Figure 6.31, is used by power utility companies. This optical cable provides a conductive ground path from end to end. Instead of a jacket, the cable has helically wrapped wires, some of which are conductors and strength members.

Tight Tube, or Stuffed, Design

This design is based on the tight tube element. The designs are common in that the core is filled, or stuffed, with flexible strength members, usually Kevlar. The design usually incorporates two or more fibers, as illustrated in Figure 6.32.

Application Considerations

Performance advantages exist for all designs, depending on what parameter is considered. For example, the tight tube design can force the ends of a broken fiber to remain in contact even after the fiber has broken. The result is that transmission may still be possible. When reliability is paramount, this feature may be important.

Loose tube designs have a different performance advantage. They offer a mechanical *dead zone*, which is not available in tight tube designs. The effect is that stress can be applied to the cable without that stress being transferred to the fiber. This dead zone exists for all mechanical forces, including tensile and crush loads, and bend strains. Tight tube designs do not have this mechanical dead zone. In the tight tube design, any force

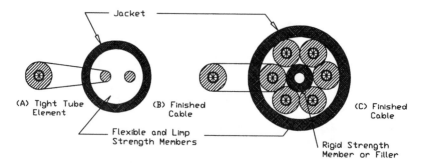

Figure 6.32 The tight tube stuffed design relies on a core filled with flexible strength members, usually Kevlar. Typically, two or more fibers are contained within the cable.

applied to the cable is also applied to the fibers. Loose tube designs also offer smaller size, lower cost, and smaller bend radii than tight tube designs.

When cable cost alone is considered, loose tube designs have the advantage over tight tube, breakout designs in long-length applications. However, when total installation cost is considered, the loose tube designs may or may not have a cost advantage. This is because loose tube designs have higher connector installation costs. The cost factor is composed of two parts:

- Labor cost

- Equipment cost

All designs, other than the breakout design, require handling of bare fibers or fibers with tight tubes. During this handling, fibers can be broken, especially where inexperienced personnel are involved.

6.6.4 Specifying Fiber-Optic Cable

In order to completely specify a fiber-optic cable, four primary performance categories must be quantified:

- Installation specifications

- Environmental specifications

- Fiber specifications

- Optical specifications

These criteria are outlined in Table 6.3. Note that not all specifications apply to all situations. The system engineer must review the specific application to determine which of the specifications are applicable. For example, cable installed in conduit or in protected locations will not need to meet a crush load specification.

Installation Specifications

The installation specifications are those that must be met to ensure successful cable installation. There are six:

- Maximum installation load in kilograms-force or pounds-force. This is the maximum tensile load that can be applied to a cable without causing fiber breakage or a permanent change in attenuation. This characteristic must always be specified. Load values for some typical installations are shown in Table 6.4. If the application requires a strength higher than those listed, specify a higher-strength cable. The increased cost of specifying a higher-strength cable is small, typically 5 to 10 percent of the cable cost.

- Minimum installation bend radius in inches or millimeters. This is the minimum radius to which the cable can be bent while loaded at the maximum installation load. This bending can be done without causing a permanent change in attenuation, fiber breakage, or breakage of any portion of the cable structure. The bend radius is usually specified as no less than 20 times the cable diameter. To determine this value, examine the locations where the cable will be installed, and identify the smallest bend the cable will encounter. Conversely, the system engineer can choose the cable and then specify that the conduits or ducts not violate this radius. The radius is actually limited more by the cabling materials than by the bend radius of the fiber.

- Diameter of the cable. Despite the space-effective nature of FO cable, it still must reside in the available space. This is especially true if the cable is to be installed in a partially filled conduit.

- Diameter of subcables or elements. The diameter of the subcable or the cable elements can become a limiting factor. In the case of a breakout-style cable, the diameter of the subcable must be smaller than the maximum diameter of the connector boot so that the boot will fit on the subcable. Also, the diameter of the element must be less than the maximum diameter acceptable to the backshell of the connector. Most breakout cables have tight-tube elements, usually with a diameter of 1 mm or less.

- Recommended temperature range for installation (°C). All cables have a temperature range within which they can be installed without damage to either the cable materials or the fibers. Generally, the temperature range is affected more by the cable materials than the fibers. Not all cable manufacturers include this parameter in their data sheets. If the parameter is not specified, select a conservative temperature range of operation.

- Recommended temperature range for storage (°C). In severe climates, such as deserts and the Arctic, the system engineer must specify a recommended temperature range for storage in °C. This range will strongly influence the materials used in the cable.

Table 6.3 Fiber Cable Specification Considerations

Installation Specifications:
    Maximum recommended installation load
    Minimum installation bend radius
    Cable diameter
    Diameter of subcables
    Maximum installation temperature range
    Maximum storage temperature range
Environmental Specifications:
    Temperature range of operation
    Minimum recommended unloaded bend radius
    Minimum long-term bend radius
    Maximum long-term use load
    Vertical rise
    National Electric Code or local electrical code requirements
    Flame resistance
    UV stability or UV resistance
    Resistance to rodent damage
    Resistance to water damage
    Crushing characteristics
    Resistance to conduction under high-voltage fields
    Toxicity
    High flexibility: static vs. dynamic applications
    Abrasion resistance
    Resistance to solvents, petrochemicals, and other substances
    Hermetically sealed fiber
    Radiation resistance
    Impact resistance
    Gas permeability
    Stability of filling compounds
Fiber Specifications:
Dimensional considerations:
    Core diameter
    Clad diameter
    Buffer coating diameter
    Mode field diameter
Optical Specifications:
Power considerations:
    Core diameter
    Numerical aperture
    Attenuation rate
    Cut-off wavelength
    Capacity Considerations:
    Bandwidth-distance product (dispersion)
    Zero-dispersion wavelength

Table 6.4 Maximum Installation Loads that Fiber Cable can be Exposed to in Various Applications

Application	Typical Maximum Recommended Installation Load Pounds force
1 fiber in raceway or tray	67 lb
1 fiber in duct or conduit	125 lb
2 fibers in duct or conduit	250–500 lb
Multi-fiber (6–12) cables	500 lb
Direct burial cables	600–800 lb
Lashed aerial cables	300 lb
Self-supported aerial cables	600 lb

Environmental Specifications

Environmental specifications are those that must be met to ensure successful long-term cable operation. Most of the items listed in Table 6.3 are self-explanatory. However, some environmental specifications deserve special attention.

The temperature range of operation is that range in which the attenuation remains less than the specified value. There are few applications where FO cable cannot be used because of temperature considerations. FO cables composed of plastic materials have maximum and minimum temperature points. If these are exceeded, the materials will not maintain their mechanical properties. After long exposures to high temperatures, plastics deteriorate and become soft. Some materials will begin to crack. After exposures to low temperatures, plastics become brittle and crack when flexed or moved. Under such conditions, the cable coverings will cease to protect the fiber.

Another reason for considering the temperature range of operation is the increase in attenuation that occurs when fibers are exposed to temperature extremes. This sensitivity occurs when the fibers are bent. When a cable is subject to extreme temperatures, the plastic materials will expand and contract. The rates at which the expansion and contraction take place are much greater (perhaps 100 times) than the rates of glass fibers. This movement results in the fiber being bent at a microscopic level. The fiber is either forced against the inside of the plastic tube as the plastic contracts, or the fiber is stretched against the inside of the tube as the plastic expands. In either case, the fiber is forced to conform to the microscopically uneven surface of the plastic. On a microscopic level, this is similar to placing the fiber against sandpaper. The bending results in light escaping from the core of the fiber. The result is referred to as a *microbend-induced* increase in attenuation.

6.7   Cabling Hardware

Permanent installation of interconnected equipment requires that some orderly means be employed for routing and supporting the cables. To provide strain relief on the connectors and a neat installation, wires and cables should be bound or harnessed into bundles running between system components and secured to some type of supporting structure or frame. Cables carrying like or related signals should be grouped together to prevent crosstalk.

6.7.1 Cable Ties

Individual cable ties provide the easiest and most flexible means of binding cable bundles. There are many factors to consider when selecting the proper cable tie for each application. Generally, the environmental concerns are limited to the effects of extreme heat and, in the case of mobile installations, extreme cold, moisture, and ultraviolet light (from sunlight). The physical properties of plastic materials normally degrade during exposure to high temperature because of oxidation. The maximum temperature for successful service depends on the material used and environmental conditions. Initially, plastics become more flexible and weaker when exposed to high temperatures. After a period of time, oxidation may occur, which will cause the material to become brittle, making plastic cable ties more susceptible to failure from impact and vibration. Low-temperature exposure will also make most plastics more brittle during exposure, but little permanent degradation of the properties remains when the material is returned to room temperature.

Mechanical stress also affects the life of a cable tie. As the bundle diameter is reduced, the tie experiences more bending stress. A thick strap on a small diameter represents a high-stress condition. If the tie is under high load, this will add more stress to the tie body. A thinner tie will have shortened life because surface cracks will penetrate the thickness of the tie faster. Applications subject to high vibration will result in impact stresses, which can cause surface cracks to propagate.

Several other external factors affect the life of a cable tie. Chemical exposure can degrade the tie material. This is the most detrimental factor to the life of a tie. Direct sunshine, high altitudes, and high temperature also decrease the life of the cable tie. Dry environments cause nylon 6/6 ties to become more brittle. High humidity and temperatures can result in degradation because of hydrolysis in nylon.

Hand-tensioning of a cable tie can result in too little or too much tension. Tie-wrap tensioning tools are available from tie manufacturers. These can be adjusted to apply the proper tension for each type and size of tie. One operation applies tension to the tie and cuts the tail of the tie when the proper tension has been reached.

Figure 6.33 shows some of the tie-wrap products commonly available from several suppliers. Use of a tensioning tool is illustrated in Figure 6.34.

A wide variety of cabling methods may be applied in a given facility. The type used will be dictated by the following:

- Number and size of cables to be secured

- Type of cable support used (cable tray, conduit, or clamps)

- Exposure to vibration or harsh environmental conditions

- Voltages and currents being transported

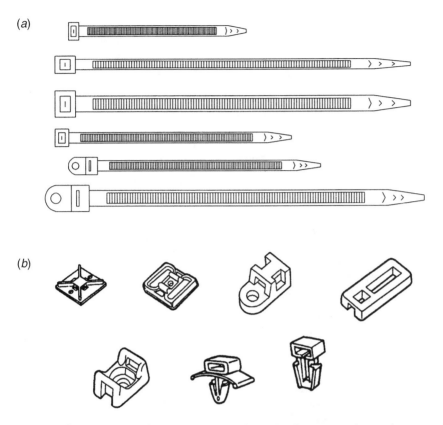

Figure 6.33 A selection of tie-wrap products available from several manufacturers: (*a*) tie-wraps of various sizes, (*b*) mounting clamps.

Figure 6.35 shows some of the more common cabling products available for small- to medium-sized runs.

Twist-on cable wrap protects cables against rough surfaces. It absorbs vibration and reduces impact damage, insulates cables, and resists abrasion. The product twists on like tape, making it easy to apply. This also facilitates breakouts of single or multiple wires and rerouting of replacement wires. Twist-on cable wrap products are available in polyethylene, nylon, and flame-retardant polyethylene for bundle sizes from 1/16-in (1.6 mm) to 7-in (178 mm) diameters. The polyethylene and nylon material is available in black and natural colors. Flame-retardant polyethylene is typically available in its natural color only. Products are also available that conform to military specifications.

6.7.2 Braided Sleeving

Braided sleeving is designed to protect wire bundles, harnesses, and cable assemblies from mechanical and environmental damage. The high-tensile-strength, damage-re-

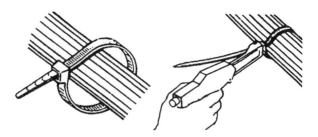

Figure 6.34 Use of a tie-wrap tensioning tool on a cable bundle.

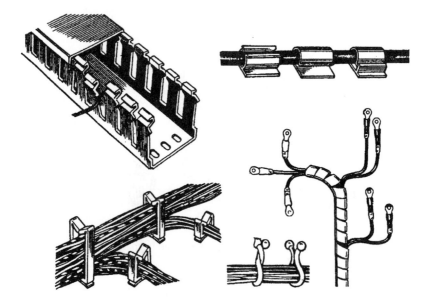

Figure 6.35 Some of the common methods of cabling small- to medium-sized runs in a facility.

sistant filament braid protects against rough surfaces, sharp edges or corners. The product offers the following benefits:

- Resists scuffing and abrasion

- Cushions against vibration and damage from impact

- Prevents condensation while allowing complete drainage

- Dissipates heat and moisture

- Does not degrade in most fluid environments

Sleeved cable bundles are more organized and more attractive than most other cabling techniques. The expandable, open-weave construction works like a Chinese finger puzzle. When pushed over a cable bundle, the weave expands. When pulled taught, it tightens around the bundle. Braided sleeving, woven from polyester and flame retardant polyester yarn, slips on quickly and is self-fitting over irregular shapes and contours. It fits a wide range of diameters, simplifying inventory requirements. The open weave allows for easy inspection of sleeved components and easy break out of individual cables. In spite of its high tensile strength, braided sleeving maintains full cable flexibility, even at low temperatures. It also does not kink like tubing when bent. It can be flexed repeatedly without damage. Because of its low surface friction, wire-pulling compounds are not required for installation. Applying slight tension after application makes for an even tighter fit.

6.7.3 Cable Identification and Marking

Cable-jacketing materials are available in a variety of colors, which can be used to identify different signal types. In large quantities, and for a fee, suppliers will apply colored striping onto the cable. Using this approach, many more color codes are possible for cable identification. Because of the limited number of stock colors, they can only be used to identify the type of signal certain cables are carrying, for example, video, sync, or data. Red, green, and blue cables might be used to identify the three cables in an RGB bundle used to carry component video signals. Colored tape, available in many widths, can also be used to distinguish particular cables.

Wire Markers

There are several types of wire markers available for placing identification information onto cables and wires. They include heat-shrinkable tubing or sleeving, wrap-around adhesive tape labels, and write-on cable ties. Each type has its own advantages and disadvantages.

Heat-shrinkable sleeving is available in several materials (PVC and polyolefin), diameters, and colors. When heated with a heating tool, the sleeving can be reduced (recovered) to one-third its supplied (expanded) size. Identification information is written onto the sleeving by hand using a pen and permanent ink. Automatic and hand-operated tools are available for printing white or black number and letter codes on dark- or light-colored sleeving, respectively. Print wheels must be set manually for each code. The manual tool is acceptable for small quantities, but it is impractical when large quantities are required. The ink used must be smear-resistant and permanent in all cases. Use high-carbon, noncorrectable fabric ribbon with impact printers. Depending on the sleeve material, the markings may be permanent upon printing before shrinking. The markings on some sleeving become permanent only after shrinking.

A computer can to be used to print sequential numbers or more elaborate labels from a cable schedule database. Software for this purpose is usually available from the manufacturers of the guide-mounted sleeving. A flat-file or relational-type database program can be used to maintain a cable schedule, and can easily be formatted to print wire markers. With the appropriate software, large typefaces can be used to make the labels more readable.

At the very least, the cable marker should include a number that identifies the cable on the cable-pull schedule. Descriptive information, such as signal type, source, and destination can also be included on the label. Cables to be used on temporary remote field applications should also be labeled with the company name.

Wrap-Around Adhesive Tape Wire Markers

Wrap-around adhesive tape wire markers are available in pre-printed and write-on strips, mounted on cards, from several manufacturers. Pre-printed numbered markers are also available on small spools for easy dispensing from a marker dispenser. Write-on markers are available in many sizes. One end of each marker is transparent and, when wrapped, is intended to cover the white printed portion to protect the ink from smudging.

Adhesive wire markers are easy to install, but they are not as durable as heat-shrinkable sleeve markers. They require meticulous care during installation for proper alignment and to avoid contaminating the adhesive. If oil gets on the adhesive, the label will eventually unwrap from the cable. Though advertised as permanent, these markers will eventually begin to peel at the end. If handled frequently or flexed, they may come off. If bent, the marker will kink, which can cause it to begin to peel off.

Write-On Cable Ties

Cable ties are available that have an enlarged flat write-on area near the female end. They are used for marking cables and cable bundles. The cable identification information is written using permanent ink. Such marker ties are sturdy and easy to install. However, the information has to be individually written or stamped onto each marker. Ties cannot be printed in bulk using a computer software program.

6.8   Cable Connectors

Many different types of connectors are used to integrate the various elements of a facility. In general, however, connectors for signal-carrying lines can be grouped into one of the following categories:

- Connectors used for coaxial cable

- Connectors used for twisted pair cables, shielded and unshielded

- Connectors for fiber optic cables

6.8.1   BNC Connector

BNC connectors are used on professional video equipment, computer network interfaces, and low-power RF systems. BNC connectors are available in a number of styles, as shown in Figure 6.36.

Select the right connector to fit the dimensions of the cable being terminated. When using crimp-type connectors, avoid using the wrong-size crimp ring. It may distort and not provide the correct pressure on the cable when crimped. The center pin should be gold plated. Gold is less susceptible to corrosion and is a good conductor. The outer contact should be made of beryllium copper, instead of phosphor bronze or spring brass. Because of its higher elasticity, beryllium copper is more reliable and can withstand more flexing cycles before mechanical failure occurs.

Strip coaxial cables using a good coaxial cable stripper. Such tools cut all of the cable elements to the specified length for the connector being used. Follow the connector manufacturer's instructions for the proper cut dimensions. Make sure the shield braid is cut clean and is not dragged along by the cutting blade. All of the strands of the braid should be the same length.

Dual Crimp-Type Connectors

Assemble a dual crimp-type connector using the following steps:

- Push the center pin onto the center conductor of the cable.

- Make sure the end of the conductor can be seen through the inspection hole in the pin.

- Crimp the pin onto the conductor using the recommended crimping tool.

- Slip the crimp sleeve onto the cable for later use.

- Insert the pin and cable into the connector until it snaps into place.

- Flair the braid to allow it to pass over the connector body easily.

- Push the connector onto the cable, guiding the shield braid over the knurled portion of the connector. This will keep the braid as neat as possible. The shielding should just touch the connector body or have a slight clearance.

- Slide the crimp sleeve forward over the shield braid and up against the connector body. No shield braid should be visible or protrude around the sleeve.

- Crimp the sleeve using the recommended crimping tool.

Screw-Type Connector

Assemble a screw-type connector using the following steps:

- Push the center pin onto the center conductor of the cable.

- Solder the pin to the center conductor so there is no gap between the pin and the insulation.

- Make sure that no solder protrudes above the surface of the pin to hinder insertion into the connector.

- Slide the compression fitting onto the cable.

(a)

(b)

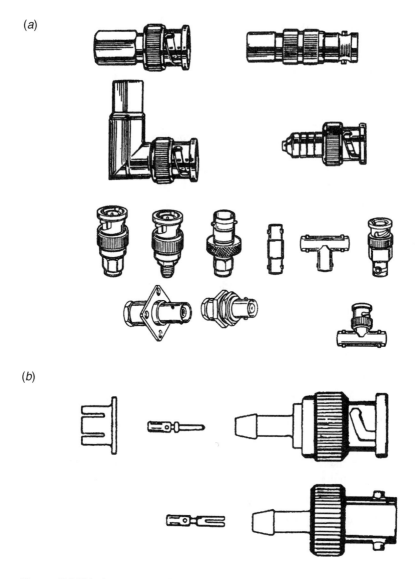

Figure 6.36 Various types of BNC connectors available for video applications: (a) general connector types, (b) typical connector construction.

- Slip the ferrule over the braid, up against the cable outer jacket.
- Fold the braid back over the ferrule and smooth it out.
- Cut off any excess braid.
- Insert the pin and cable into the connector until it snaps into place.

- Screw on the compression fitting, using a wrench to tighten.

After installing the connector onto the cable, inspect it. Make sure the center pin is secure and straight and protrudes by the correct amount. It should snap into place when the connector is pushed into the cable. If the cable was not pushed in far enough before crimping, the center pin will not be protruding sufficiently to make contact with the center pin of the mating connector. Hold the cable, and pull the connector as hard as you can. It should not slip or come off of the cable. Check to make sure that none of the shield braid is sticking out around either edge of the connector crimp ring.

As an added assurance, the continuity of the finished cable should be tested using an ohm meter.

6.8.2 Twisted Pair Connectors

A variety of connector types are available for use with low frequency signals carried over twisted pair lines. For audio applications, connectors that have become standardized in the industry for professional use include XLR and phone. Phono (or RCA) and miniphone connectors are used on industrial-grade and consumer equipment. Quality, construction, and workmanship of connectors vary. Quality and reliability are determined largely by the materials used. This is especially true for the contact and contact spring elements used in connectors. These elements, in turn, determine the cost of the connector. The method of fastening component parts together also affects the durability and cost of the product.

Audio Connectors

XLR connectors are the standard connector used for professional audio. This connector uses three pins. When using a shielded-pair cable, the shield is connected to pin 1, the high side is connected to pin 2, and the low side is connected to pin 3. (See Figure 6.37.)

Phone plugs are used for quick-connect applications, such as headphones, microphones (for consumer or industrial applications), and speakers. Mono and stereo versions are available. The mono version has two elements: *tip* and *sleeve*. The tip is the high side, and the sleeve is the low or ground side. The stereo phone connector has three elements: *tip*, *ring*, and *sleeve*. The sleeve is used as the common and the left and right channels are connected to the tip and ring, respectively. This is illustrated in Figure 6.38. Phone connectors are used in audio patch panels. Using shielded-pair cable, the shield is connected to the sleeve, the high side to the tip, and the low side to the ring.

Data Connectors

A considerable variety of connector types exist for computer applications. Some of the more common devices are illustrated in Figure 6.39. For reasons of cost, installation time, and reliability, virtually all computer device connectors are supplied factory-assembled. Very few connectors are applied in the field.

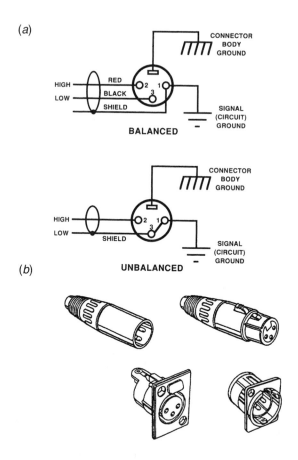

Figure 6.37 XLR connector for audio: (*a*) standard wiring for balanced and unbalanced circuits, (*b*) illustration of common XLR jacks and plugs.

6.8.3 Terminal Blocks

Three basic types of terminals are available on terminal blocks:

- Solder

- Twist-on, including barrier strips

- Push-on, including *insulation displacement* types

Solder-type terminals provide reliable connection but require more work to connect and make changes. Push-on-type connectors provide the easiest and fastest connections. Figure 6.40 shows the basic design of an insulation displacement terminal block.

In a facility designed around a central distribution scheme, all inputs and outputs are brought together at one location for interconnection by way of patch panels or a signal

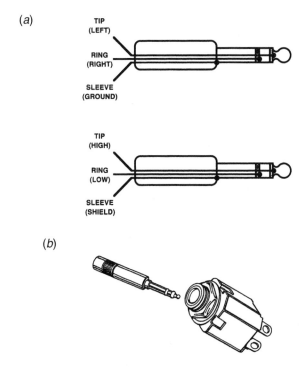

(a)

TIP
(LEFT)

RING
(RIGHT)

SLEEVE
(GROUND)

TIP
(HIGH)

RING
(LOW)

SLEEVE
(SHIELD)

(b)

Figure 6.38 1/4 in stereo phone connector: (a) standard wiring diagram for an unbalanced (top) and balanced circuit, (b) illustration of a jack and mating plug.

router. Jacks from the individual patch panels are wired to terminals on terminal blocks. All interconnections to equipment are also brought to terminal blocks at the central location. Interconnection between equipment and patch panels is made by connecting short jumpers from point to point on the terminal blocks. Changes are simply a matter of removing one end of a particular wire and connecting it to a new set of terminals on the same or another terminal block.

When designing a central equipment interconnection point, incorporate wire forms or hangers to support the cables entering and leaving the terminal point. This approach will reduce or eliminate the need for cable ties or other cable bundling in the distribution area. A typical wire form installation is shown in Figure 6.41.

6.8.4 Fiber Optic Connectors

The purpose of a fiber optic connector is to efficiently convey the optical signal from one link or element to the next. Most connectors share a design similar to the assembly shown in Figure 6.42. Typically, connectors are plugs (male) and are mated to precision couplers or sleeves (female). While the specific mechanical design of each

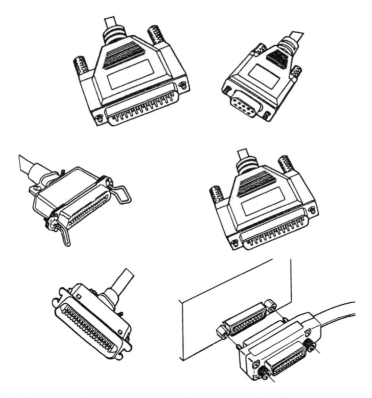

Figure 6.39 Some of the more common computer peripheral connectors.

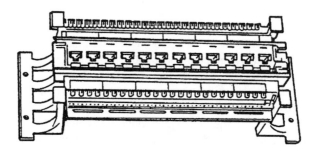

Figure 6.40 A standard insulation displacement terminal block similar to the type used by telephone companies in the U.S.

connector type varies from one manufacturer to the next, the basic concept is the same: provide precise alignment of the optical fiber cores through a ferrule in the coupler. Some connectors are designed to keep the fiber ends separated, while other

(a)　　　　　　　　(b)

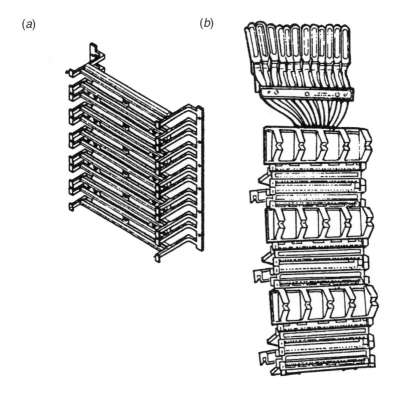

Figure 6.41 Use of wire forms to organize a terminal center: (a) basic wiring block, (b) blocks formed into a wire center.

designs permit the fiber ends to touch in order to reduce reflections resulting from the glass-to-air-to-glass transition.

The fiber is prepared and attached to the connector, usually with an adhesive or epoxy cement, and polished flush with the connector tip. Factory-installed connectors typically use heat-cured epoxy and hand or machine polishing. Field installable connectors include epoxy-and-polish types, and crimp-on types. The crimp-on connector simplifies field assembly considerably.

The ferrule is a critical element of the connector. The ferrule functions to hold the fiber in place for optimum transmission of light energy. Ferrule materials include ceramics, stainless steel alloys, and glass.

Connector Properties

There are many types of fiber optic connectors. Each design has evolved to fill a specific application, or class of applications. Figure 6.43 shows three common fiber optic connectors.

The selection of a connector should take into consideration the following issues:

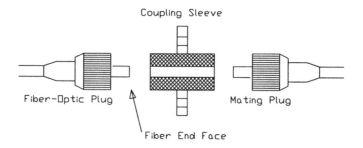

Figure 6.42 The mechanical arrangement of a simple fiber-optic connector.

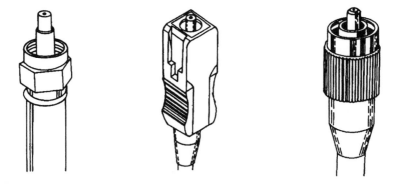

Figure 6.43 Common types of fiber-optic connectors.

- Insertion loss
- Allowable loss budget for the fiber system
- Consistent loss characteristics over a minimum number of connect/disconnect cycles
- Sufficiently high return loss for proper system operation
- Ruggedness
- Compatibility with fiber connectors of the same type
- High tensile strength
- Stable thermal characteristics

As with any system that transports energy, the fewer number of connectors and/or splices, the better. Pigtail leads are often required between a fiber termination panel and the transmission/reception system; however, keep such links to a minimum. Figure

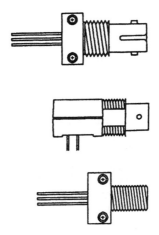

Figure 6.44 Circuit board mounted LED optical sources with connector terminations.

6.44 shows how the fiber-optic light source may be terminated as a panel-mounted connector in order to minimize the number of pigtail links.

Performance Considerations

The optical loss in a fiber optic connector is the primary measure of device quality. Connector loss specifications are derived by measuring the optical power through a length of fiber. Next, the fiber is cut in the center of its length and the connectors are attached and mated with a coupler. The power is then measured again at the end of the fiber. The additional loss in the link represents the loss in the connector.

Return loss is another important measurement of connector quality. Return loss is the optical power that is reflected toward the source by a connector. Connector return loss in a single-mode link, for example, can diffuse back into the laser cavity, degrading its stability. In a multi-mode link, return loss can cause extraneous signals, reducing overall performance.

6.9 Bibliography

Ajemian, Ronald G., "Fiber Optic Connector Considerations for Professional Audio," *Journal of the Audio Engineering Society*, Audio Engineering Society, New York, NY, June 1992.

Benson, K. B., and J. Whitaker: *Television and Audio Handbook for Engineers and Technicians*, McGraw-Hill, New York, NY, 1989.

Block, Roger: "The Grounds for Lightning and EMP Protection," PolyPhaser Corporation, Gardnerville, NV, 1987.

Crutchfield, E. B. (ed.), *NAB Engineering Handbook*, 8th ed., National Association of Broadcasters, Washington, D.C., 1992.

Dahlgren, Michael W., "Servicing Local Area Networks," *Broadcast Engineering*, Intertec Publishing, Overland Park, KS, November 1989.

Davis, Gary, and Ralph Jones: *Sound Reinforcement Handbook*, Yamaha Music Corporation, Hal Leonard Publishing, Milwaukee, WI, 1987.

Fardo, S., and D. Patrick: *Electrical Power Systems Technology*, Prentice-Hall, Englewood Cliffs, NJ, 1985.

Federal Information Processing Standards Publication No. 94, *Guideline on Electrical Power for ADP Installations*, U.S. Department of Commerce, National Bureau of Standards, Washington, DC, 1983.

International Organization for Standardization, "Information Processing Systems—Open Systems Interconnection—Basic Reference Model," ISO 7498, 1984.

Lanphere, John: "Establishing a Clean Ground," *Sound & Video Contractor*, Intertec Publishing, Overland Park, KS., August 1987.

Lawrie, Robert: *Electrical Systems for Computer Installations*, McGraw-Hill, New York, NY, 1988.

Morrison, Ralph, and Warren Lewis: *Grounding and Shielding in Facilities*, John Wiley & Sons, New York, NY, 1990.

Mullinack, Howard G.: "Grounding for Safety and Performance," *Broadcast Engineering*, Intertec Publishing, Overland Park, KS., October 1986.

Pearson, Eric: *How to Specify and Choose Fiber-Optic Cables,* Pearson Technologies, Acworth, GA, 1991.

Whitaker, Jerry C, *AC Power Systems*, 2nd ed., CRC Press, Boca Raton, FL, 1998.

Whitaker, Jerry C., *Maintaining Electronic Systems*, CRC Press, Boca Raton, FL, 1991.

System Reliability

7.1 Introduction

The ultimate goal of any design engineer or maintenance department is zero downtime. This is an elusive goal, but one that can be approximated by examining the vulnerable areas of plant operation and taking steps to prevent a sequence of events that could result in system failure. In cases where failure prevention is not practical, a reliability assessment should encompass the stocking of spare parts, circuit boards, or even entire systems. A large facility may be able to cost-justify the purchase of backup gear that can be used as spares for the entire complex. Backup hardware is expensive, but so is downtime.

Failures can, and do, occur in electronic systems. The goal of product quality assurance at every step in the manufacturing and operating chain is to ensure that failures do not produce a systematic or repeatable pattern. The ideal is to eliminate failures altogether. Short of that, the secondary goal is to end up with a random distribution of failure modes. This indicates that the design of the system is fundamentally optimized and that failures are caused by random events that cannot be predicted. In an imperfect world, this is often the best that end users can hope for. Reliability and maintainability must be built into products or systems at every step in the design, construction, and maintenance process. They cannot be treated as an afterthought.

7.1.1 Terminology

To understand the principles of reliability engineering, the following basic terms must be defined:

- **Availability**. The probability that a system subject to repair will operate satisfactorily on demand.

- **Average life**. The mean value for a normal distribution of product or component lives. This term is generally applied to mechanical failures resulting from "wear-out."

- **Burn-in**. The initially high failure rate encountered when a component is placed on test. Burn-in failures usually are associated with manufacturing defects and the debugging phase of early service.

- **Defect**. Any deviation of a unit or product from specified requirements. A unit or product may contain more than one defect.

- **Degradation failure**. A failure that results from a gradual change, over time, in the performance characteristics of a system or part.

- **Downtime**. Time during which equipment is not capable of doing useful work because of malfunction. This does not include preventive maintenance time. Downtime is measured from the occurrence of a malfunction to its correction.

- **Failure**. A detected cessation of ability to perform a specified function or functions within previously established limits. A failure is beyond adjustment by the operator by means of controls normally accessible during routine operation of the system.

- **Failure mode and effects analysis (FMEA)**. An iterative documented process performed to identify basic faults at the component level and determine their effects at higher levels of assembly.

- **Failure rate**. The rate at which failure occurs during an interval of time as a function of the total interval length.

- **Fault tree analysis (FTA)**. An iterative documented process of a systematic nature performed to identify basic faults, determine their causes and effects, and establish their probabilities of occurrence.

- **Lot size**. A specific quantity of similar material or a collection of similar units from a common source; in inspection work, the quantity offered for inspection and acceptance at any one time. This may be a collection of raw material, parts, subassemblies inspected during production, or a consignment of finished products to be sent out for service.

- **Maintainability**. The probability that a failure will be repaired within a specified time after it occurs.

- **Mean time between failure (MTBF)**. The measured operating time of a single piece of equipment divided by the total number of failures during the measured period of time. This measurement normally is made during that period between early life and wear-out failures.

- **Mean time to repair (MTTR)**. The measured repair time divided by the total number of failures of the equipment.

- **Mode of failure**. The physical description of the manner in which a failure occurs and the operating condition of the equipment or part at the time of the failure.

- **Part failure rate**. The rate at which a part fails to perform its intended function.

- **Quality assurance (QA).** All those activities, including surveillance, inspection, control, and documentation, aimed at ensuring that a product will meet its performance specifications.

- **Reliability.** The probability that an item will perform satisfactorily for a specified period of time under a stated set of use conditions.

- **Reliability growth.** Actions taken to move a hardware item toward its reliability potential, during development, subsequent manufacturing, or operation.

- **Reliability predictions.** Compiled failure rates for parts, components, subassemblies, assemblies, and systems. These generic failure rates are used as basic data to predict a value for reliability.

- **Sample.** One or more units selected at random from a quantity of product to represent that product for inspection purposes.

- **Sequential sampling.** Sampling inspection in which, after each unit is inspected, the decision is made to accept, reject, or inspect another unit. (Note: Sequential sampling as defined here is sometimes called *unit sequential sampling* or *multiple sampling*.)

- **System.** A combination of parts, assemblies, and sets joined together to perform a specific operational function or functions.

- **Test to failure.** Testing conducted on one or more items until a predetermined number of failures have been observed. Failures are induced by increasing electrical, mechanical, and/or environmental stress levels, usually in contrast to *life tests*, in which failures occur after extended exposure to predetermined stress levels. A life test can be considered a test to failure using age as the stress.

7.1.2 Quality Assurance

Electronic component and system manufacturers design and implement quality assurance procedures for one fundamental reason: Nothing is perfect. The goal of a QA program is to ensure, for both the manufacturer and the customer, that all but some small, mutually acceptable percentage of devices or systems shipped will be as close to perfection as economics and the state of the art allow. There are tradeoffs in this process. It would be unrealistic, for example, to perform extensive testing to identify potential failures if the cost of that testing exceeded the cost savings that would be realized by not having to replace the devices later in the field.

The focal points of any QA effort are *quality* and *reliability*. These terms are not synonymous. They are related, but they do not provide the same measure of a product:

- Quality is the measure of a product's performance relative to some established criteria.

- Reliability is the measure of a product's life expectancy.

Stated from a different perspective, quality answers the question of whether the product meets applicable specifications *now*; reliability answers the question of *how long* the product will continue to meet its specifications.

7.1.3 Inspection Process

Quality assurance for components normally is performed through sampling rather than through 100 percent inspection. The primary means used by QA departments for controlling product quality at the various processing steps include:

- *Gates.* A mandatory sampling of every lot passing through a critical production stage. Material cannot move on to the next operation until QA has inspected and accepted the lot.

- *Monitor points.* A periodic sampling of some attribute of the component. QA personnel sample devices at a predetermined frequency to verify that machines and operators are producing material that meets preestablished criteria.

- *Quality audit.* An audit carried out by a separate group within the QA department. This group is charged with ensuring that all production steps throughout the manufacturer's facility are in accordance with current specifications.

- *Statistical quality control.* A technique, based on computer modeling, that incorporates data accumulated at each gate and monitor point to construct statistical profiles for each product, operation, and piece of equipment within the plant. Analysis of this data over time allows QA engineers to assess trends in product performance and failure rates.

Quality assurance for a finished subassembly or system may range from a simple go/no-go test to a thorough operational checkout that may take days to complete.

7.1.4 Reliability Evaluation

Reliability prediction is the process of quantitatively assessing the reliability of a component or system during development, before large-scale fabrication and field operation. During product development, predictions serve as a guide by which design alternatives can be judged for reliability. To be effective, the prediction technique must relate engineering variables to reliability variables.

A prediction of reliability is obtained by determining the reliability of each critical item at the lowest system level and proceeding through intermediate levels until an estimate of overall reliability is obtained. This prediction method depends on the availability of accurate evaluation models that reflect the reliability of lower-level components. Various formal prediction procedures are used, based on theoretical and statistical concepts.

Parts-Count Method

The parts-count approach to reliability prediction provides an estimate of reliability based on a count by part type (ICs, transistors, vacuum tube devices, resistors, capacitors, and other components). This method is useful during the early design stage of a product, when the amount of available data is limited. The technique involves counting the number of components of each type, multiplying that number by a generic failure rate for each part type, and summing the products to obtain the failure rate of each functional circuit, subassembly, assembly, and/or block depicted in the system block diagram. The parts-count method is useful in the design phase because it provides rapid estimates of reliability, permitting assessment of the feasibility of a given concept.

Stress-Analysis Method

The stress-analysis technique is similar to the parts-count method, but utilizes a detailed parts model plus calculation of circuit stress values for each part before determining the failure rate. Each part is evaluated in its electric circuit and mechanical assembly application based on an electrical and thermal stress analysis. After part failure rates have been established, a combined failure rate for each functional block is determined.

7.1.5 Failure Analysis

Failure mode and effects analysis can be performed with data taken from actual failure modes observed in the field, or from hypothetical failure modes derived from one of the following:

- Design analysis
- Reliability prediction activities
- Experience with how specific parts fail

In the most complete form of FMEA, failure modes are identified at the component level. Failures are induced analytically into each component, and failure effects are evaluated and noted, including the severity and frequency (or probability) of occurrence. Using this approach, the probability of various system failures can be calculated, based on the probability of lower-level failure modes.

Fault tree analysis is a tool commonly used to analyze failure modes found during design, factory test, or field operations. The approach involves several steps, including the development of a detailed logic diagram that depicts basic faults and events that can lead to system failure and/or safety hazards. These data are used to formulate corrective suggestions that, when implemented, will eliminate or minimize faults considered critical. An example FTA chart is shown in Figure 7.1.

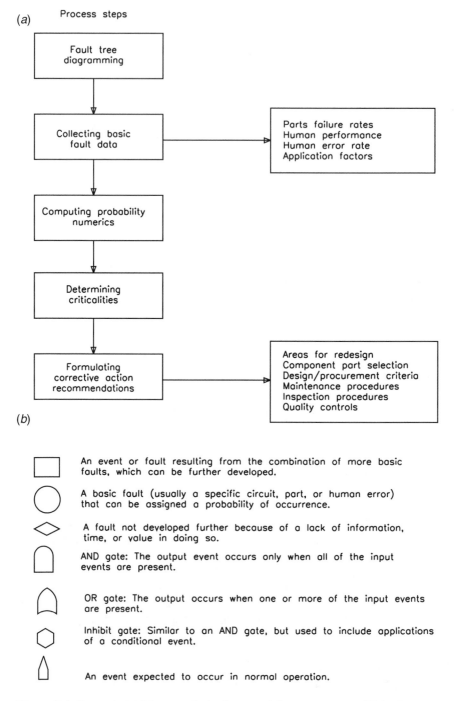

Figure 7.1 Example fault tree analysis diagram: (*a*) process steps, (*b*) fault tree symbols, (*c*, next page) example diagram.

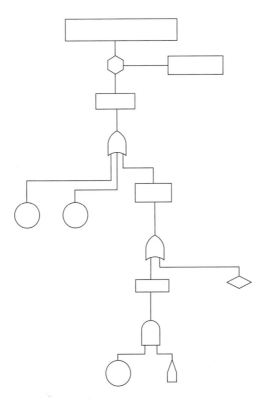

Figure 7.1c

7.1.6 Standardization

Standardization and reliability go hand in hand. Standardization of electronic components began with military applications in mind; the first recorded work was performed by the U.S. Navy with vacuum tubes. The navy recognized that some control
at the component level was essential to the successful incorporation of electronics
into naval systems.

Standardization and reliability are closely related, although there are many aspects
of standardization whose reliability implications are subtle. The primary advantages of
standardization include:

- *Total product interchangeability.* Standardization ensures that products of the
  same part number provide the same physical and electrical characteristics. There
  have been innumerable instances of a replacement device bearing the same part
  number as a failed device, but not functioning identically to it. In some cases, the
  differences in performance were so great that the system would not function at all
  with the new device.

- *Consistency and configuration control.* Component manufacturers constantly re-define their products to improve yields and performance. Consistency and con-figuration control assure the user that product changes will not affect the interchangeability of the part.

- *Efficiencies of volume production.* Standardization programs usually result in production efficiencies that reduce the costs of parts, relative to components with the same level of reliability screening and control.

- *Effective spares management.* The use of standardized components makes the stocking of spare parts a much easier task. This aspect of standardization is not a minor consideration. For example, the costs of placing, expediting, and receiving material against one Department of Defense purchase order may range from $300 to $1100 (or more). Accepting the lowest estimate, the conversion of 10 separate part numbers to one standardized component could effect immediate savings of $3000 just in purchasing and receiving costs.

- *Multiple product sources.* Standardization encourages second-sourcing. Multi-ple sources help hold down product costs and encourage manufacturers to strive for better product performance.

7.1.7 Reliability Analysis

The science of reliability and maintainability matured during the 1960s with the de-velopment of sophisticated computer systems and complex military and spacecraft electronics. Components and systems never fail without a reason. That reason may be difficult to find, but determination of failure modes and weak areas in system design or installation is fundamental to increasing the reliability of any component or sys-tem, whether it is a power vacuum tube, integrated circuit, aircraft autopilot, or radio transmitter.

All equipment failures are logical; some are predictable. A system failure usually is related to poor-quality components or to abuse of the system or a part within, either be-cause of underrating or environmental stress. Even the best-designed components can be badly manufactured. A process can go awry, or a step involving operator interven-tion may result in an occasional device that is substandard or likely to fail under normal stress. Hence, the process of screening and/or *burn-in* to weed out problem parts is a universally accepted quality control tool for achieving high reliability.

7.1.8 Statistical Reliability

Figure 7.2 illustrates what is commonly known as the *bathtub curve.* It divides the ex-pected lifetime of a class of parts into three segments: *infant mortality, useful life,* and *wear-out.* A typical burn-in procedure consists of the following steps:

- The parts are electrically biased and loaded; that is, they are connected in a circuit representing a typical application.

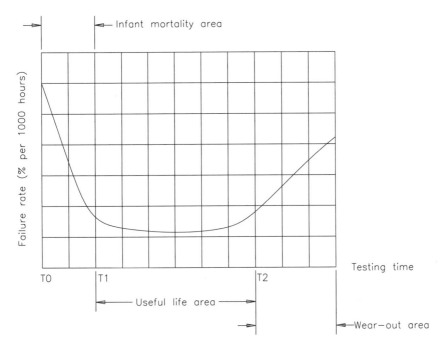

Figure 7.2 The statistical distribution of equipment or component failures vs. time for electronic systems and devices.

- The parts are cycled on and off (power applied, then removed) for a predetermined number of times. The number of cycles may range from 10 to several thousand during the burn-in period, depending on the component under test.

- The components under load are exposed to high operating temperatures for a selected time (typically 72 to 168 hours). This constitutes an accelerated life test for the part.

An alternative approach involves temperature shock testing, in which the component product is subjected to temperature extremes, with rapid changes between the *hot-soak* and *cold-soak* conditions. After the stress period, the components are tested for adherence to specifications. Parts meeting the established specifications are accepted for shipment to customers. Parts that fail to meet them are discarded.

Figure 7.3 illustrates the benefits of temperature cycling to product reliability. The charts compare the distribution of component failures identified through steady-state high-temperature burn-in vs. temperature cycling. Note that cycling screened out a significant number of failures. The distribution of failures under temperature cycling usually resembles the distribution of field failures. Temperature cycling simulates real-world conditions more closely than steady-state burn-in. The goal of burn-in testing is to ensure that the component lot is advanced beyond the infant mortality stage (*T1*

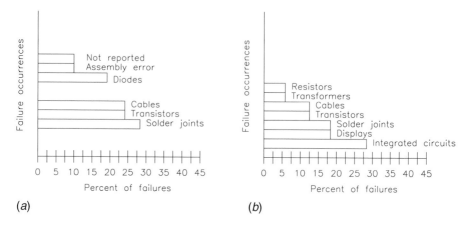

Figure 7.3 Distribution of component failures identified through burn-in testing: (*a*) steady-state high-temperature burn-in, (*b*) temperature cycling.

on the bathtub curve). This process is used not only for individual components, but for entire systems as well.

Such a systems approach to reliability is effective, but not foolproof. The burn-in period is a function of statistical analysis; there are no absolute guarantees. The natural enemies of electronic parts are heat, vibration, and excessive voltage. Figure 7.4 documents failures vs. hours in the field for a piece of avionics equipment. The conclusion is made that a burn-in period of 200 hours or more will eliminate 60 percent of the expected failures. However, the burn-in period for another system using different components may well be a different number of hours.

The goal of burn-in testing is to catch system problems and potential faults before the device or unit leaves the manufacturer. The longer the burn-in period, the greater the likelihood of catching additional failures. The problems with extended burn-in, however, are time and money. Longer burn-in translates to longer delivery delays and additional costs for the equipment manufacturer, which are likely to be passed on to the end user. The point at which a product is shipped is based largely on experience with similar components or systems and the financial requirement to move products to customers.

Roller-Coaster Hazard Rate

The bathtub curve has been used for decades to represent the failure rate of an electronic system. More recent data, however, has raised questions regarding the accuracy of the curve shape. A growing number of reliability scientists now believe that the probability of failure, known in the trade as the *hazard rate*, is more accurately represented as a roller-coaster track, as illustrated in Figure 7.5. Hazard rate calculations require analysis of the number of failures of the system under test, as well as the num-

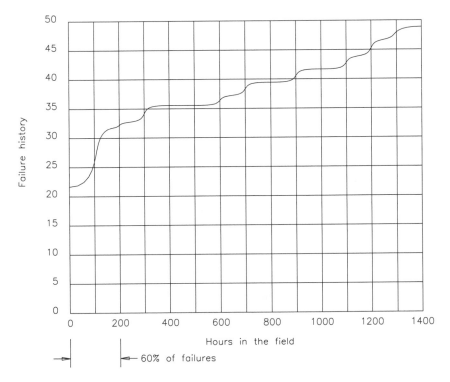

Figure 7.4 The failure history of a piece of avionics equipment vs. time. Note that 60 percent of the failures occurred within the first 200 hours of service. (*After* [1].)

ber of survivors. Advocates of this approach point out that previous estimating processes and averaging tended to smooth the roller-coaster curve so that the humps were less pronounced, leading to an incorrect conclusion insofar as the hazard rate was concerned. The testing environment also has a significant effect on the shape of the hazard curve, as illustrated in Figure 7.6. Note that at the higher operating temperature (greater environmental stress), the roller-coaster hump has moved to an earlier age.

7.1.9 Environmental Stress Screening

The science of reliability analysis is rooted in the understanding that there is no such thing as a random failure; every failure has a cause. For reasonably designed and constructed electronic equipment, failures not caused by outside forces result from built-in flaws or *latent defects*. Because different flaws are sensitive to different stresses, a variety of environmental forces must be applied to a unit under test to identify any latent defects. This is the underlying concept behind *environmental stress screening* (ESS).

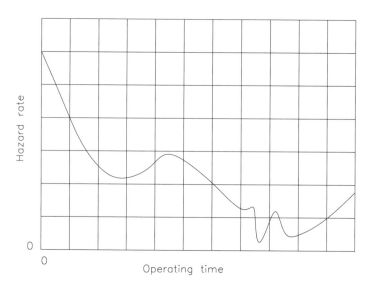

Figure 7.5 The roller-coaster hazard rate curve for electronic systems. (*After* [2].)

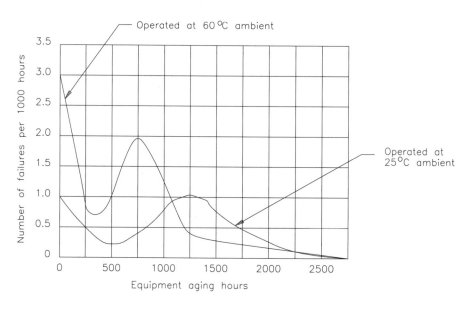

Figure 7.6 The effects of environmental conditions on the roller-coaster hazard rate curve. (*After* [2].)

ESS, which has come into widespread use for aeronautics and military products, takes the burn-in process a step further by combining two of the major environmental

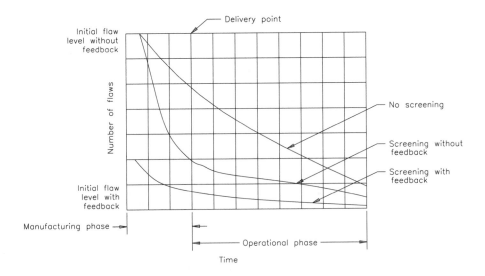

Figure 7.7 The effects of environmental stress screening on the reliability *bathtub* curve. (*After* [3].)

factors that cause parts or units to fail: heat and vibration. Qualification testing for products at a factory practicing ESS involves a carefully planned series of checks for each unit off the assembly line. Units are subjected to random vibration and temperature cycling during production (for subassemblies and discrete components) and upon completion (for systems). The goal is to catch product defects at the earliest possible stage of production. ESS also can lead to product improvements in design and manufacture if feedback from the qualification stage to the design and manufacturing stages is implemented. Figure 7.7 illustrates the improvement in reliability that typically can be achieved through ESS over simple burn-in screening, and through ESS with feedback to earlier production stages. Significant reductions in equipment failures in the field can be gained. Table 7.1 compares the merits of conventional reliability testing and ESS.

Designing an ESS procedure for a given product is no easy task. The environmental stresses imposed on the product must be great enough to cause fallout of marginal components during qualification testing. The stresses must not be so great, however, as to cause failures in good products. Any unit that is stressed beyond its design limits eventually will fail. The proper selection of stress parameters—generally, random vibration on a vibration generator and temperature cycling in an environmental chamber—can, in minutes, uncover product flaws that might take weeks or months to manifest themselves in the field. The result is greater product reliability for the user.

The ESS concept requires that every product undergo qualification testing before integration into a larger system for shipment to an end user. The flaws uncovered by

Table 7.1 Comparison of Conventional Reliability Testing and Environmental Stress Screening (*After* [2].)

Parameter	Conventional Testing	Environmental Stress Screening
Test condition	Simulates operational equipment profile	Accelerated stress condition
Test sample size	Small	100 percent of production
Total test time	Limited	High
Number of failures	Small	Large
Reliability growth	Potential for gathering useful data small	Potential for gathering useful data good

ESS vary from one unit to the next, but types of failures tend to respond to particular environmental stresses. Available data clearly demonstrate that the burn-in screens must match the flaws sought; otherwise, the flaws will probably not be found.

The concept of flaw-stimulus relationships can be presented in Venn diagram form. Figure 7.8 shows a Venn diagram for a hypothetical, but specific, product. The required screen would be different for a different product. For clarity, not all stimuli are shown. Note that there are many latent defects that will not be uncovered by any one stimulus. For example, a solder splash that is just barely clinging to a circuit element probably would not be broken loose by high-temperature burn-in or voltage cycling, but vibration or thermal cycling probably would break the particle loose. Remember also that the defect may be observable only during stimulation and not during a static bench test.

The levels of stress imposed on a product during ESS should be greater than the stress to which the product will be subjected during its operational lifetime, but still be below the maximum design parameters. This rule of thumb is pushed to the limits under an *enhanced screening* process. Enhanced screening places the component or system at well above the expected field environmental levels. This process has been found to be useful and cost-effective for many programs and products. Enhanced screening, however, requires the development of screens that are carefully identified during product design and development so that the product can survive the qualification tests. Enhanced screening techniques often are required for cost-effective products on a cradle-to-grave basis; that is, early design changes for screenability save tremendous costs over the lifetime of the product.

The types of products that can be checked economically through ESS break down into two categories: high-dollar items and mass-produced items. Units that are physically large in size, such as RF generators, are difficult to test in the finished state. Still, qualification tests using more primitive methods, such as cam-driven truck-bed shakers, are practical. Because most large systems generate a large amount of heat, subjecting the equipment to temperature extremes also may be accomplished. Sophisticated

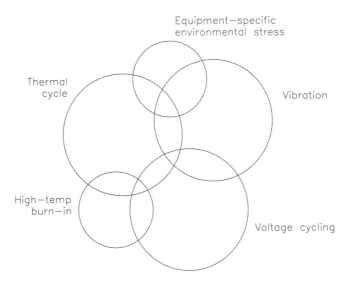

Figure 7.8 Venn diagram representation of the relationship between flaw precipitation and applied environmental stress. (*After* [4].)

ESS for large systems, however, must rely on qualification testing at the subassembly stage.

The basic hardware complement for an ESS test station includes a thermal chamber shaker and controller/monitor. A typical test sequence includes 10 minutes of exposure to random vibration, followed by 10 cycles between temperature minimum and maximum. To save time, the two tests may be performed simultaneously.

7.1.10 Latent Defects

The cumulative failure rate observed during the early life of an electronic system is dominated by the latent defect content of the product, not its inherent failure rate. Product design is the major determinant of inherent failure rate. A product design will show a higher-than-expected inherent rate if the system contains components that are marginally overstressed, have inadequate functional margin, or contain a subpopulation of components that exhibit a wear-out life shorter than the useful life of the product. Industry has grown to expect the high instantaneous failure rate observed when a new product is placed into service. The burn-in process, whether ESS or conventional, is aimed at shielding customers from the detrimental effects of infant mortality. The key to reducing early-product-life failures lies in reducing the number of latent defects.

A latent defect is some abnormal characteristic of the product or its parts that is likely to result in failure at some point, depending on:

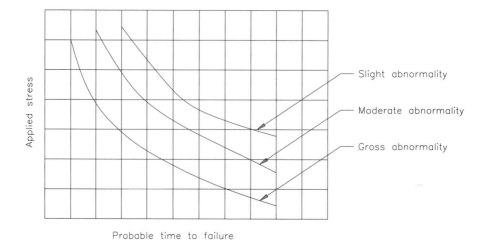

Figure 7.9 Estimation of the probable time to failure from an abnormal solder joint. (*After* [5].)

- The degree of abnormality
- The magnitude of applied stress
- The duration of applied stress

For example, consider a solder joint on the connecting pin of a vacuum tube device. One characteristic of the joint is the degree to which the pin hole is filled with solder, characterized as "percent fill." All other characteristics being acceptable, a joint that is 100 percent filled offers the maximum mechanical strength, minimum resistance, and greatest current carrying capacity. Conversely, a joint that is zero percent filled has no mechanical strength, and only if the lead is touching the barrel does it have any significant electrical properties. Between these two extreme cases are degrees of abnormality. For a fixed magnitude of applied stress:

- A grossly abnormal solder joint probably will fail in a short time.

- A moderately abnormal solder joint probably will fail, but after a longer period of time than a grossly abnormal joint.

- A mildly abnormal solder joint probably will fail, but after a much longer period of time than in either of the preceding conditions.

Figure 7.9 illustrates this concept. A similar time-stress relationship holds for a fixed degree of abnormality and variable applied stress.

A latent defect eventually will advance to a *patent defect* when exposed to environmental, or other, stimuli. A patent defect is a flaw that has advanced to the point at which an abnormality actually exists. To carry on the solder example, a cold solder joint represents a flaw (latent defect). After vibration and/or thermal cycling, the joint (it is assumed) will crack. The joint will now have become a detectable (patent) defect. Some latent defects can be stimulated into patent defects by thermal cycling, some by vibration, and some by voltage cycling. Not all flaws respond to all stimuli.

There is strong correlation between the total number of physical and functional defects found per unit of product during the manufacturing process, and the average latent defect content of shipping product. Product- and process-design changes aimed at reducing latent defects not only improve the reliability of shipping product, but also result in substantial manufacturing cost savings.

7.1.11 Operating Environment

The operating environment of an electronic system, either because of external environmental conditions or unintentional component underrating, may be significantly more stressful than the system manufacturer or the component supplier anticipated. Unintentional component underrating represents a design fault, but unexpected environmental conditions are possible for many applications, particularly in remote locations.

Conditions of extreme low or high temperatures, high humidity, and vibration during transportation may have a significant impact on long-term reliability of the system. For example, it is possible—and more likely, probable—that the vibration stress of the truck ride to a remote transmitting site will represent the worst-case vibration exposure of the transmitter and all components within it during the lifetime of the product.

Manufacturers report that most of the significant vibration and shock problems for land-based products arise from the shipping and handling environment. Shipping tends to be an order of magnitude more severe than the operating environment with respect to vibration and shock. Early testing for these problems involved simulation of actual shipping and handling events, such as end-drops, truck trips, side impacts, and rolls over curbs and cobblestones. Although unsophisticated by today's standards, these tests are capable of improving product resistance to shipping-induced damage.

7.1.12 Failure Modes

Latent failures aside, the circuit elements most vulnerable to failure in any piece of electronic hardware are those exposed to the outside world. In most systems, the greatest threat typically involves one or more of the following components or subsystems:

- The ac-to-dc power supply

- Sensitive signal-input circuitry

- High-power output stages and devices

- Circuitry operating into an unpredictable load, or into a load that may be exposed to lightning and other transient effects (such as an antenna)

Derating of individual components is a key factor in improving the overall reliability of a given system. The goal of derating is the reduction of electrical, mechanical, thermal, and other environmental stresses on a component to decrease the degradation rate and prolong expected life. Through derating, the margin of safety between the operating stress level and the permissible stress level for a given part is increased. This adjustment provides added protection from system overstress, unforeseen during design.

7.1.13 Maintenance Considerations

The reliability and operating costs over the lifetime of a device or system can be affected significantly by the effectiveness of the preventive maintenance program designed and implemented by the engineering staff. In the case of a *critical-system* unit that must be operational continuously or during certain periods, maintenance can have a major impact—either positive or negative—on downtime.

The reliability of any electronic system may be compromised by an *enabling event phenomenon*. This is an event that does not cause a failure by itself, but sets up (or enables) a second event that can lead to failure of the system. Such a phenomenon is insidious because the enabling event may not be self-revealing. Examples include the following:

- A warning system that has failed or has been disabled for maintenance

- One or more controls that are set incorrectly, providing false readouts for operations personnel

- Redundant hardware that is out of service for maintenance

- Remote metering that is out of calibration

Common-Mode Failure

A *common-mode failure* is one that can lead to the failure of all paths in a redundant configuration. In the design of redundant systems, therefore, it is important to identify and eliminate sources of common-mode failures, or to increase their reliability to at least an order of magnitude above the reliability of the redundant system. Common-mode failure points include the following:

- Switching circuits that activate standby or redundant hardware

- Sensors that detect a hardware failure

- Indicators that alert personnel to a hardware failure

- Software that is common to all paths in a redundant system

The concept of software reliability in control and monitoring has limited meaning in that a good program will always run, and copies of a good program will always run. On the other hand, a program with one or more errors will always fail, and so will the copies, given the same input data. The reliability of software, unlike hardware, cannot be improved through redundancy if the software in the parallel path is identical to that in the primary path.

Spare Parts

The spare parts inventory is a key aspect of any successful equipment maintenance program. Having adequate replacement components on hand is important not only to correct equipment failures, but to identify those failures as well. Many parts—particularly in the high-voltage power supply and RF chain—are difficult to test under static conditions. The only reliable way to test the component may be to substitute one of known quality. If the system returns to normal operation, then the original component is defective. Substitution is also a valuable tool in troubleshooting intermittent failures caused by component breakdown under peak power conditions.

7.1.14 ISO 9000 Series

At its core, the ISO 9000 Series defines what a total quality system should do in order to guarantee product and service consistency. To that end, the ISO 9000 Series philosophically supports the age old argument that form follows function; if a system's processes are defined and held within limits, consistent products and services will follow.

The ISO 9000 Series are documents that pertain to quality management standards. Individually titled and defined, they are listed in Table 7.2.

7.2   Disaster Preparedness Issues

Preparing for the unpredictable is an important part of engineering. Whether facing a natural disaster or an accident, thorough planning will help you get through the event with the least threat to life and property. A time-dependent business cannot shut down its operations to effect repairs without suffering a loss of income. Developing detailed contingency plans is the key to minimizing the disruption of operations during an emergency.

7.2.1 Emergency Situations

Major natural disasters help us focus on assuring that information systems will work during emergencies. These events hammer home the weaknesses in the system. They can also be rare windows of opportunity to learn from past mistakes and make improvements for the future.

A plan that has not been fully tested under realistic conditions is not a plan. Emergency management experts specialize in not only building plans, and building exercises to test them. Sometimes called a *tabletop*, plan exercises can be quite realistic. The exercise manager writes a scenario for a likely emergency. Messages are written to stress the plan and the team assignments it makes. One or more referees watch as the ex-

Table 7.2 ISO 9000 Series Levels

Standard	Use
ISO 9000: quality management and assurance standards, guidelines for selection and use	Like a road map, this standard is to be used as a guideline to facilitate decisions with respect to selection and use of the other standards in the ISO 9000 Series
ISO 9001: model for quality assurance in design, development, production, installation, and servicing	This standard is to be used when conformance to specified requirements are to be assured by the supplier during several stages: design, development, production, installation, and servicing
ISO 9002: model for quality assurance in production and installation	This standard is to be used when conformance to specified requirements are to be assured by the supplier during production and installation
ISO 9003: model for quality assurance in final inspection and testing	This standard is to be used when conformance to specified requirements are to be assured by the supplier solely at final inspection and test
ISO 9004: quality management and quality system elements, guidelines	This standard is to used as a model to develop and implement a quality management system. Basic elements of a quality management system are described. There is a heavy emphasis on meeting customer needs.

ercise unfolds. There is a debriefing immediately after the exercise. The emergency plan is adjusted if necessary. A tabletop is, in microcosm, a safe way to close the emergency preparedness loop before a real emergency strikes.

7.2.2 The Planning Process

It is impossible to separate emergency planning from the facility where the plan will be put into action. Emergency planning must be integral to a functional facility. It must support the main mission and the people who must carry it out. It must work when all else fails. Designers first must obtain firm commitment and backing from top management. Commitment is always easier to get if management has experienced first-hand a major earthquake or powerful storm. Fear is a powerful source of motivation.

Disaster planning and recovery is an art, a science and a technology. Like entities such as the Institute of Electrical and Electronics Engineers (IEEE) or the Society of Broadcast Engineers (SBE), disaster planners have their own professional groups and certification standards.

Many states provide year-round classroom training for government disaster planners. Some planners work full time for the military or in the public safety arenas of government, others have found homes in businesses who recognize that staying in business after a major disaster is smart business. Still others offer their skills and services as consultants to firms who need to jump-start their disaster planning process.

The technical support group must have responsibility or supervision over the environmental infrastructure of the facility. Without oversight, electronic systems are at the mercy of whomever controls that environment. Local emergencies can be triggered by preventable failures in air supply systems, preventable roof leaks, or uncoordinated telephone, computer or ac wiring changes. Seemingly harmless acts such as employees plugging electric heaters into the wrong ac outlet have downed entire facilities. Successful practitioners of systems design and support must take daily emergencies into account in the overall planning process.

7.2.3 Identifying Realistic Risks

Before you can plan for future problems, you must identify what those problems may be. To this end, a list of realistic risks should be developed, based on specific hazards identified by local conditions. Such risks include the following:

- Regional high water marks for the 100 and 150 year storms
- Regional social, political, and governmental conditions
- Regional commercial electrical power reliability
- Regional weather conditions
- Regional geography
- Regional geology

Next, assess specific local hazards that could be triggered by the following:

- Threats from present or former employees who may hold grudges
- External parties that may wish to damage the facility for whatever reason
- Other factors that could make the facility an easy target
- Nearby man-made hazards
- Construction of the facility
- Electrical power
- Other utilities, including buried pipelines

If possible, seek aid from emergency planning professionals when compiling this list. They can help devise a well-written and comprehensive emergency plan. They can also help with detailed research on factors such as geology and dealing hazardous materials such as stored diesel fuel. After there is agreement on the major goals for operations under emergency conditions, you will have a clear direction for refining the emergency plan.

Perform a realistic assessment of the risks that this plan suggests. Do not overlook the obvious. If computers, transmitters, or routing equipment depend on cool air, how can they continue to operate during a heat wave when the facility's one air conditioner has malfunctioned?

What level of reliability should a designer build into the facility? Emergencies introduce chaos into the reliability equation. Most engineers are quite happy when a system achieves 99.9999 percent reliability. While the glass is certainly more than half full, *four nines* reliability still means eight minutes of outage over a one year period. Reliability is an educated prediction based on a number of factors that may or may not be directly applicable to a particular installation.

7.2.4 Alternate Sites

No matter how well you plan, something still could happen that will require you to abandon your facility for some period of time. Governmental entities, hospitals, broadcasters, and other organizations usually have standing mutual aid agreements whereby certain facilities are made available to the affected facility for a specified period of time. Sometimes this is the only way to resume service to the public in the event of a major disaster. If management shows reluctance to share their facilities with a competitor, respectfully ask what they would do if their own facility is rendered useless.

7.2.5 Standby Power Options

Of all the failures that a facility is likely to experience in any given year, the loss of ac power is clearly the most common. To ensure the continuity of power, most facilities depend upon some form of on-site generation.

The engine-generator shown in Figure 7.10 is the classic standby power system. An automatic transfer switch monitors the ac voltage coming from the utility company line for power failure conditions. Upon detection of an outage for a predetermined period of time (generally 1 to 10 s), the standby generator is started; after the generator is up to speed, the load is transferred from the utility to the local generator. Upon return of the utility feed, the load is switched back after some pre-determined "safe time-delay", and the generator is stopped. This basic type of system is used widely at government and commercial facilities and provides economical protection against prolonged power outages (5 min or more).

In some areas, usually metropolitan centers, two utility company power drops can be brought into a facility as a means of providing a source of standby power. As shown in Figure 7.11, two separate utility service drops—from separate power-distribution systems—are brought into the plant, and an *automatic transfer switch* changes the load to the backup line in the event of a main-line failure. The dual feeder system provides an advantage over the auxiliary diesel arrangement in that power transfer from main to standby can be made in a fraction of a second if a *static transfer switch* is used. Time de-

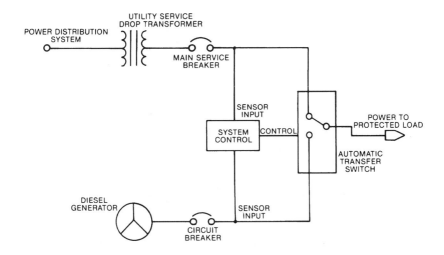

Figure 7.10 The classic standby power system using an engine-generator set. This system protects a facility from prolonged utility company power failures.

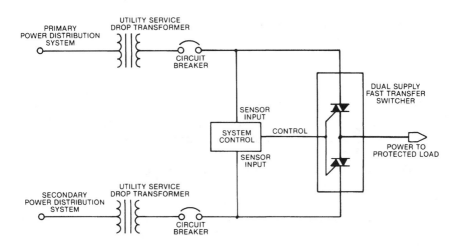

Figure 7.11 The dual utility feeder system of ac power loss protection. An automatic transfer switch changes the load from the main utility line to the standby line in the event of a power interruption.

lays are involved in the diesel generator system that limit its usefulness to power failures lasting more than several minutes.

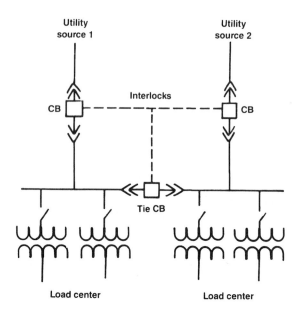

Figure 7.12 A dual utility feeder system with interlocked circuit breakers.

The dual feeder system of protection is based on the assumption that each of the service drops brought into the facility is routed via different paths. This being the case, the likelihood of a failure on both power lines simultaneously is remote. The dual feeder system will not, however, protect against area-wide power failures, which can occur from time to time.

The dual feeder system is limited primarily to urban areas. Rural or mountainous regions generally are not equipped for dual redundant utility company operation. Even in urban areas, the cost of bringing a second power line into a facility can be high, particularly if special lines must be installed for the feed. If two separate utility services are available at or near the site, redundant feeds generally will be less expensive than engine-driven generators of equivalent capacity.

Figure 7.12 illustrates a dual feeder system that utilizes both utility inputs simultaneously at the facility. Notice that during normal operation, both ac lines feed loads, and the tie circuit-breaker is open. In the event of a loss of either line, the circuit-breaker switches reconfigure the load to place the entire facility on the single remaining ac feed. Switching is performed automatically; manual control is provided in the event of a planned shutdown on one of the lines.

A more sophisticated power-control system is shown in Figure 7.13, where a dual feeder supply is coupled with a *motor-generator set* to provide clean, undisturbed ac power to the load. The m-g set will smooth over the transition from the main utility feed to the standby, often making a commercial power failure unnoticed by on-site person-

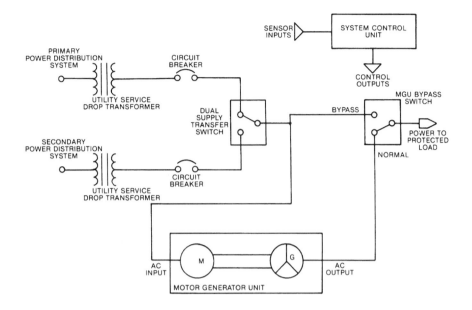

Figure 7.13 A dual feeder standby power system using a motor-generator set to provide power fail ride-through and transient-disturbance protection. Switching circuits allow the m-g set to be bypassed if necessary.

nel. An m-g typically will give up to ½ s of power fail ride-through, more than enough to accomplish a transfer from one utility feed to the other. Special-purpose m-g sets can provide a power fail ride-through of one minute or greater, under specified conditions. This standby power system is further refined in the application illustrated in Figure 7.14, where a diesel generator has been added to the system. With the automatic overlap transfer switch shown at the generator output, this arrangement also can be used for peak-demand *power shaving*.

Generators are available for power levels ranging from less than 1 kVA to several thousand kVA or more. Machines also may be paralleled to provide greater capacity. Generator sets typically are divided by the type of power plant used:

- **Diesel**. Advantages: rugged and dependable, low fuel costs, low fire and/or explosion hazard. Disadvantages: somewhat more costly than other engines, heavier in smaller sizes.

- **Natural and liquefied-petroleum gas**. Advantages: quick starting after long shutdown periods, long life, low maintenance . Disadvantage: availability of natural gas during area-wide power failure subject to question.

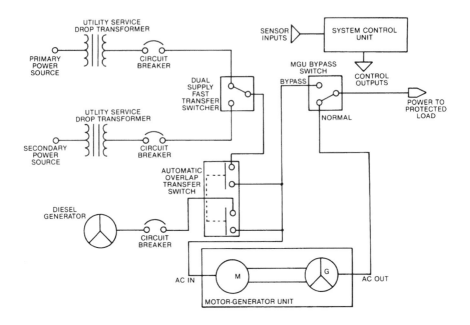

Figure 7.14 A premium power-supply backup and conditioning system using dual utility company feeds, a diesel generator, and a motor-generator set.

- **Gasoline**. Advantages: rapid starting, low initial cost. Disadvantages: greater hazard associated with storing and handling gasoline, generally shorter mean time between overhaul.

- **Gas turbine**. Advantages: smaller and lighter than piston engines of comparable horsepower, rooftop installations practical, rapid response to load changes. Disadvantages: longer time required to start and reach operating speed, sensitive to high input air temperature .

The type of power plant chosen usually is determined primarily by the environment in which the system will be operated and by the cost of ownership. For example, a standby generator located in an urban area office complex may be best suited to the use of an engine powered by natural gas, because of the problems inherent in storing large amounts of fuel. State or local building codes may place expensive restrictions on fuel-storage tanks and make the use of a gasoline- or diesel-powered engine impractical. The use of propane usually is restricted to rural areas. The availability of propane during periods of bad weather (when most power failures occur) also must be considered.

The nature of most power outages requires a sophisticated monitoring system for the engine-generator set. Most power failures occur during periods of bad weather. Most standby generators are unattended. More often than not, the standby system will start, run, and shut down without any human intervention or supervision. For reliable operation, the monitoring system must check the status of the machine continually to ensure that all parameters are within normal limits. Time-delay periods usually are provided by the controller that require an outage to last from 5 to 10 s before the generator is started and the load is transferred. This prevents false starts that needlessly exercise the system. A time delay of 5 to 30 min usually is allowed between the restoration of utility power and return of the load. This delay permits the utility ac lines to stabilize before the load is reapplied.

7.2.6 Batteries

Batteries are the lifeblood of most low-power portable devices. Batteries also play a key role in the operation of *uninterruptible power systems* (UPS), which have become critical to the proper operation of most computers and computer-based systems. Recent research has brought about a number of different battery chemistries, each offering distinct advantages over the others but none providing a fully satisfactory solution to all common applications. Today's most common and promising chemistries available include:

- *Nickel cadmium* (NiCd)—used for portable radios, cellular phones, video cameras, laptop computers and power tools. NiCds have a good load characteristics, are economically priced, and are simple to use.

- *Nickel metal hydride* (NiMH)—used for cellular phones and laptop computers where high-energy is of importance and cost is secondary.

- *Sealed lead acid* (SLA)—used for UPS systems and other demanding applications where energy-to-weight ratio is not critical and low battery cost is desirable.

- *Lithium ion* (Li-Ion)—used for video cameras and other portable electronic devices. This chemistry is replacing some NiCds for high energy-density applications, but at a higher cost.

- *Lithium polymer* (Li-Polymer)—this battery offers the highest energy density and lowest self-discharge rate of the devices compared here. The load characteristics, however, currently suit mainly low current applications.

- *Reusable alkaline*—used for light duty applications. Because of its low self-discharge, this battery is suitable for portable entertainment devices and other non-critical appliances that are used only occasionally.

No single battery offers all the answers, rather, each chemistry is based on a number of compromises, optimized to a particular application or group of applications.

It is interesting to observe that the NiCd has the shortest charge time, delivers the highest load current, and offers the lowest cost-per-cycle, but is most demanding on exercise requirements. For applications where high energy density is critical, regular exercise is impractical and cost is secondary, the NiMH is usually considered the best choice. Not without problems, NiMH batteries have a cycle life one-third that of NiCds. Furthermore, field use has revealed that the NiMH also needs some level of exercise to maximize service life, but to a lesser extent than the NiCd. In comparison, the SLA needs little or no maintenance but has a low energy density.

Among rechargeable batteries, the NiCd has been around the longest (since 1950). It is also one of the best understood chemistries and has become a standard against which other batteries are compared.

7.2.7 Plan Ahead

The bottom line in dealing with a disaster is preparation. First, develop a realistic plan to deal with a wide range of possible occurrences. Second, gain management and staff support for the plan. Third, test the plan with practice drills and revise the procedures as necessary.

While it may be impossible to prevent certain disasters, at least you will be able to deal with them.

7.3   References

1. Capitano, J., and J. Feinstein, "Environmental Stress Screening Demonstrates Its Value in the Field," *Proceedings of the IEEE Reliability and Maintainability Symposium*, IEEE, New York, 1986.
2. Wong, Kam L., "Demonstrating Reliability and Reliability Growth with Environmental Stress Screening Data," *Proceedings of the IEEE Reliability and Maintainability Symposium*, IEEE, New York, 1990.
3. Tustin, Wayne, "Recipe for Reliability: Shake and Bake," *IEEE Spectrum*, IEEE, New York, December 1986.
4. Hobbs, Gregg K., "Development of Stress Screens," *Proceedings of the IEEE Reliability and Maintainability Symposium*, IEEE, New York, 1987.
5. Smith, William B., "Integrated Product and Process Design to Achieve High Reliability in Both Early and Useful Life of the Product," *Proceedings of the IEEE Reliability and Maintainability Symposium*, IEEE, New York, 1987.

7.4   Bibliography

Buckmann, Isidor, "Batteries," in *The Electronics Handbook*, Jerry C. Whitaker (ed.), CRC Press, Boca Raton, FL, 1996.
Rudman, Richard, "Disaster Planning and Recovery," in *The Electronics Handbook*, Jerry C. Whitaker (ed.), CRC Press, Boca Raton, FL, 1996.

Safety Considerations

8.1   Introduction

Safety is critically important to engineering personnel who work around powered hardware, especially if they work under time pressures. Safety should not be taken lightly. *Life safety* systems are those designed to protect life and property. Such systems include emergency lighting, fire alarms, smoke exhaust and ventilating fans, and site security.

8.1.1   Facility Safety Equipment

Personnel safety is the responsibility of the facility manager. Proper life safety procedures and equipment must be installed. Safety-related hardware includes the following:

- *Emergency power off (EPO) button.* EPO push buttons are required by safety codes for data processing (DP) centers. One must be located at each principal exit from the DP room. Other EPO buttons may be located near operator workstations. The EPO system, intended only for emergencies, disconnects all power to the room, except for lighting.

- *Smoke detector.* Two basic types of smoke detectors are common. The first compares the transmission of light through air in the room with light through a sealed optical path into which smoke cannot penetrate. Smoke causes a differential or *backscattering* effect that, when detected, triggers an alarm after a preset threshold has been exceeded. The second type senses the ionization of combustion products, rather than visible smoke. A mildly radioactive source, usually nickel, ionizes the air passing through a screened chamber. A charged probe captures ions and detects the small current that is proportional to the rate of capture. When combustion products or material other than air molecules enter the probe area, the rate of ion production changes abruptly, generating a signal that triggers the alarm.

- *Flame detector*. The flame sensor responds not to heated surfaces or objects, but to infrared, when it flickers with the unique characteristics of a fire. Such detectors, for example, will respond to a lighted match, but not to a cigarette. The ultraviolet light from a flame also is used to distinguish between hot, glowing objects and open flame.

- *Halon*. The Halon fire-extinguishing agent is a low-toxicity, compressed gas that is contained in pressurized vessels. Discharge nozzles in DP and other types of equipment rooms are arranged to dispense the entire contents of a central container or of multiple smaller containers of Halon when actuated by a command from the fire control system. The discharge is sufficient to extinguish flame and stop combustion of most flammable substances. Halon is one of the more common fire-extinguishing agents used for DP applications. Halon systems are usually not practical, however, in large, open-space computer or communications centers.

- *Water sprinkler*. Although water is an effective agent against a fire, activation of a sprinkler system will cause damage to the equipment it is meant to protect. Interlock systems must drop all power (except for emergency lighting) before the water system is discharged. Most water systems use a two-stage alarm. Two or more fire sensors, often of different design, must signal an alarm condition before water is discharged into the protected area. Where sprinklers are used, floor drains and EPO controls must be provided.

- *Fire damper*. When a fire is detected, dampers are used to block ventilating passages in strategic parts of the system. This prevents fire from spreading through the passages and keeps fresh air from fanning the flames. A fire damper system, combined with the shutdown of cooling and ventilating air, enables Halon to be retained in the protected space until the fire is extinguished.

Many life safety system functions can be automated. The decision of what to automate and what to operate manually requires considerable thought. If the life safety control panels are accessible to a large number of site employees, most functions should be automatic. Alarm-silencing controls should be kept locked away. A mimic board can be used to identify problem areas readily. Figure 8.1 illustrates a well-organized life safety control system. Note that fire, HVAC (heating, ventilation, and air conditioning), security, and EPO controls all are readily accessible. Note also that operating instructions are posted for life safety equipment, and an evacuation route is shown. Important telephone numbers are posted, and a direct-line telephone (not via the building switchboard) is provided. All equipment is located adjacent to a lighted emergency exit door.

Life safety equipment must be maintained just as diligently as the electronic hardware that it protects. Conduct regular tests and drills. It is, obviously, not necessary or advisable to discharge Halon or water during a drill.

Configure the life safety control system to monitor the premises for dangerous conditions and also to monitor the equipment designed to protect the facility. Important monitoring points include HVAC machine parameters, water and/or Halon pressure, emergency battery-supply status, and other elements of the system that could compro-

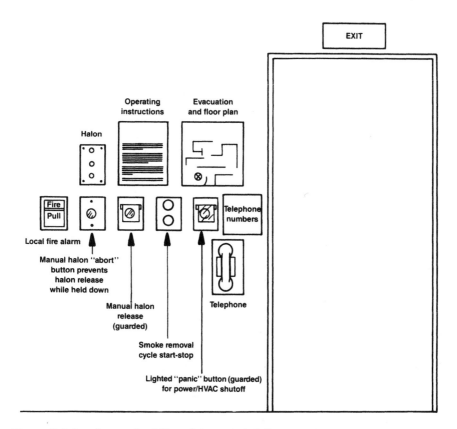

Figure 8.1 A well-organized life safety control station.

mise the ability of life safety equipment to function properly. Basic guidelines for life safety systems include the following:

- Carefully analyze the primary threats to life and property within the facility. Develop contingency plans to meet each threat.

- Prepare a life safety manual, and distribute it to all employees at the facility. Require them to read it.

- Conduct drills for employees at random times without notice. Require acceptable performance from employees.

- Prepare simple, step-by-step instructions on what to do in an emergency. Post the instructions in a conspicuous place.

- Assign after-hours responsibility for emergency situations. Prepare a list of supervisors that operators should contact if problems arise. Post the list with phone numbers. Keep the list accurate and up-to-date. Always provide the names of three individuals who may be contacted in an emergency.

- Work with a life safety consultant to develop a coordinated control and monitoring system for the facility. Such hardware will be expensive, but it must be provided. The facility may be able to secure a reduction in insurance rates if comprehensive safety efforts can be demonstrated.

- Interface the life safety system with automatic data-logging equipment so that documentation can be assembled on any event.

- Insist on complete, up-to-date schematic diagrams for all facility hardware. Insist that the diagrams include any changes made during installation or subsequent modification.

- Provide sufficient emergency lighting.

- Provide easy-access emergency exits.

The importance of providing standby power for sensitive loads at commercial and industrial facilities is obvious. It is equally important to provide standby power for life safety systems. A lack of ac power must not render the life safety system inoperative. Sensors and alarm control units should include their own backup battery supplies. In a properly designed system, all life safety equipment will be fully operational despite the loss of all ac power to the facility, including backup power for sensitive loads.

Place cables linking the life safety control system with remote sensors and actuators in a separate conduit containing only life safety conductors. Study the National Electrical Code and all applicable local and federal codes relating to safety. Follow them strictly.

8.1.2   A Systems Approach to Safety

Electrical safety is important when working with any type of electronic hardware. Because vacuum tubes and many other devices operate at high voltages and currents, safety is doubly important. The primary areas of concern, from a safety standpoint, include:

- Electric shock

- Nonionizing radiation

- Beryllium oxide (BeO) ceramic dust

- Hot surfaces of vacuum tube devices

- Polychlorinated biphenyls (PCBs)

8.2   Electric Shock

Surprisingly little current is required to injure a person. Studies at Underwriters Laboratories (UL) show that the electrical resistance of the human body varies with the amount of moisture on the skin, the muscular structure of the body, and the applied voltage. The typical hand-to-hand resistance ranges between 500 Ω and 600 kΩ, de-

Table 8.1 The Effects of Current on the Human Body

Current	Effect
1 mA or less	No sensation, not felt
More than 3 mA	Painful shock
More than 10 mA	Local muscle contractions, sufficient to cause "freezing" to the circuit for 2.5 percent of the population
More than 15 mA	Local muscle contractions, sufficient to cause "freezing" to the circuit for 50 percent of the population
More than 30 mA	Breathing is difficult, can cause unconsciousness
50 mA to 100 mA	Possible ventricular fibrillation
100 mA to 200 mA	Certain ventricular fibrillation
More than 200 mA	Severe burns and muscular contractions; heart more apt to stop than to go into fibrillation
More than a few amperes	Irreparable damage to body tissue

pending on the conditions. Higher voltages have the capability to break down the outer layers of the skin, which can reduce the overall resistance value. UL uses the lower value, 500 Ω, as the standard resistance between major extremities, such as from the hand to the foot. This value is generally considered the minimum that would be encountered and, in fact, may not be unusual because wet conditions or a cut or other break in the skin significantly reduces human body resistance.

8.2.1 Effects on the Human Body

Table 8.1 lists some effects that typically result when a person is connected across a current source with a hand-to-hand resistance of 2.4 kΩ. The table shows that a current of approximately 50 mA will flow between the hands, if one hand is in contact with a 120 V ac source and the other hand is grounded. The table indicates that even the relatively small current of 50 mA can produce *ventricular fibrillation* of the heart, and perhaps death. Medical literature describes ventricular fibrillation as rapid, uncoordinated contractions of the ventricles of the heart, resulting in loss of synchronization between heartbeat and pulse beat. The electrocardiograms shown in Figure 8.2 compare a healthy heart rhythm with one in ventricular fibrillation. Unfortunately, once ventricular fibrillation occurs, it will continue. Barring resuscitation techniques, death will ensue within a few minutes.

The route taken by the current through the body has a significant effect on the degree of injury. Even a small current, passing from one extremity through the heart to another extremity, is dangerous and capable of causing severe injury or electrocution. There are cases where a person has contacted extremely high current levels and lived to tell about it. However, usually when this happens, the current passes only through a single limb and not through the body. In these instances, the limb is often lost, but the person survives.

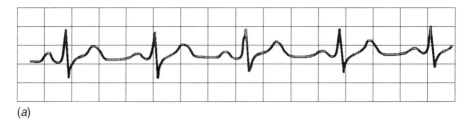

(a)

(b)

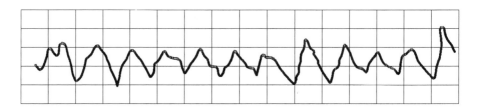

Figure 8.2 Electrocardiogram of a human heartbeat: (a) healthy rhythm, (b) ventricular fibrillation.

Current is not the only factor in electrocution. Figure 8.3 summarizes the relationship between current and time on the human body. The graph shows that 100 mA flowing through a human adult body for 2 s will cause death by electrocution. An important factor in electrocution, the *let-go range*, also is shown on the graph. This range is described as the amount of current that causes "freezing", or the inability to let go of the conductor. At 10 mA, 2.5 percent of the population will be unable to let go of a "live" conductor. At 15 mA, 50 percent of the population will be unable to let go of an energized conductor. It is apparent from the graph that even a small amount of current can "freeze" someone to a conductor. The objective for those who must work around electric equipment is how to protect themselves from electric shock. Table 8.2 lists required precautions for personnel working around high voltages.

8.2.2 Circuit Protection Hardware

The typical primary panel or equipment circuit breaker or fuse will not protect a person from electrocution. In the time it takes a fuse or circuit breaker to blow, someone could die. However, there are protection devices that, properly used, may help prevent electrocution. The *ground-fault current interrupter* (GFCI), shown in Figure 8.4, works by monitoring the current being applied to the load. The GFI uses a differential transformer and looks for an imbalance in load current. If a current (5 mA, ±1 mA) begins to flow between the neutral and ground or between the hot and ground leads,

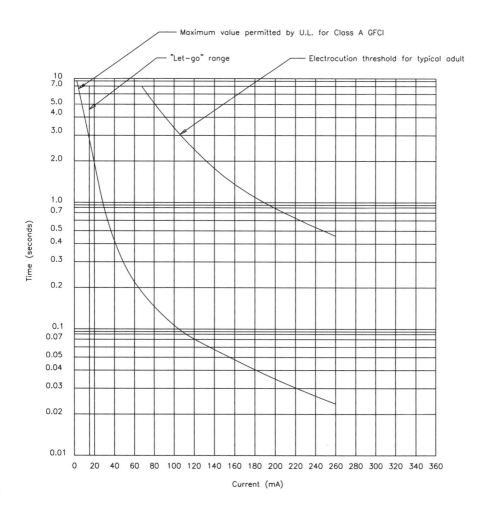

Figure 8.3 Effects of electric current and time on the human body. Note the "let-go" range.

the differential transformer detects the leakage and opens up the primary circuit within 2.5 ms.

GFIs will not protect a person from every type of electrocution. If the victim becomes connected to both the neutral and the hot wire, the GFI will not detect an imbalance.

Three-Phase Systems

For large, three-phase loads, detecting ground currents and interrupting the circuit before injury or damage can occur is a more complicated proposition. The classic

Table 8.2 Required Safety Practices for Engineers Working Around High-Voltage Equipment

High-Voltage Precautions
✓ Remove all ac power from the equipment. Do not rely on internal contactors or SCRs to remove dangerous ac.
✓ Trip the appropriate power distribution circuit breakers at the main breaker panel.
✓ Place signs as needed to indicate that the circuit is being serviced.
✓ Switch the equipment being serviced to the *local control* mode as provided.
✓ Discharge all capacitors using the discharge stick provided by the manufacturer.
✓ Do not remove, short circuit, or tamper with interlock switches on access covers, doors, enclosures, gates, panels, or shields.
✓ Keep away from live circuits.

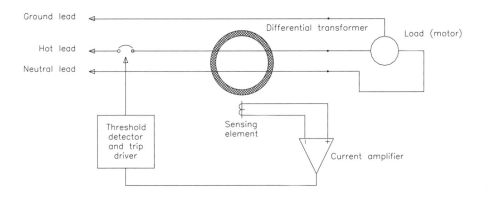

Figure 8.4 Basic design of a ground-fault interrupter (GFI).

method of protection involves the use of a zero-sequence *current transformer* (CT). Such devices are basically an extension of the single-phase GFI circuit, shown in Figure 8.4. Three-phase CTs have been developed to fit over bus ducts, switchboard buses, and circuit-breaker studs. Rectangular core-balanced CTs are able to detect leakage currents as small as several milliamperes when the system carries as much as 4 kA. "Doughnut-type" toroidal zero-sequence CTs also are available in varying diameters.

The zero-sequence current transformer is designed to detect the magnetic field surrounding a group of conductors. As shown in Figure 8.5, in a properly operating three-phase system, the current flowing through the conductors of the system—includ-

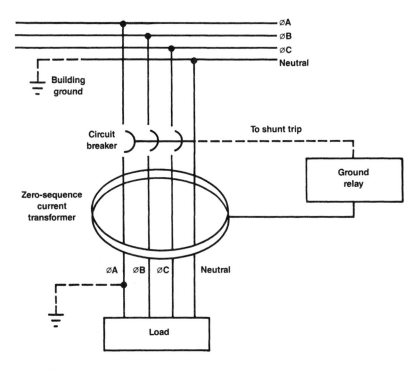

Figure 8.5 Ground-fault detection in a three-phase ac system.

ing the neutral—travels out and returns along those same conductors. The net magnetic flux detected by the CT is zero. No signal is generated in the transformer winding, regardless of current magnitudes—symmetrical or asymmetrical. If one phase conductor is faulted to ground, however, the current balance will be upset. The ground-fault detection circuit then will trip the breaker and open the line.

For optimum protection in a large facility, GFI units are placed at natural branch points of the ac power system. Obviously, it is preferable to lose only a small portion of a facility in the event of a ground fault than it is to have the entire plant dropped. Figure 8.6 illustrates such a distributed system. Sensors are placed at major branch points to isolate any ground fault from the remainder of the distribution network. In this way, the individual GFI units can be set for higher sensitivity and shorter time delays than would be practical with a large, distributed load. The technology of GFI devices has improved significantly in the past few years. New integrated circuit devices and improved CT designs have provided improved protection components at a lower cost.

Sophisticated GFI monitoring *systems* are available that analyze ground-fault currents and isolate the faulty branch circuit. This feature prevents needless tripping of GFI units up the line toward the utility service entrance. For example, if a ground fault is sensed in a fourth-level branch circuit, the GFI system controller automatically locks

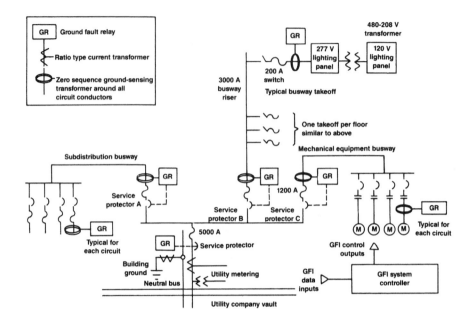

Figure 8.6 Ground-fault protection system for a large, multistory building.

out first-, second-, and third-level devices from operating to clear the fault. The problem, therefore, is safely confined to the fourth-level branch. The GFI control system is designed to operate in a fail-safe mode. In the event of a control-system shutdown, the individual GFI trip relays would operate independently to clear whatever fault currents may exist.

8.2.3 Working with High Voltage

Rubber gloves are commonly used by engineers working on high-voltage equipment. These gloves are designed to provide protection from hazardous voltages or RF when the wearer is working on "hot" ac or RF circuits. Although the gloves may provide some protection from these hazards, placing too much reliance on them can have disastrous consequences. There are several reasons why gloves should be used with a great deal of caution and respect. A common mistake made by engineers is to assume that the gloves always provide complete protection. The gloves found in many facilities may be old or untested. Some may show signs of user repair, perhaps with electrical tape. Few tools could be more hazardous than such a pair of gloves.

Another mistake is not knowing the voltage rating of the gloves. Gloves are rated differently for both ac and dc voltages. For example, a *class 0* glove has a minimum dc

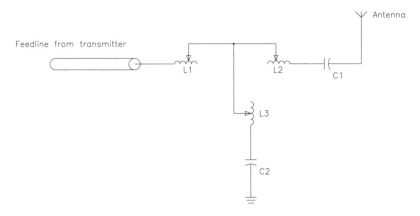

Figure 8.7 Example of how high voltages can be generated in an RF load matching unit.

breakdown voltage of 35 kV; the minimum ac breakdown voltage, however, is only 6 kV. Furthermore, high-voltage rubber gloves are not usually tested at RF frequencies, and RF can burn a hole in the best of them. It is possible to develop dangerous working habits by assuming that gloves will offer the required protection.

Gloves alone may not be enough to protect an individual in certain situations. Recall the axiom of keeping one hand in a pocket while working around a device with current flowing? That advice is actually based on simple electricity. It is not the "hot" connection that causes the problem, but the ground connection that lets the current begin to flow. Studies have shown that more than 90 percent of electric equipment fatalities occurred when the grounded person contacted a live conductor. Line-to-line electrocution accounted for less than 10 percent of the deaths.

When working around high voltages, always look for grounded surfaces. Keep hands, feet, and other parts of the body away from any grounded surface. Even concrete can act as a ground if the voltage is sufficiently high. If work must be performed in "live" cabinets, then consider using, in addition to rubber gloves, a rubber floor mat, rubber vest, and rubber sleeves. Although this may seem to be a lot of trouble, consider the consequences of making a mistake. Of course, the best troubleshooting methodology is never to work on any circuit without being certain that no hazardous voltages are present. In addition, any circuits or contactors that normally contain hazardous voltages should be firmly grounded before work begins.

RF Considerations

Engineers often rely on electrical gloves when making adjustments to live RF circuits. This practice, however, can be extremely dangerous. Consider the typical load matching unit shown in Figure 8.7. In this configuration, disconnecting the coil from either L2 or L3 places the full RF output literally at the engineer's fingertips. Depending on the impedances involved, the voltages can become quite high, even in a circuit that normally is relatively tame.

In the Figure 8.7 example, assume that the load impedance is approximately 106 $+j202\,\Omega$. With 1 kW feeding into the load, the rms voltage at the matching output will be approximately 700 V. The peak voltage (which determines insulating requirements) will be close to 1 kV, and perhaps more than twice that if the carrier is being amplitude-modulated. At the instant the output coil clip is disconnected, the current in the shunt leg will increase rapidly, and the voltage easily could more than double.

8.2.4 First Aid Procedures

All engineers working around high-voltage equipment should be familiar with first aid treatment for electric shock and burns. Always keep a first aid kit on hand at the facility. Figure 8.8 illustrates the basic treatment for victims of electric shock. Copy the information, and post it in a prominent location. Better yet, obtain more detailed information from the local heart association or Red Cross chapter. Personalized instruction on first aid usually is available locally.

8.3 Operating Hazards

A number of potential hazards exist in the operation and maintenance of high-power equipment. Maintenance personnel must exercise extreme care around such hardware. Consider the following guidelines:

- Use caution around the high-voltage stages of the equipment. Many power tubes operate at voltages high enough to kill through electrocution. Always break the primary ac circuit of the power supply, and discharge all high-voltage capacitors.

- Minimize exposure to RF radiation. Do not permit personnel to be in the vicinity of open, energized RF generating circuits, RF transmission systems (waveguides, cables, or connectors), or energized antennas. High levels of radiation can result in severe bodily injury, including blindness. Cardiac pacemakers may also be affected.

- Avoid contact with beryllium oxide (BeO) ceramic dust and fumes. BeO ceramic material may be used as a thermal link to carry heat from a tube to the heat sink. Do not perform any operation on any BeO ceramic that might produce dust or fumes, such as grinding, grit blasting, or acid cleaning. Beryllium oxide dust and fumes are highly toxic, and breathing them can result in serious injury or death. BeO ceramics must be disposed of as prescribed by the device manufacturer.

- Avoid contact with hot surfaces within the equipment. The anode portion of most power tubes is air-cooled. The external surface normally operates at a high temperature (up to 250°C). Other portions of the tube also may reach high temperatures, especially the cathode insulator and the cathode/heater surfaces. All hot surfaces may remain hot for an extended time after the system is shut off. To prevent serious burns, avoid bodily contact with these surfaces during operation and

If the victim is not responsive, follow the A—B—Cs of basic life support.

A   AIRWAY: If the victim is unconscious, open airway.

1. Lift up neck
2. Push forehead back
3. Clear out mouth if necessary
4. Observe for breathing

B   BREATHING: If the victim is not breathing, begin artificial breathing.

1. Tilt head
2. Pinch nostrils
3. Make airtight seal
4. Provide four quick full breaths

Check carotid pulse. If pulse is absent, begin artificial circulation. Remember that mouth—to—mouth resuscitation must be commenced as soon as possible.

C   CIRCULATION: Depress the sternum 1.2 to 2 inches.

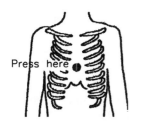

Press here

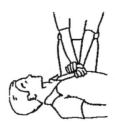

For situations in which there is one rescuer, provide 15 compressions and then 2 quick breaths. The approximate rate of compressions should be 80 per minute.
For situations in which there are two rescuers, provide 5 compressions and then 1 breath. The approximate rate of compressions should be 60 per minute.
Do not interrupt the rhythm of compressions when a second person is giving breaths.

If the victim is responsive, keep warm and quiet, loosen clothing, and place in a reclining position. Call for medical assistance as soon as possible.

Figure 8.8 Basic first aid treatment for electric shock.

Table 8.3 Basic First Aid Procedures for Burns (More detailed information can be obtained from any Red Cross office.)

Extensively Burned and Broken Skin	
✓	Cover affected area with a clean sheet or cloth.
✓	Do not break blisters, remove tissue, remove adhered particles of clothing, or apply any salve or ointment.
✓	Treat victim for shock as required.
✓	Arrange for transportation to a hospital as quickly as possible.
✓	If arms or legs are affected, keep them elevated.
✓	If medical help will not be available within an hour and the victim is conscious and not vomiting, prepare a weak solution of salt and soda: 1 level teaspoon of salt and 1/2 level teaspoon of baking soda to each quart of tepid water. Allow the victim to sip slowly about 4 ounces (half a glass) over a period of 15 minutes. Discontinue fluid intake if vomiting occurs. (Do not offer alcohol.)
Less Severe Burns (First and Second Degree)	
✓	Apply cool (not ice-cold) compresses using the cleanest available cloth article.
✓	Do not break blisters, remove tissue, remove adhered particles of clothing, or apply salve or ointment.
✓	Apply clean, dry dressing if necessary.
✓	Treat victim for shock as required.
✓	Arrange for transportation to a hospital as quickly as possible.
✓	If arms or legs are affected, keep them elevated.

for a reasonable cool-down period afterward. Table 8.3 lists basic first aid procedures for burns.

8.3.1 OSHA Safety Considerations

The U.S. government has taken a number of steps to help improve safety within the workplace under the auspices of the Occupational Safety and Health Administration (OSHA). The agency helps industries monitor and correct safety practices. OSHA has developed a number of guidelines designed to help prevent accidents. OSHA records show that electrical standards are among the most frequently violated of all safety standards. Table 8.4 lists 16 of the most common electrical violations, including exposure of live conductors, improperly labeled equipment, and faulty grounding.

Protective Covers

Exposure of live conductors is a common safety violation. All potentially dangerous electric conductors should be covered with protective panels. The danger is that someone may come into contact with the exposed current-carrying conductors. It is also possible for metallic objects such as ladders, cable, or tools to contact a hazard-

Table 8.4 Sixteen Common OSHA Violations (*After* [1].)

Fact Sheet	Subject	NEC[1] Reference
1	Guarding of live parts	110-17
2	Identification	110-22
3	Uses allowed for flexible cord	400-7
4	Prohibited uses of flexible cord	400-8
5	Pull at joints and terminals must be prevented	400-10
6.1	Effective grounding, Part 1	250-51
6.2	Effective grounding, Part 2	250-51
7	Grounding of fixed equipment, general	250-42
8	Grounding of fixed equipment, specific	250-43
9	Grounding of equipment connected by cord and plug	250-45
10	Methods of grounding, cord and plug-connected equipment	250-59
11	AC circuits and systems to be grounded	250-5
12	Location of overcurrent devices	240-24
13	Splices in flexible cords	400-9
14	Electrical connections	110-14
15	Marking equipment	110-21
16	Working clearances about electric equipment	110-16
[1] National Electrical Code		

ous voltage, creating a life-threatening condition. Open panels also present a fire hazard.

Identification and Marking

Circuit breakers and switch panels should be properly identified and labeled. Labels on breakers and equipment switches may be many years old and may no longer reflect the equipment actually in use. This is a safety hazard. Casualties or unnecessary damage can be the result of an improperly labeled circuit panel if no one who understands the system is available in an emergency. If a number of devices are connected to a single disconnect switch or breaker, a diagram should be provided for clarification. Label with brief phrases, and use clear, permanent, and legible markings.

Equipment marking is a closely related area of concern. This is not the same thing as equipment identification. Marking equipment means labeling the equipment breaker panels and ac disconnect switches according to device rating. Breaker boxes should contain a nameplate showing the manufacturer, rating, and other pertinent electrical factors. The intent is to prevent devices from being subjected to excessive loads or voltages.

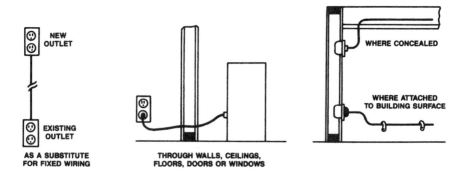

Figure 8.9 Flexible cord uses prohibited under NEC rules.

Extension Cords

Extension (flexible) cords often are misused. Although it may be easy to connect a new piece of equipment with a flexible cord, be careful. The National Electrical Code lists only eight approved uses for flexible cords.

The use of a flexible cord where the cable passes through a hole in the wall, ceiling, or floor is a common violation. Running the cord through doorways, windows, or similar openings also is prohibited. A flexible cord should not be attached to building surfaces or concealed behind building walls or ceilings. These common violations are illustrated in Figure 8.9.

Failure to provide adequate strain relief on connectors is another common problem. Whenever possible, use manufactured cable connections.

Grounding

OSHA regulations describe two types of grounding: *system grounding* and *equipment grounding*. System grounding actually connects one of the current-carrying conductors (such as the terminals of a supply transformer) to ground. (See Figure 8.10.) Equipment grounding connects all of the noncurrent-carrying metal surfaces together and to ground. From a grounding standpoint, the only difference between a grounded electrical system and an ungrounded electrical system is that the *main bonding jumper* from the service equipment ground to a current-carrying conductor is omitted in the ungrounded system. The system ground performs two tasks:

- It provides the final connection from equipment-grounding conductors to the grounded circuit conductor, thus completing the ground-fault loop.

- It solidly ties the electrical system and its enclosures to their surroundings (usually earth, structural steel, and plumbing). This prevents voltages at any source from rising to harmfully high voltage-to-ground levels.

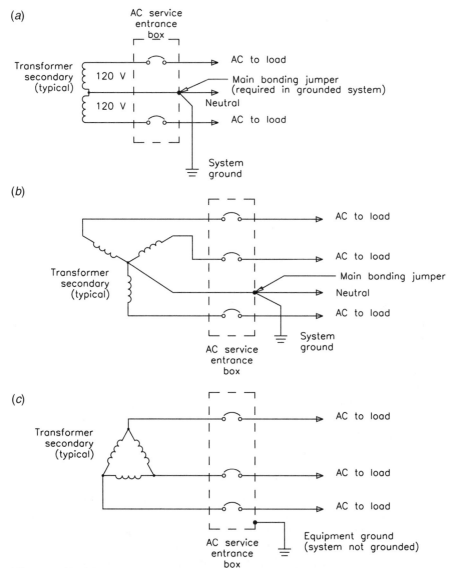

Figure 8.10 AC service entrance bonding requirements: (a) 120 V phase-to-neutral (240 V phase-to-phase), (b) 3-phase 208 V wye (120 V phase-to-neutral), (c) 3-phase 240 V (or 480 V) delta. Note that the main bonding jumper is required in only two of the designs.

Note that equipment grounding—bonding all electric equipment to ground—is required whether or not the system is grounded. Equipment grounding serves two important tasks:

- It bonds all surfaces together so that there can be no voltage difference among them.

- It provides a ground-fault current path from a fault location back to the electrical source, so that if a fault current develops, it will rise to a level high enough to operate the breaker or fuse.

The National Electrical Code (NEC) is complex and contains numerous requirements concerning electrical safety. The fact sheets listed in Table 8.4 are available from OSHA.

8.3.2 Beryllium Oxide Ceramics

Some vacuum tubes, both power grid and microwave, contain beryllium oxide (BeO) ceramics, typically at the output waveguide window or around the cathode. Never perform any operations on BeO ceramics that produce dust or fumes, such as grinding, grit blasting, or acid cleaning. Beryllium oxide dust and fumes are highly toxic, and breathing them can result in serious personal injury or death.

If a broken window is suspected on a microwave tube, carefully remove the device from its waveguide, and seal the output flange of the tube with tape. Because BeO warning labels may be obliterated or missing, maintenance personnel should contact the tube manufacturer before performing any work on the device. Some tubes have BeO internal to the vacuum envelope.

Take precautions to protect personnel working in the disposal or salvage of tubes containing BeO. All such personnel should be made aware of the deadly hazards involved and the necessity for great care and attention to safety precautions. Some tube manufacturers will dispose of tubes without charge, provided they are returned to the manufacturer prepaid, with a written request for disposal.

8.3.3 Corrosive and Poisonous Compounds

The external output waveguides and cathode high-voltage bushings of microwave tubes are sometimes operated in systems that use a dielectric gas to impede microwave or high-voltage breakdown. If breakdown does occur, the gas may decompose and combine with impurities, such as air or water vapor, to form highly toxic and corrosive compounds. Examples include Freon gas, which may form lethal *phosgene*, and sulfur hexafluoride (SF_6) gas, which may form highly toxic and corrosive sulfur or fluorine compounds such as *beryllium fluoride*. When breakdown does occur in the presence of these gases, proceed as follows:

- Ventilate the area to outside air

- Avoid breathing any fumes or touching any liquids that develop

- Take precautions appropriate for beryllium compounds and for other highly toxic and corrosive substances

If a coolant other than pure water is used, follow the precautions supplied by the coolant manufacturer.

8.3.4 FC-75 Toxic Vapor

The decomposition products of FC-75 are highly toxic. Decomposition may occur as a result of any of the following:

- Exposure to temperatures above 200°C

- Exposure to liquid fluorine or alkali metals (lithium, potassium, or sodium)

- Exposure to ionizing radiation

Known thermal decomposition products include *perfluoroisobutylene* (PFIB; $[CF_3]_2$ $C = CF_2$), which is highly toxic in small concentrations.

If FC-75 has been exposed to temperatures above 200°C through fire, electric heating, or prolonged electric arcs, or has been exposed to alkali metals or strong ionizing radiation, take the following steps:

- Strictly avoid breathing any fumes or vapors.

- Thoroughly ventilate the area.

- Strictly avoid any contact with the FC-75.

Under such conditions, promptly replace the FC-75 and handle and dispose of the contaminated FC-75 as a toxic waste.

8.3.5 Nonionizing Radiation

Nonionizing radio frequency radiation (RFR) resulting from high-intensity RF fields is a serious concern to engineers who must work around high-power transmission equipment. The principal medical issue regarding nonionizing radiation involves heating of various body tissues, which can have serious effects, particularly if there is no mechanism for heat removal. Recent research has also noted, in some cases, subtle psychological and physiological changes at radiation levels below the threshold for heat-induced biological effects. However, the consensus is that most effects are thermal in nature.

High levels of RFR can affect one or more body systems or organs. Areas identified as potentially sensitive include the ocular (eye) system, reproductive system, and the immune system. Nonionizing radiation also is thought to be responsible for metabolic effects on the central nervous system and cardiac system.

In spite of these studies, many of which are ongoing, there is still no clear evidence in Western literature that exposure to medium-level nonionizing radiation results in detrimental effects. Russian findings, on the other hand, suggest that occupational exposure to RFR at power densities above $1.0 \, \text{mW/cm}^2$ does result in symptoms, particularly in the central nervous system.

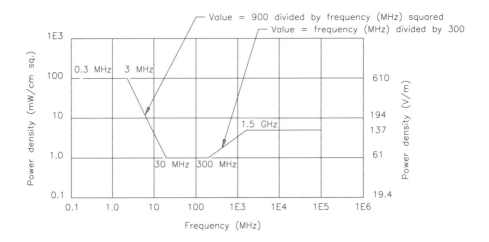

Figure 8.11 The power density limits for nonionizing radiation exposure for humans.

Clearly, the jury is still out as to the ultimate biological effects of RFR. Until the situation is better defined, however, the assumption must be made that potentially serious effects can result from excessive exposure. Compliance with existing standards should be the minimum goal, to protect members of the public as well as facility employees.

NEPA Mandate

The National Environmental Policy Act of 1969 required the Federal Communications Commission to place controls on nonionizing radiation. The purpose was to prevent possible harm to the public at large and to those who must work near sources of the radiation. Action was delayed because no hard and fast evidence existed that low- and medium-level RF energy is harmful to human life. Also, there was no evidence showing that radio waves from radio and TV stations did not constitute a health hazard.

During the delay, many studies were carried out in an attempt to identify those levels of radiation that might be harmful. From the research, suggested limits were developed by the American National Standards Institute (ANSI) and stated in the document known as ANSI C95.1-1982. The protection criteria outlined in the standard are shown in Figure 8.11.

The energy-level criteria were developed by representatives from a number of industries and educational institutions after performing research on the possible effects of nonionizing radiation. The projects focused on absorption of RF energy by the human body, based upon simulated human body models. In preparing the document, ANSI attempted to determine those levels of incident radiation that would cause the body to absorb less than 0.4 W/kg of mass (averaged over the whole body) or peak absorption values of 8 W/kg over any 1 gram of body tissue.

From the data, the researchers found that energy would be absorbed more readily at some frequencies than at others. The absorption rates were found to be functions of the size of a specific individual and the frequency of the signal being evaluated. It was the result of these absorption rates that culminated in the shape of the *safe curve* shown in the figure. ANSI concluded that no harm would come to individuals exposed to radio energy fields, as long as specific values were not exceeded when averaged over a period of 0.1 hour. It was also concluded that higher values for a brief period would not pose difficulties if the levels shown in the standard document were not exceeded when averaged over the 0.1-hour time period.

The FCC adopted ANSI C95.1-1982 as a standard that would ensure adequate protection to the public and to industry personnel who are involved in working around RF equipment and antenna structures.

Revised Guidelines

The ANSI C95.1-1982 standard was intended to be reviewed at 5-year intervals. Accordingly, the 1982 standard was due for reaffirmation or revision in 1987. The process was indeed begun by ANSI, but was handed off to the Institute of Electrical and Electronics Engineers (IEEE) for completion. In 1991, the revised document was completed and submitted to ANSI for acceptance as ANSI/IEEE C95.1-1992.

The IEEE standard incorporated changes from the 1982 ANSI document in four major areas:

- An additional safety factor was provided in certain situations. The most significant change was the introduction of new *uncontrolled* (public) exposure guidelines, generally established at one-fifth of the *controlled* (occupational) exposure guidelines. Figure 8.12 illustrates the concept for the microwave frequency band.

- For the first time, guidelines were included for body currents; examination of the electric and magnetic fields were determined to be insufficient to determine compliance.

- Minor adjustments were made to occupational guidelines, including relaxation of the guidelines at certain frequencies and the introduction of *breakpoints* at new frequencies.

- Measurement procedures were changed in several aspects, most notably with respect to spatial averaging and to minimum separation from reradiating objects and structures at the site.

The revised guidelines are complex and beyond the scope of this handbook. Refer to the ANSI/IEEE document for details.

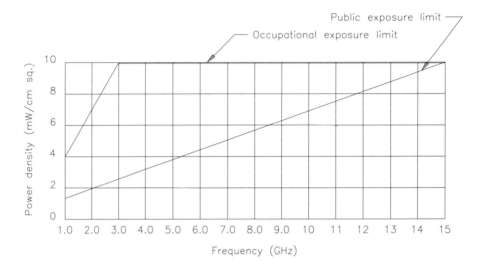

Figure 8.12 ANSI/IEEE exposure guidelines for microwave frequencies.

Multiple-User Sites

At a multiple-user site, the responsibility for assessing the RFR situation—although officially triggered by either a new user or the license renewal of all site tenants—is, in reality, the joint responsibility of all the site tenants. In a multiple-user environment involving various frequencies, and various protection criteria, compliance is indicated when the fraction of the RFR limit within each pertinent frequency band is established and added to the sum of all the other fractional contributions. The sum must not be greater than 1.0. Evaluating the multiple-user environment is not a simple matter, and corrective actions, if indicated, may be quite complex.

Operator Safety Considerations

RF energy must be contained properly by shielding and transmission lines. All input and output RF connections, cables, flanges, and gaskets must be RF leakproof. The following guidelines should be followed at all times:

- Never operate a power tube without a properly matched RF energy absorbing load attached.

- Never look into or expose any part of the body to an antenna or open RF generating tube, circuit, or RF transmission system that is energized.

- Monitor the RF system for radiation leakage at regular intervals and after servicing.

8.3.6 X-Ray Radiation Hazard

The voltages typically used in microwave tubes are capable of producing dangerous X rays. As voltages increase beyond 15 kV, metal-body tubes are capable of producing progressively more dangerous radiation. Adequate X-ray shielding must be provided on all sides of such tubes, particularly at the cathode and collector ends, as well as at the modulator and pulse transformer tanks (as appropriate). High-voltage tubes never should be operated without adequate X-ray shielding in place. The X-ray radiation of the device should be checked at regular intervals and after servicing.

8.3.7 Implosion Hazard

Because of the high internal vacuum in power grid, microwave, and cathode ray tubes, the glass or ceramic output window or envelope can shatter inward (implode) if struck with sufficient force or exposed to sufficient mechanical shock. Flying debris could result in bodily injury, including cuts and puncture wounds. If the device is made of beryllium oxide ceramic, implosion may produce highly toxic dust or fumes.

In the event of such an implosion, assume that toxic BeO ceramic is involved unless confirmed otherwise.

8.3.8 Hot Coolant and Surfaces

Extreme heat occurs in the electron collector of a microwave tube and the anode of a power grid tube during operation. Coolant channels used for water or vapor cooling also can reach high temperatures (boiling—100°C—and above), and the coolant is typically under pressure (as high as 100 psi). Some devices are cooled by boiling the coolant to form steam.

Contact with hot portions of the tube or its cooling system can scald or burn. Carefully check that all fittings and connections are secure, and monitor back pressure for changes in cooling system performance. If back pressure is increased above normal operating values, shut the system down and clear the restriction.

For a device whose anode or collector is air-cooled, the external surface normally operates at a temperature of 200 to 300°C. Other parts of the tube also may reach high temperatures, particularly the cathode insulator and the cathode/heater surfaces. All hot surfaces remain hot for an extended time after the tube is shut off. To prevent serious burns, take care to avoid bodily contact with these surfaces during operation and for a reasonable cool-down period afterward.

8.3.9 Polychlorinated Biphenyls

PCBs belong to a family of organic compounds known as *chlorinated hydrocarbons*. Virtually all PCBs in existence today have been synthetically manufactured. PCBs have a heavy oil-like consistency, high boiling point, a high degree of chemical stability, low flammability, and low electrical conductivity. These characteristics resulted in the widespread use of PCBs in high-voltage capacitors and transformers. Commercial

Table 8.5 Commonly Used Names for PCB Insulating Material

PCB Trade Names					
Apirolio	Abestol	Askarel[1]	Aroclor B	Chlorexto	Chlophen
Chlorinol	Clorphon	Diaclor	DK	Dykanol	EEC-18
Elemex	Eucarel	Fenclor	Hyvol	Inclor	Inerteen
Kanechlor	No-Flamol	Phenodlor	Pydraul	Pyralene	Pyranol
Pyroclor	Sal-T-Kuhl	Santothern FR	Santovac	Solvol	Thermin

[1] Generic name used for nonflammable dielectric fluids containing PCBs.

products containing PCBs were widely distributed between 1957 and 1977 under several trade names including:

- Aroclor

- Pyroclor

- Sanotherm

- Pyranol

- Askarel

Askarel is also a generic name used for nonflammable dielectric fluids containing PCBs. Table 8.5 lists some common trade names used for Askarel. These trade names typically will be listed on the nameplate of a PCB transformer or capacitor.

PCBs are harmful because once they are released into the environment, they tend not to break apart into other substances. Instead, PCBs persist, taking several decades to slowly decompose. By remaining in the environment, they can be taken up and stored in the fatty tissues of all organisms, from which they are slowly released into the bloodstream. Therefore, because of the storage in fat, the concentration of PCBs in body tissues can increase with time, even though PCB exposure levels may be quite low. This process is called *bioaccumulation*. Furthermore, as PCBs accumulate in the tissues of simple organisms, and as they are consumed by progressively higher organisms, the concentration increases. This process is called *biomagnification*. These two factors are especially significant because PCBs are harmful even at low levels. Specifically, PCBs have been shown to cause chronic (long-term) toxic effects in some species of animals and aquatic life. Well-documented tests on laboratory animals show that various levels of PCBs can cause reproductive effects, gastric disorders, skin lesions, and cancerous tumors.

PCBs may enter the body through the lungs, the gastrointestinal tract, and the skin. After absorption, PCBs are circulated in the blood throughout the body and stored in fatty tissues and a variety of organs, including the liver, kidneys, lungs, adrenal glands, brain, heart, and skin.

The health risk from PCBs lies not only in the PCB itself, but also in the chemicals that develop when PCBs are heated. Laboratory studies have confirmed that PCB by-products, including *polychlorinated dibenzofurans* (PCDFs) and *polychlorinated dibenzo-p-dioxins* (PCDDs), are formed when PCBs or *chlorobenzenes* are heated to temperatures ranging from approximately 900 to 1300°F. Unfortunately, these products are more toxic than PCBs themselves.

Governmental Action

The U.S. Congress took action to control PCBs in October 1975 by passing the Toxic Substances Control Act (TSCA). A section of this law specifically directed the EPA to regulate PCBs. Three years later the Environmental Protection Agency (EPA) issued regulations to implement the congressional ban on the manufacture, processing, distribution, and disposal of PCBs. Since that time, several revisions and updates have been issued by the EPA. One of these revisions, issued in 1982, specifically addressed the type of equipment used in industrial plants and transmitting stations. Failure to properly follow the rules regarding the use and disposal of PCBs has resulted in high fines and even jail sentences.

Although PCBs are no longer being produced for electrical products in the United States, there are thousands of PCB transformers and PCB capacitors still in use or in storage. The threat of widespread contamination from PCB fire-related incidents is one reason behind the EPA's efforts to reduce the number of PCB products in the environment. The users of high-power equipment are affected by the regulations primarily because of the widespread use of PCB transformers and capacitors. These components usually are located in older (pre-1979) systems, so this is the first place to look for them. However, some facilities also maintain their own primary power transformers. Unless these transformers are of recent vintage, it is possible that they too contain a PCB dielectric. Table 8.6 lists the primary classifications of PCB devices.

PCB Components

The two most common PCB components are transformers and capacitors. A PCB transformer is one containing at least 500 ppm (parts per million) PCBs in the dielectric fluid. An Askarel transformer generally has 600,000 ppm or more. A PCB transformer may be converted to a *PCB-contaminated device* (50 to 500 ppm) or a *non-PCB device* (less than 50 ppm) by having it drained, refilled, and tested. The testing must not take place until the transformer has been in service for a minimum of 90 days. Note that this is *not* something a maintenance technician can do. It is the exclusive domain of specialized remanufacturing companies.

PCB transformers must be inspected quarterly for leaks. If an impervious dike is built around the transformer sufficient to contain all of the liquid material, the inspections can be conducted yearly. Similarly, if the transformer is tested and found to contain less than 60,000 ppm, a yearly inspection is sufficient. Failed PCB transformers cannot be repaired; they must be properly disposed of.

Table 8.6 Definition of PCB Terms as Identified by the EPA

Term	Definition	Examples
PCB	Any chemical substance that is limited to the biphenyl molecule that has been chlorinated to varying degrees, or any combination of substances that contain such substances.	PCB dielectric fluids, PCB heat-transfer fluids, PCB hydraulic fluids, 2,2',4-trichlorobiphenyl
PCB article	Any manufactured article, other than a PCB container, that contains PCBs and whose surface has been in direct contact with PCBs.	Capacitors, transformers, electric motors, pumps, pipes
PCB container	A device used to contain PCBs or PCB articles, and whose surface has been in direct contact with PCBs.	Packages, cans, bottles, bags, barrels, drums, tanks
PCB article container	A device used to contain PCB articles or equipment, and whose surface has not been in direct contact with PCBs.	Packages, cans, bottles, bags, barrels, drums, tanks
PCB equipment	Any manufactured item, other than a PCB container or PCB article container, that contains a PCB article or other PCB equipment.	Microwave systems, fluorescent light ballasts, electronic equipment
PCB item	Any PCB article, PCB article container, PCB container, or PCB equipment that deliberately or unintentionally contains, or has as a part of it, any PCBs.	
PCB transformer	Any transformer that contains PCBs in concentrations of 500 ppm or greater.	
PCB contaminated	Any electric equipment that contains more than 50, but less than 500 ppm of PCBs. (Oil-filled electric equipment other than circuit breakers, reclosers, and cable whose PCB concentration is unknown must be assumed to be PCB-contaminated electric equipment.)	Transformers, capacitors, contaminated circuit breakers, reclosers, voltage regulators, switches, cable, electromagnets

If a leak develops, it must be contained and daily inspections begun. A cleanup must be initiated as soon as possible, but no later than 48 hours after the leak is discovered. Adequate records must be kept of all inspections, leaks, and actions taken for 3 years after disposal of the component. Combustible materials must be kept a minimum of 5 m from a PCB transformer and its enclosure.

As of October 1, 1990, the use of PCB transformers (500 ppm or greater) was prohibited in or near commercial buildings when the secondary voltages are 480 V ac or higher.

The EPA regulations also require that the operator notify others of the possible dangers. All PCB transformers (including PCB transformers in storage for reuse) must be registered with the local fire department. The following information must be supplied:

- The location of the PCB transformer(s).

- Address(es) of the building(s) and, for outdoor PCB transformers, the location.

- Principal constituent of the dielectric fluid in the transformer(s).

- Name and telephone number of the contact person in the event of a fire involving the equipment.

Any PCB transformers used in a commercial building must be registered with the building owner. All building owners within 30 m of such PCB transformers also must be notified. In the event of a fire-related incident involving the release of PCBs, the Coast Guard National Spill Response Center (800-424-8802) must be notified immediately. Appropriate measures also must be taken to contain and control any possible PCB release into water.

Capacitors are divided into two size classes, *large* and *small*. A PCB small capacitor contains less than 1.36 kg (3 lbs) of dielectric fluid. A capacitor having less than 100 in^3 also is considered to contain less than 3 lb of dielectric fluid. A PCB large capacitor has a volume of more than 200 in^3 and is considered to contain more than 3 lb of dielectric fluid. Any capacitor having a volume between 100 and 200 in^3 is considered to contain 3 lb of dielectric, provided the total weight is less than 9 lb. A PCB *large high-voltage capacitor* contains 3 lb or more of dielectric fluid and operates at voltages of 2 kV or greater. A *large low-voltage capacitor* also contains 3 lb or more of dielectric fluid but operates below 2 kV.

The use and servicing of PCB small capacitors is not restricted by the EPA unless there is a leak. In that event, the leak must be repaired or the capacitor disposed of. Disposal may be handled by an approved incineration facility, or the component may be placed in a specified container and buried in an approved chemical waste landfill. Items such as capacitors that are leaking oil greater than 500 ppm PCBs should be taken to an EPA-approved PCB disposal facility.

PCB Liability Management

Properly managing the PCB risk is not particularly difficult; the keys are understanding the regulations and following them carefully. Any program should include the following steps:

Table 8.7 Major Points of a Facility Safety Program

Management assumes the leadership role regarding safety policies.
Responsibility for safety- and health-related activities is clearly assigned.
Hazards are identified, and steps are taken to eliminate them.
Employees at all levels are trained in proper safety procedures.
Thorough accident/injury records are maintained.
Medical attention and first aid is readily available.
Employee awareness and participation is fostered through incentives and an ongoing, high-profile approach to workplace safety.

- Locate and identify all PCB devices. Check all stored or spare devices.

- Properly label PCB transformers and capacitors according to EPA requirements.

- Perform the required inspections and maintain an accurate log of PCB items, their location, inspection results, and actions taken. These records must be maintained for 3 years after disposal of the PCB component.

- Complete the annual report of PCBs and PCB items by July 1 of each year. This report must be retained for 5 years.

- Arrange for necessary disposal through a company licensed to handle PCBs. If there are any doubts about the company's license, contact the EPA.

- Report the location of all PCB transformers to the local fire department and to the owners of any nearby buildings.

The importance of following the EPA regulations cannot be overstated.

8.4   Management Responsibility

The key to operating a safe facility is diligent management. A carefully thought-out plan ensures a coordinated approach to protecting staff members from injury, and the facility from potential litigation. Although the details and overall organization may vary from workplace to workplace, some general guidelines can be stated. Common practices are summarized in Table 8.7.

If managers are concerned about safety, employees probably also will be. Display safety pamphlets, and recruit employee help in identifying hazards. Reward workers for good safety performance. Often, an incentive program will help to encourage safe work practices. Eliminate any hazards that have been identified, and obtain OSHA forms and any first-aid supplies that would be needed in an emergency. The OSHA "Handbook for Small Business" outlines the legal requirements imposed by the Occupational Safety and Health Act of 1970. The handbook, which is available from OSHA, also suggests ways in which a company can develop an effective safety program.

Free on-site consultations also are available from OSHA. A consultant will tour the facility and offer practical advice about safety. These consultants do not issue citations,

Table 8.8 Safety Program Checklist

Refer regularly to this checklist to maintain a safe facility. For each category shown, be sure that:

Electrical Safety:
Fuses of the proper size have been installed.
All ac switches are mounted in clean, tightly closed metal boxes.
Each electrical switch is marked to show its purpose.
Motors are clean and free of excessive grease and oil.
Motors are maintained properly and provided with adequate overcurrent protection.
Bearings are in good condition.
Portable lights are equipped with proper guards.
All portable equipment is double-insulated or properly grounded.
The facility electrical system is checked periodically by a contractor competent in the NEC.
The equipment-grounding conductor or separate ground wire has been carried all the way back to the supply conductor.
All extension cords are in good condition, and the grounding pin is not missing or bent.
Ground-fault interrupters are installed as required.

Exits and Access:
All exits are visible and unobstructed.
All exits are marked with a readily visible, properly illuminated sign.
There are sufficient exits to ensure prompt escape in the event of an emergency.

Fire Protection:
Portable fire extinguishers of the appropriate type are provided in adequate numbers.
All remote vehicles have proper fire extinguishers.
Fire extinguishers are inspected monthly for general condition and operability, which is noted on the inspection tag.
Fire extinguishers are mounted in readily accessible locations.
The fire alarm system is tested annually.

---

propose penalties, or routinely provide information about workplace conditions to the federal inspection staff. Contact the nearest OSHA office for additional information. Table 8.8 provides a basic checklist of safety points for consideration.

8.5   References

1. National Electrical Code, NFPA #70.

8.6   Bibliography

Code of Federal Regulations, 40, Part 761.

"Current Intelligence Bulletin #45," National Institute for Occupational Safety and Health, Division of Standards Development and Technology Transfer, February 24, 1986.

"Electrical Standards Reference Manual," U.S. Department of Labor, Washington, DC.

Hammar, Willie, *Occupational Safety Management and Engineering*, Prentice Hall, New York.

Hammett, William F., "Meeting IEEE C95.1-1991 Requirements," *NAB 1993 Broadcast Engineering Conference Proceedings*, National Association of Broadcasters, Washington, D.C., pp. 471–476, April 1993.

Markley, Donald, "Complying with RF Emission Standards," *Broadcast Engineering*, Intertec Publishing, Overland Park, KS, May 1986.

"Occupational Injuries and Illnesses in the United States by Industry," OSHA Bulletin 2278, U.S. Department of Labor, Washington, DC, 1985.

OSHA, "Electrical Hazard Fact Sheets," U.S. Department of Labor, Washington, DC, January 1987.

OSHA, "Handbook for Small Business," U.S. Department of Labor, Washington, DC.

Pfrimmer, Jack, "Identifying and Managing PCBs in Broadcast Facilities," *1987 NAB Engineering Conference Proceedings*, National Association of Broadcasters, Washington, DC, 1987.

"Safety Precautions," Publication no. 3386A, Varian Associates, Palo Alto, CA, March 1985.

Smith, Milford K., Jr., "RF Radiation Compliance," *Proceedings of the Broadcast Engineering Conference*, Society of Broadcast Engineers, Indianapolis, IN, 1989.

"Toxics Information Series," Office of Toxic Substances, July 1983.

Whitaker, Jerry C., *AC Power Systems*, 2nd Ed., CRC Press, Boca Raton, FL, 1998.

Whitaker, Jerry C., G. DeSantis, and C. Paulson, *Interconnecting Electronic Systems*, CRC Press, Boca Raton, FL, 1993.

Whitaker, Jerry C., *Maintaining Electronic Systems*, CRC Press, Boca Raton, FL, 1991.

Whitaker, Jerry C., *Power Vacuum Tubes Handbook*, 2nd ed., CRC Press, Boca Raton, FL, 1999.

Whitaker, Jerry C., *Radio Frequency Transmission Systems: Design and Operation*, McGraw-Hill, New York, 1990.

9

Dictionary

A

absolute delay The amount of time a signal is delayed. The delay may be expressed in time or number of pulse events.

absolute zero The lowest temperature theoretically possible, –273.16°C. *Absolute zero* is equal to zero degrees Kelvin.

absorption The transference of some or all of the energy contained in an electromagnetic wave to the substance or medium in which it is propagating or upon which it is incident.

absorption auroral The loss of energy in a radio wave passing through an area affected by solar auroral activity.

ac coupling A method of coupling one circuit to another through a capacitor or transformer so as to transmit the varying (ac) characteristics of the signal while blocking the static (dc) characteristics.

ac/dc coupling Coupling between circuits that accommodates the passing of both ac and dc signals (may also be referred to as simply dc coupling).

accelerated life test A special form of reliability testing performed by an equipment manufacturer. The unit under test is subjected to stresses that exceed those typically experienced in normal operation. The goal of an *accelerated life test* is to improve the reliability of products shipped by forcing latent failures in components to become evident before the unit leaves the factory.

accelerating electrode The electrode that causes electrons emitted from an electron gun to accelerate in their journey to the screen of a cathode ray tube.

accelerating voltage The voltage applied to an electrode that accelerates a beam of electrons or other charged particles.

acceptable reliability level The maximum number of failures allowed per thousand operating hours of a given component or system.

acceptance test The process of testing newly purchased equipment to ensure that it is fully compliant with contractual specifications.

access The point at which entry is gained to a circuit or facility.

acquisition time In a communication system, the amount of time required to attain synchronism.

active Any device or circuit that introduces gain or uses a source of energy other than that inherent in the signal to perform its function.

adapter A fitting or electrical connector that links equipment that cannot be connected directly.

adaptive A device able to adjust or react to a condition or application, as an *adaptive circuit*. This term usually refers to filter circuits.

adaptive system A general name for a system that is capable of reconfiguring itself to meet new requirements.

adder A device whose output represents the sum of its inputs.

adjacent channel interference Interference to communications caused by a transmitter operating on an adjacent radio channel. The sidebands of the transmitter mix with the carrier being received on the desired channel, resulting in noise.

admittance A measure of how well alternating current flows in a conductor. It is the reciprocal of *impedance* and is expressed in *siemens*. The real part of admittance is *conductance*; the imaginary part is *susceptance*.

AFC (automatic frequency control) A circuit that automatically keeps an oscillator on frequency by comparing the output of the oscillator with a standard frequency source or signal.

air core An inductor with no magnetic material in its core.

algorithm A prescribed finite set of well-defined rules or processes for the solution of a problem in a finite number of steps.

alignment The adjustment of circuit components so that an entire system meets minimum performance values. For example, the stages in a radio are aligned to ensure proper reception.

allocation The planned use of certain facilities and equipment to meet current, pending, and/or forecasted circuit- and carrier-system requirements.

alternating current (ac) A continuously variable current, rising to a maximum in one direction, falling to zero, then reversing direction and rising to a maximum in the other direction, then falling to zero and repeating the cycle. Alternating current usually follows a sinusoidal growth and decay curve. Note that the correct usage of the term *ac* is lower case.

alternator A generator that produces alternating current electric power.

ambient electromagnetic environment The radiated or conducted electromagnetic signals and noise at a specific location and time.

ambient level The magnitude of radiated or conducted electromagnetic signals and noise at a specific test location when equipment-under-test is not powered.

ambient temperature The temperature of the surrounding medium, typically air, that comes into contact with an apparatus. Ambient temperature may also refer simply to room temperature.

American National Standards Institute (ANSI) A nonprofit organization that coordinates voluntary standards activities in the U.S.

American Wire Gauge (AWG) The standard American method of classifying wire diameter.

ammeter An instrument that measures and records the amount of current in amperes flowing in a circuit.

amp (A) An abbreviation of the term *ampere*.

ampacity A measure of the current carrying capacity of a power cable. *Ampacity* is determined by the maximum continuous-performance temperature of the insulation, by the heat generated in the cable (as a result of conductor and insulation losses), and by the heat-dissipating properties of the cable and its environment.

ampere (amp) The standard unit of electric current.

ampere per meter The standard unit of magnetic field strength.

ampere-hour The energy that is consumed when a current of one ampere flows for a period of one hour.

ampere-turns The product of the number of turns of a coil and the current in amperes flowing through the coil.

amplification The process that results when the output of a circuit is an enlarged reproduction of the input signal. Amplifiers may be designed to provide amplification of voltage, current, or power, or a combination of these quantities.

amplification factor In a vacuum tube, the ratio of the change in plate voltage to the change in grid voltage that causes a corresponding change in plate current. Amplification factor is expressed by the Greek letter μ (*mu*).

amplifier (1—general) A device that receives an input signal and provides as an output a magnified replica of the input waveform. **(2—audio)** An amplifier designed to cover the normal audio frequency range (20 Hz to 20 kHz). **(3—balanced)** A circuit with two identical connected signal branches that operate in phase opposition, with input and output connections each balanced to ground. **(4—bridging)** An amplifying circuit featuring high input impedance to prevent loading of the source. **(5—broadband)** An amplifier capable of operating over a specified broad band of frequencies with acceptably small amplitude variations as a function of frequency. **(6—buffer)** An amplifier stage used to isolate a frequency-sensitive circuit from variations in the load presented by following stages. **(7—linear)** An amplifier in which the instantaneous output signal is a linear function of the corresponding input signal. **(8—magnetic)** An amplifier incorporating a control device dependent on magnetic saturation. A small dc signal applied to a control circuit triggers a large change in operating impedance and, hence, in the output of the circuit. **(9—microphone)** A circuit that amplifies the low level output from a microphone to make it sufficient to be used as an input signal to a power amplifier or another stage in a modulation circuit. Such a circuit is commonly known as a *preamplifier*. **(10—push-pull)** A balanced amplifier with two similar amplifying units connected in phase opposition in order to cancel undesired harmonics and minimize distortion. **(11—tuned radio frequency)** An amplifier tuned to a particular radio frequency or band so that only selected frequencies are amplified.

amplifier operating class (1—general) The operating point of an amplifying stage. The operating point, termed the operating *class*, determines the period during which current flows in the output. **(2—class A)** An amplifier in which output current flows during the whole of the input current cycle. **(3—class AB)** An amplifier in which the output current flows for more than half but less than the whole of the input cycle. **(4—class B)** An amplifier in which output current is cut off at zero input signal; a half-wave rectified output is produced. **(5—class C)** An amplifier in which output current flows for less than half the input cycle. **(6—class D)** An amplifier operating in a pulse-only mode.

amplitude The magnitude of a signal in voltage or current, frequently expressed in terms of *peak*, *peak-to-peak*, or *root-mean-square* (RMS). The actual amplitude of a quantity at a particular instant often varies in a sinusoidal manner.

amplitude distortion A distortion mechanism occurring in an amplifier or other device when the output amplitude is not a linear function of the input amplitude under specified conditions.

amplitude equalizer A corrective network that is designed to modify the amplitude characteristics of a circuit or system over a desired frequency range.

amplitude-versus-frequency distortion The distortion in a transmission system caused by the nonuniform attenuation or gain of the system with respect to frequency under specified conditions.

analog carrier system A carrier system whose signal amplitude, frequency, or phase is varied continuously as a function of a modulating input.

anode (1 — general) A positive pole or element. **(2—vacuum tube)** The outermost positive element in a vacuum tube, also called the *plate*. **(3—battery)** The positive element of a battery or cell.

anodize The formation of a thin film of oxide on a metallic surface, usually to produce an insulating layer.

antenna (1—general) A device used to transmit or receive a radio signal. An antenna is usually designed for a specified frequency range and serves to couple electromagnetic energy from a transmission line to and/or from the free space through which it travels. Directional antennas concentrate the energy in a particular horizontal or vertical direction. **(2—aperiodic)** An antenna that is not periodic or resonant at particular frequencies, and so can be used over a wide band of frequencies. **(3—artificial)** A device that behaves, so far as the transmitter is concerned, like a proper antenna, but does not radiate any power at radio frequencies. **(4—broadband)** An antenna that operates within specified performance limits over a wide band of frequencies, without requiring retuning for each individual frequency. **(5—Cassegrain)** A double reflecting antenna, often used for ground stations in satellite systems. **(6—coaxial)** A dipole antenna made by folding back on itself a quarter wavelength of the outer conductor of a coaxial line, leaving a quarter wavelength of the inner conductor exposed. **(7—corner)** An antenna within the angle formed by two plane-reflecting surfaces. **(8—dipole)** A center-fed antenna, one half-wavelength long. **(9—directional)** An antenna designed to receive or emit radiation more efficiently in a particular direction. **(10—dummy)** An artificial antenna, designed to accept power from the transmitter but not to radiate it. **(11—ferrite)** A common AM broadcast receive antenna that uses a small coil mounted on a short rod of ferrite material. **(12—flat top)** An antenna in which all the horizontal components are in the same horizontal plane. **(13—folded dipole)** A radiating device consisting of two ordinary half-wave dipoles joined at their outer ends and fed at the center of one of the dipoles. **(14—horn reflector)** A radiator in which the feed horn extends into a parabolic reflector, and the power is radiated through a window in the horn. **(15—isotropic)** A theoretical antenna in free space that transmits or receives with the same efficiency in all directions. **(16—log-periodic)** A broadband directional antenna incorporating an array of dipoles of different lengths, the length and spacing between dipoles increasing logarithmically away from the feeder element. **(17—long wire)** An antenna made up of one or more conductors in a straight line pointing in the required direction with a total length of several wavelengths at the operating frequency. **(18—loop)** An antenna consisting of one or more turns of wire in the same or parallel planes. **(19—nested rhombic)** An assembly of two rhombic antennas, one smaller than the other, so that the complete diamond-shaped antenna fits inside the area occupied by the larger unit. **(20—omnidirectional)** An antenna whose radiating or receiving properties are the same in all horizontal plane directions. **(21—periodic)** A resonant antenna designed for use at a particular frequency. **(22—quarter-wave)** A dipole antenna whose length is equal to one quarter of a wavelength at the operating frequency. **(23—rhombic)** A large diamond-shaped antenna, with sides of the diamond several wavelengths long. The rhombic antenna is fed at one of the corners, with directional efficiency in the direction of the diagonal. **(24—series fed)** A vertical antenna that is fed at its lower end. **(25—shunt fed)** A vertical antenna whose base is grounded, and is fed at a specified point above ground. The point at which the antenna is fed above ground determines the operating impedance. **(26—steerable)** An antenna so constructed that its major lobe may readily be changed in direction. **(27—top-loaded)** A vertical antenna capacitively loaded at its upper end, often by simple enlargement or the attachment of a disc or plate. **(28—turnstile)** An antenna with one or more tiers of horizontal dipoles, crossed at right angles to each other and with excitation of the dipoles in phase quadrature. **(29—whip)** An antenna constructed of a thin semiflexible metal rod or tube, fed at its base. **(30—Yagi)** A directional antenna constructed of a series of dipoles cut to specific lengths. *Director* elements are placed in front of the active dipole and *reflector* elements are placed behind the active element.

antenna array A group of several antennas coupled together to yield a required degree of directivity.

antenna beamwidth The angle between the *half-power* points (3 dB points) of the main lobe of the antenna pattern when referenced to the peak power point of the antenna pattern. *Antenna beamwidth* is measured in degrees and normally refers to the horizontal radiation pattern.

antenna directivity factor The ratio of the power flux density in the desired direction to the average value of power flux density at crests in the antenna directivity pattern in the interference section.

antenna factor A factor that, when applied to the voltage appearing at the terminals of measurement equipment, yields the electrical field strength at an antenna. The unit of antenna factor is volts per meter per measured volt.

antenna gain The ratio of the power required at the input of a theoretically perfect omnidirectional reference antenna to the power supplied to the input of the given antenna to produce the same field at the same distance. When not specified otherwise, the figure expressing the gain of an antenna refers to the gain in the direction of the radiation main lobe. In services using *scattering* modes of propagation, the full gain of an antenna may not be realizable in practice and the apparent gain may vary with time.

antenna gain-to-noise temperature For a satellite earth terminal receiving system, a figure of merit that equals G/T, where G is the gain in dB of the earth terminal antenna at the receive frequency, and T is the equivalent noise temperature of the receiving system in Kelvins.

antenna matching The process of adjusting an antenna matching circuit (or the antenna itself) so that the input impedance of the antenna is equal to the characteristic impedance of the transmission line.

antenna monitor A device used to measure the ratio and phase between the currents flowing in the towers of a directional AM broadcast station.

antenna noise temperature The temperature of a resistor having an available noise power per unit bandwidth equal to that at the antenna output at a specified frequency.

antenna pattern A diagram showing the efficiency of radiation in all directions from the antenna.

antenna power rating The maximum continuous-wave power that can be applied to an antenna without degrading its performance.

antenna preamplifier A small amplifier, usually mast-mounted, for amplifying weak signals to a level sufficient to compensate for down-lead losses.

apparent power The product of the root-mean-square values of the voltage and current in an alternating-current circuit without a correction for the phase difference between the voltage and current.

arc A sustained luminous discharge between two or more electrodes.

arithmetic mean The sum of the values of several quantities divided by the number of quantities, also referred to as the *average.*

armature winding The winding of an electrical machine, either a motor or generator, in which current is induced.

array (1—antenna) An assembly of several directional antennas so placed and interconnected that directivity may be enhanced. **(2—broadside)** An antenna array whose elements are all in the same plane, producing a major lobe perpendicular to the plane. **(3—colinear)** An antenna array whose elements are in the same line, either horizontal or vertical. **(4—end-fire)** An antenna array whose elements are in parallel rows, one behind the other, producing a major lobe perpendicular to the plane in which individual elements are placed. **(5—linear)** An antenna array whose elements are arranged end-to-end. **(6—stacked)** An antenna array whose elements are stacked, one above the other.

artificial line An assembly of resistors, inductors, and capacitors that simulates the electrical characteristics of a transmission line.

assembly A manufactured part made by combining several other parts or subassemblies.

assumed values A range of values, parameters, levels, and other elements assumed for a mathematical model, hypothetical circuit, or network, from which analysis, additional estimates, or calculations will be made. The range of values, while not measured, represents the best engineering judgment and is generally derived from values found or measured in real circuits or networks of the same generic type, and includes projected improvements.

atmosphere The gaseous envelope surrounding the earth, composed largely of oxygen, carbon dioxide, and water vapor. The atmosphere is divided into four primary layers: *troposphere, stratosphere, ionosphere,* and *exosphere.*

atmospheric noise Radio noise caused by natural atmospheric processes, such as lightning.

attack time The time interval in seconds required for a device to respond to a control stimulus.

attenuation The decrease in amplitude of an electrical signal traveling through a transmission medium caused by dielectric and conductor losses.

attenuation coefficient The rate of decrease in the amplitude of an electrical signal caused by attenuation. The *attenuation coefficient* can be expressed in decibels or nepers per unit length. It may also be referred to as the *attenuation constant.*

attenuation distortion The distortion caused by attenuation that varies over the frequency range of a signal.

attenuation-limited operation The condition prevailing when the received signal amplitude (rather than distortion) limits overall system performance.

attenuator A fixed or adjustable component that reduces the amplitude of an electrical signal without causing distortion.

atto A prefix meaning one *quintillionth.*

attraction The attractive force between two unlike magnetic poles (N/S) or electrically charged bodies (+/–).

attributes The characteristics of equipment that aid planning and circuit design.

automatic frequency control (AFC) A system designed to maintain the correct operating frequency of a receiver. Any drift in tuning results in the production of a control voltage, which is used to adjust the frequency of a local oscillator so as to minimize the tuning error.

automatic gain control (AGC) An electronic circuit that compares the level of an incoming signal with a previously defined standard and automatically amplifies or attenuates the signal so it arrives at its destination at the correct level.

autotransformer A transformer in which both the primary and secondary currents flow through one common part of the coil.

auxiliary power An alternate source of electric power, serving as a back-up for the primary utility company ac power.

availability A measure of the degree to which a system, subsystem, or equipment is operable and not in a stage of congestion or failure at any given point in time.

avalanche effect The effect obtained when the electric field across a barrier region is sufficiently strong for electrons to collide with *valence electrons*, thereby releasing more electrons and giving a cumulative multiplication effect in a semiconductor.

average life The mean value for a normal distribution of product or component lives, generally applied to mechanical failures resulting from "wear-out."

B

back emf A voltage induced in the reverse direction when current flows through an inductance. *Back emf* is also known as *counter-emf*.

back scattering A form of wave scattering in which at least one component of the scattered wave is deflected opposite to the direction of propagation of the incident wave.

background noise The total system noise in the absence of information transmission, independent of the presence or absence of a signal.

backscatter The deflection or reflection of radiant energy through angles greater than 90° with respect to the original angle of travel.

backscatter range The maximum distance from which backscattered radiant energy can be measured.

backup A circuit element or facility used to replace an element that has failed.

backup supply A redundant power supply that takes over if the primary power supply fails.

balance The process of equalizing the voltage, current, or other parameter between two or more circuits or systems.

balanced A circuit having two sides (conductors) carrying voltages that are symmetrical about a common reference point, typically ground.

balanced circuit A circuit whose two sides are electrically equal in all transmission respects.

balanced line A transmission line consisting of two conductors in the presence of ground capable of being operated in such a way that when the voltages of the two conductors at all transverse planes are equal in magnitude and opposite in polarity with respect to ground, the currents in the two conductors are equal in magnitude and opposite in direction.

balanced modulator A modulator that combines the information signal and the carrier so that the output contains the two sidebands without the carrier.

balanced three-wire system A power distribution system using three conductors, one of which is balanced to have a potential midway between the potentials of the other two.

balanced-to-ground The condition when the impedance to ground on one wire of a two-wire circuit is equal to the impedance to ground on the other wire.

balun (balanced/unbalanced) A device used to connect balanced circuits with unbalanced circuits.

band A range of frequencies between a specified upper and lower limit.

band elimination filter A filter having a single continuous attenuation band, with neither the upper nor lower cut-off frequencies being zero or infinite. A *band elimination filter* may also be referred to as a *band-stop, notch,* or *band reject* filter.

bandpass filter A filter having a single continuous transmission band with neither the upper nor the lower cut-off frequencies being zero or infinite. A bandpass filter permits only a specific band of frequencies to pass; frequencies above or below are attenuated.

bandwidth The range of signal frequencies that can be transmitted by a communications channel with a defined maximum loss or distortion. Bandwidth indicates the information-carrying capacity of a channel.

bandwidth expansion ratio The ratio of the necessary bandwidth to the baseband bandwidth.

bandwidth-limited operation The condition prevailing when the frequency spectrum or bandwidth, rather than the amplitude (or power) of the signal, is the limiting factor in communication capability. This condition is reached when the system distorts the shape of the waveform beyond tolerable limits.

bank A group of similar items connected together in a specified manner and used in conjunction with one another.

bare A wire conductor that is not enameled or enclosed in an insulating sheath.

baseband The band of frequencies occupied by a signal before it modulates a carrier wave to form a transmitted radio or line signal.

baseband channel A channel that carries a signal without modulation, in contrast to a *passband* channel.

baseband signal The original form of a signal, unchanged by modulation.

bath tub The shape of a typical graph of component failure rates: high during an initial period of operation, falling to an acceptable low level during the normal usage period, and then rising again as the components become time-expired.

battery A group of several cells connected together to furnish current by conversion of chemical, thermal, solar, or nuclear energy into electrical energy. A single cell is itself sometimes also called a battery.

bay A row or suite of racks on which transmission, switching, and/or processing equipment is mounted.

Bel A unit of power measurement, named in honor of Alexander Graham Bell. The commonly used unit is one tenth of a Bel, or a decibel (dB). One Bel is defined as a tenfold increase in power. If an amplifier increases the power of a signal by 10 times, the power gain of the amplifier is equal to 1 Bel or 10 *decibels* (dB). If power is increased by 100 times, the power gain is 2 Bels or 20 decibels.

bend A transition component between two elements of a transmission waveguide.

bending radius The smallest bend that may be put into a cable under a stated pulling force. The bending radius is typically expressed in inches.

bias A dc voltage difference applied between two elements of an active electronic device, such as a vacuum tube, transistor, or integrated circuit. Bias currents may or may not be drawn, depending on the device and circuit type.

bidirectional An operational qualification which implies that the transmission of information occurs in both directions.

bifilar winding A type of winding in which two insulated wires are placed side by side. In some components, bifilar winding is used to produce balanced circuits.

bipolar A signal that contains both positive-going and negative-going amplitude components. A bipolar signal may also contain a zero amplitude state.

bleeder A high resistance connected in parallel with one or more filter capacitors in a high voltage dc system. If the power supply load is disconnected, the capacitors discharge through the bleeder.

block diagram An overview diagram that uses geometric figures to represent the principal divisions or sections of a circuit, and lines and arrows to show the path of a signal, or to show program functionalities. It is not a *schematic*, which provides greater detail.

blocking capacitor A capacitor included in a circuit to stop the passage of direct current.

BNC An abbreviation for *bayonet Neill-Concelman*, a type of cable connector used extensively in RF applications (named for its inventor).

Boltzmann's constant 1.38×10^{-23} joules.

bridge A type of network circuit used to match different circuits to each other, ensuring minimum transmission impairment.

bridging The shunting or paralleling of one circuit with another.

broadband The quality of a communications link having essentially uniform response over a given range of frequencies. A communications link is said to be *broadband* if it offers no perceptible degradation to the signal being transported.

buffer A circuit or component that isolates one electrical circuit from another.

burn-in The operation of a device, sometimes under extreme conditions, to stabilize its characteristics and identify latent component failures before bringing the device into normal service.

bus A central conductor for the primary signal path. The term bus may also refer to a signal path to which a number of inputs may be connected for feed to one or more outputs.

busbar A main dc power bus.

bypass capacitor A capacitor that provides a signal path that effectively shunts or bypasses other components.

bypass relay A switch used to bypass the normal electrical route of a signal or current in the event of power, signal, or equipment failure.

C

cable An electrically and/or optically conductive interconnecting device.

cable loss Signal loss caused by passing a signal through a coaxial cable. Losses are the result of resistance, capacitance, and inductance in the cable.

cable splice The connection of two pieces of cable by joining them mechanically and closing the joint with a weather-tight case or sleeve.

cabling The wiring used to interconnect electronic equipment.

calibrate The process of checking, and adjusting if necessary, a test instrument against one known to be set correctly.

calibration The process of identifying and measuring errors in instruments and/or procedures.

capacitance The property of a device or component that enables it to store energy in an electrostatic field and to release it later. A capacitor consists of two conductors separated by an insulating material. When the conductors have a voltage difference between them, a charge will be stored in the electrostatic field between the conductors.

capacitor A device that stores electrical energy. A capacitor allows the apparent flow of alternating current, while blocking the flow of direct current. The degree to which the device permits ac current flow depends on the frequency of the signal and the size of the capacitor. Capacitors are used in filters, delay-line components, couplers, frequency selectors, timing elements, voltage transient suppression, and other applications.

carrier A single frequency wave that, prior to transmission, is modulated by another wave containing information. A carrier may be modulated by manipulating its amplitude and/or frequency in direct relation to one or more applied signals.

carrier frequency The frequency of an unmodulated oscillator or transmitter. Also, the average frequency of a transmitter when a signal is frequency modulated by a symmetrical signal.

cascade connection A tandem arrangement of two or more similar component devices or circuits, with the output of one connected to the input of the next.

cascaded An arrangement of two or more circuits in which the output of one circuit is connected to the input of the next circuit.

cathode ray tube (CRT) A vacuum tube device, usually glass, that is narrow at one end and widens at the other to create a surface onto which images can be projected. The narrow end contains the necessary circuits to generate and focus an electron beam on the luminescent screen at the other end. CRTs are used to display pictures in TV receivers, video monitors, oscilloscopes, computers, and other systems.

cell An elementary unit of communication, of power supply, or of equipment.

Celsius A temperature measurement scale, expressed in degrees C, in which water freezes at 0°C and boils at 100°C. To convert to degrees Fahrenheit, multiply by 0.555 and add 32. To convert to Kelvins add 273 (approximately).

center frequency In frequency modulation, the resting frequency or initial frequency of the carrier before modulation.

center tap A connection made at the electrical center of a coil.

channel The smallest subdivision of a circuit that provides a single type of communication service.

channel decoder A device that converts an incoming modulated signal on a given channel back into the source-encoded signal.

channel encoder A device that takes a given signal and converts it into a form suitable for transmission over the communications channel.

channel noise level The ratio of the channel noise at any point in a transmission system to some arbitrary amount of circuit noise chosen as a reference. This ratio is usually expressed in *decibels above reference noise*, abbreviated *dBrn*.

channel reliability The percent of time a channel is available for use in a specific direction during a specified period.

channelization The allocation of communication circuits to channels and the forming of these channels into groups for higher order multiplexing.

characteristic The property of a circuit or component.

characteristic impedance The impedance of a transmission line, as measured at the driving point, if the line were of infinite length. In such a line, there would be no standing waves. The *characteristic impedance* may also be referred to as the *surge impedance*.

charge The process of replenishing or replacing the electrical charge in a secondary cell or storage battery.

charger A device used to recharge a battery. Types of charging include: (1) *constant voltage charge*, (2) *equalizing charge*, and (3) *trickle charge*.

chassis ground A connection to the metal frame of an electronic system that holds the components in a place. The chassis ground connection serves as the ground return or electrical common for the system.

circuit Any closed path through which an electrical current can flow. In a *parallel circuit*, components are connected between common inputs and outputs such that all paths are parallel to each other. The same voltage appears across all paths. In a *series circuit*, the same current flows through all components.

circuit noise level The ratio of the circuit noise at some given point in a transmission system to an established reference, usually expressed in decibels above the reference.

circuit reliability The percentage of time a circuit is available to the user during a specified period of scheduled availability.

circular mil The measurement unit of the cross-sectional area of a circular conductor. A *circular mil* is the area of a circle whose diameter is one mil, or 0.001 inch.

clear channel A transmission path wherein the full bandwidth is available to the user, with no portions of the channel used for control, framing, or signaling. Can also refer to a classification of AM broadcast station.

clipper A limiting circuit which ensures that a specified output level is not exceeded by restricting the output waveform to a maximum peak amplitude.

clipping The distortion of a signal caused by removing a portion of the waveform through restriction of the amplitude of the signal by a circuit or device.

coax A short-hand expression for *coaxial cable*, which is used to transport high-frequency signals.

coaxial cable A transmission line consisting of an inner conductor surrounded first by an insulating material and then by an outer conductor, either solid or braided. The mechanical dimensions of the cable determine its *characteristic impedance*.

coherence The correlation between the phases of two or more waves.

coherent The condition characterized by a fixed phase relationship among points on an electromagnetic wave.

coherent pulse The condition in which a fixed phase relationship is maintained between consecutive pulses during pulse transmission.

cold joint A soldered connection that was inadequately heated, with the result that the wire is held in place by rosin flux, not solder. A cold joint is sometimes referred to as a *dry joint*.

comb filter An electrical filter circuit that passes a series of frequencies and rejects the frequencies in between, producing a frequency response similar to the teeth of a comb.

common A point that acts as a reference for circuits, often equal in potential to the local ground.

common mode Signals identical with respect to amplitude, frequency, and phase that are applied to both terminals of a cable and/or both the input and reference of an amplifier.

common return A return path that is common to two or more circuits, and returns currents to their source or to ground.

common return offset The dc common return potential difference of a line.

communications system A collection of individual communications networks, transmission systems, relay stations, tributary stations, and terminal equipment capable of interconnection and interoperation to form an integral whole. The individual components must serve a common purpose, be technically compatible, employ common procedures, respond to some form of control, and, in general, operate in unison.

commutation A successive switching process carried out by a commutator.

commutator A circular assembly of contacts, insulated one from another, each leading to a different portion of the circuit or machine.

compatibility The ability of diverse systems to exchange necessary information at appropriate levels of command directly and in usable form. Communications equipment items are compatible if signals can be exchanged between them without the addition of buffering or translation for the specific purpose of achieving workable interface connections, and if the equipment or systems being interconnected possess comparable performance characteristics, including the suppression of undesired radiation.

complex wave A waveform consisting of two or more sinewave components. At any instant of time, a complex wave is the algebraic sum of all its sinewave components.

compliance For mechanical systems, a property which is the reciprocal of stiffness.

component An assembly, or part thereof, that is essential to the operation of some larger circuit or system. A *component* is an immediate subdivision of the assembly to which it belongs.

COMSAT The *Communications Satellite Corporation*, an organization established by an act of Congress in 1962. COMSAT launches and operates the international satellites for the INTELSAT consortium of countries.

concentricity A measure of the deviation of the center conductor position relative to its ideal location in the exact center of the dielectric cross-section of a coaxial cable.

conditioning The adjustment of a channel in order to provide the appropriate transmission characteristics needed for data or other special services.

conditioning equipment The equipment used to match transmission levels and impedances, and to provide equalization between facilities.

conductance A measure of the capability of a material to conduct electricity. It is the reciprocal of *resistance* (ohm) and is expressed in *siemens*. (Formerly expressed as *mho*.)

conducted emission An electromagnetic energy propagated along a conductor.

conduction The transfer of energy through a medium, such as the conduction of electricity by a wire, or of heat by a metallic frame.

conduction band A partially filled or empty atomic energy band in which electrons are free to move easily, allowing the material to carry an electric current.

conductivity The conductance per unit length.

conductor Any material that is capable of carrying an electric current.

configuration A relative arrangement of parts.

connection A point at which a junction of two or more conductors is made.

connector A device mounted on the end of a wire or fiber optic cable that mates to a similar device on a specific piece of equipment or another cable.

constant-current source A source with infinitely high output impedance so that output current is independent of voltage, for a specified range of output voltages.

constant-voltage charge A method of charging a secondary cell or storage battery during which the terminal voltage is kept at a constant value.

constant-voltage source A source with low, ideally zero, internal impedance, so that voltage will remain constant, independent of current supplied.

contact The points that are brought together or separated to complete or break an electrical circuit.

contact bounce The rebound of a contact, which temporarily opens the circuit after its initial *make*.

contact form The configuration of a contact assembly on a relay. Many different configurations are possible from simple *single-make* contacts to complex arrangements involving *breaks* and *makes*.

contact noise A noise resulting from current flow through an electrical contact that has a rapidly varying resistance, as when the contacts are corroded or dirty.

contact resistance The resistance at the surface when two conductors make contact.

continuity A continuous path for the flow of current in an electrical circuit.

continuous wave An electromagnetic signal in which successive oscillations of the waves are identical.

control The supervision that an operator or device exercises over a circuit or system.

control grid The grid in an electron tube that controls the flow of current from the cathode to the anode.

convention A generally acceptable symbol, sign, or practice in a given industry.

Coordinated Universal Time (UTC) The time scale, maintained by the BIH (Bureau International de l'Heure) that forms the basis of a coordinated dissemination of standard frequencies and time signals.

copper loss The loss resulting from the heating effect of current.

corona A bluish luminous discharge resulting from ionization of the air near a conductor carrying a voltage gradient above a certain *critical level*.

corrective maintenance The necessary tests, measurements, and adjustments required to remove or correct a fault.

cosmic noise The random noise originating outside the earth's atmosphere.

coulomb The standard unit of electric quantity or charge. One *coulomb* is equal to the quantity of electricity transported in 1 second by a current of 1 ampere.

Coulomb's Law The attraction and repulsion of electric charges act on a line between them. The charges are inversely proportional to the square of the distance between them, and proportional to the product of their magnitudes. (Named for the French physicist Charles-Augustine de Coulomb, 1736–1806.)

counter-electromotive force The effective electromotive force within a system that opposes the passage of current in a specified direction.

couple The process of linking two circuits by inductance, so that energy is transferred from one circuit to another.

coupled mode The selection of either ac or dc coupling.

coupling The relationship between two components that enables the transfer of energy between them. Included are *direct coupling* through a direct electrical connection, such as a wire; *capacitive coupling* through the capacitance formed by two adjacent conductors; and *inductive coupling* in which energy is transferred through a magnetic field. Capacitive coupling is also called *electrostatic coupling*. Inductive coupling is often referred to as *electromagnetic coupling*.

coupling coefficient A measure of the electrical coupling that exists between two circuits. The *coupling coefficient* is equal to the ratio of the mutual impedance to the square root of the product of the self impedances of the coupled circuits.

cross coupling The coupling of a signal from one channel, circuit, or conductor to another, where it becomes an undesired signal.

crossover distortion A distortion that results in an amplifier when an irregularity is introduced into the signal as it crosses through a zero reference point. If an amplifier is properly designed and biased, the upper half cycle and lower half cycle of the signal coincide at the zero crossover reference.

crossover frequency The frequency at which output signals pass from one channel to the other in a *crossover network*. At the *crossover frequency* itself, the outputs to each side are equal.

crossover network A type of filter that divides an incoming signal into two or more outputs, with higher frequencies directed to one output, and lower frequencies to another.

crosstalk Undesired transmission of signals from one circuit into another circuit in the same system. Crosstalk is usually caused by unintentional capacitive (ac) coupling.

crosstalk coupling The ratio of the power in a disturbing circuit to the induced power in the disturbed circuit, observed at a particular point under specified conditions. Crosstalk coupling is typically expressed in dB.

crowbar A short-circuit or low resistance path placed across the input to a circuit, usually for protective purposes..

CRT (cathode ray tube) A vacuum tube device that produces light when energized by the electron beam generated inside the tube. A CRT includes an electron gun, deflection mechanism, and phosphor-covered faceplate.

crystal A solidified form of a substance that has atoms and molecules arranged in a symmetrical pattern.

crystal filter A filter that uses piezoelectric crystals to create resonant or antiresonant circuits.

crystal oscillator An oscillator using a piezoelectric crystal as the tuned circuit that controls the resonant frequency.

crystal-controlled oscillator An oscillator in which a piezoelectric-effect crystal is coupled to a tuned oscillator circuit in such a way that the crystal pulls the oscillator frequency to its own natural frequency and does not allow frequency drift.

current (1—general) A general term for the transfer of electricity, or the movement of electrons or *holes*. **(2—alternating)** An electric current that is constantly varying in amplitude and periodically reversing direction. **(3—average)** The arithmetic mean of the instantaneous values of current, averaged over one complete half cycle. **(4—charging)** The current that flows in to charge a capacitor when it is first connected to a source of electric potential. **(5—direct)** Electric current that flows in one direction only. **(6—eddy)** A wasteful current that flows in the core of a transformer and produces heat. *Eddy currents* are largely eliminated through the use of laminated cores. **(7—effective)** The ac current that will produce the same effective heat in a resistor as is produced by dc. If the ac is sinusoidal, the *effective current* value is 0.707 times the peak ac value. **(8—fault)** The current that flows between conductors or to ground during a fault condition. **(9—ground fault)** A fault current that flows to ground. **(10—ground return)** A current that returns through the earth. **(11—lagging)** A phenomenon observed in an inductive circuit where alternating current lags behind the voltage that produces it. **(12—leading)** A phenomenon observed in a capacitive circuit where alternating current leads the voltage that produces it. **(13—magnetizing)** The current in a transformer primary winding that is just sufficient to magnetize the core and offset iron losses. **(14—neutral)** The current that flows in the neutral conductor of an unbalanced polyphase power circuit. If correctly balanced, the neutral would carry no net current. **(15—peak)** The maximum value reached by a varying current during one cycle. **(16—pick-up)** The minimum current at which a relay just begins to operate. **(17—plate)** The anode current of an electron tube. **(18—residual)** The vector sum of the currents in the phase wires of an unbalanced polyphase power circuit. **(19—space)** The total current flowing through an electron tube.

current amplifier A low output impedance amplifier capable of providing high current output.

current probe A sensor, clamped around an electrical conductor, in which an induced current is developed from the magnetic field surrounding the conductor. For measurements, the current probe is connected to a suitable test instrument.

current transformer A transformer-type of instrument in which the primary carries the current to be measured and the secondary is in series with a low current ammeter. A current transformer is used to measure high values of alternating current.

current-carrying capacity A measure of the maximum current that can be carried continuously without damage to components or devices in a circuit.

cut-off frequency The frequency above or below which the output current in a circuit is reduced to a specified level.

cycle The interval of time or space required for a periodic signal to complete one period.

cycles per second The standard unit of frequency, expressed in Hertz (one cycle per second).

D

damped oscillation An oscillation exhibiting a progressive diminution of amplitude with time.

damping The dissipation and resultant reduction of any type of energy, such as electromagnetic waves.

dB (decibel) A measure of voltage, current, or power gain equal to 0.1 Bel. Decibels are given by the equations

$$20 \log \frac{V_{out}}{V_{in}}, \ 20 \log \frac{I_{out}}{I_{in}}, \text{ or } 10 \log \frac{P_{out}}{P_{in}}.$$

dBk A measure of power relative to 1 kilowatt. 0 dBk equals 1 kW.

dBm (decibels above 1 milliwatt) A logarithmic measure of power with respect to a reference power of one milliwatt.

dBmv A measure of voltage gain relative to 1 millivolt at 75 ohms.

dBr The power difference expressed in dB between any point and a reference point selected as the *zero relative transmission level* point. A power expressed in *dBr* does not specify the absolute power; it is a relative measurement only.

dBu A term that reflects comparison between a measured value of voltage and a reference value of 0.775 V, expressed under conditions in which the impedance at the point of measurement (and of the reference source) are not considered.

dbV A measure of voltage gain relative to 1 V.

dBW A measure of power relative to 1 watt. 0 dBW equals 1 W.

dc An abbreviation for *direct current*. Note that the preferred usage of the term *dc* is lower case.

dc amplifier A circuit capable of amplifying dc and slowly varying alternating current signals.

dc component The portion of a signal that consists of direct current. This term may also refer to the average value of a signal.

dc coupled A connection configured so that both the signal (ac component) and the constant voltage on which it is riding (dc component) are passed from one stage to the next.

dc coupling A method of coupling one circuit to another so as to transmit the static (dc) characteristics of the signal as well as the varying (ac) characteristics. Any dc offset present on the input signal is maintained and will be present in the output.

dc offset The amount that the dc component of a given signal has shifted from its correct level.

dc signal bounce Overshoot of the proper dc voltage level resulting from multiple ac couplings in a signal path.

de-energized A system from which sources of power have been disconnected.

deca A prefix meaning *ten*.

decay The reduction in amplitude of a signal on an exponential basis.

decay time The time required for a signal to fall to a certain fraction of its original value.

decibel (dB) One tenth of a Bel. The decibel is a logarithmic measure of the ratio between two powers.

decode The process of recovering information from a signal into which the information has been encoded.

decoder A device capable of deciphering encoded signals. A decoder interprets input instructions and initiates the appropriate control operations as a result.

decoupling The reduction or removal of undesired coupling between two circuits or stages.

deemphasis The reduction of the high-frequency components of a received signal to reverse the preemphasis that was placed on them to overcome attenuation and noise in the transmission process.

defect An error made during initial planning that is normally detected and corrected during the development phase. Note that a *fault* is an error that occurs in an in-service system.

deflection The control placed on electron direction and motion in CRTs and other vacuum tube devices by varying the strengths of electrostatic (electrical) or electromagnetic fields.

degradation In susceptibility testing, any undesirable change in the operational performance of a test specimen. This term does not necessarily mean malfunction or catastrophic failure.

degradation failure A failure that results from a gradual change in performance characteristics of a system or part with time.

delay The amount of time by which a signal is delayed or an event is retarded.

delay circuit A circuit designed to delay a signal passing through it by a specified amount.

delay distortion The distortion resulting from the difference in phase delays at two frequencies of interest.

delay equalizer A network that adjusts the velocity of propagation of the frequency components of a complex signal to counteract the delay distortion characteristics of a transmission channel.

delay line A transmission network that increases the propagation time of a signal traveling through it.

delta connection A common method of joining together a three-phase power supply, with each phase across a different pair of the three wires used.

delta-connected system A 3-phase power distribution system where a single-phase output can be derived from each of the adjacent pairs of an equilateral triangle formed by the service drop transformer secondary windings.

demodulator Any device that recovers the original signal after it has modulated a high-frequency carrier. The output from the unit may be in baseband composite form.

demultiplexer (demux) A device used to separate two or more signals that were previously combined by a compatible multiplexer and are transmitted over a single channel.

derating factor An operating safety margin provided for a component or system to ensure reliable performance. A *derating allowance* also is typically provided for operation under extreme environmental conditions, or under stringent reliability requirements.

desiccant A drying agent used for drying out cable splices or sensitive equipment.

design A layout of all the necessary equipment and facilities required to make a special circuit, piece of equipment, or system work.

design objective The desired electrical or mechanical performance characteristic for electronic circuits and equipment.

detection The rectification process that results in the modulating signal being separated from a modulated wave.

detectivity The reciprocal of *noise equivalent power.*

detector A device that converts one type of energy into another.

device A functional circuit, component, or network unit, such as a vacuum tube or transistor.

dewpoint The temperature at which moisture will condense out.

diagnosis The process of locating errors in software, or equipment faults in hardware.

diagnostic routine A software program designed to trace errors in software, locate hardware faults, or identify the cause of a breakdown.

dielectric An insulating material that separates the elements of various components, including capacitors and transmission lines. Dielectric materials include air, plastic, mica, ceramic, and Teflon. A dielectric material must be an insulator. (*Teflon* is a registered trademark of Du Pont.)

dielectric constant The ratio of the capacitance of a capacitor with a certain dielectric material to the capacitance with a vacuum as the dielectric. The *dielectric constant* is considered a measure of the capability of a dielectric material to store an electrostatic charge.

dielectric strength The potential gradient at which electrical breakdown occurs.

differential amplifier An input circuit that rejects voltages that are the same at both input terminals but amplifies any voltage difference between the inputs. Use of a differential amplifier causes any signal present on both terminals, such as common mode hum, to cancel itself.

differential dc The maximum dc voltage that can be applied between the differential inputs of an amplifier while maintaining linear operation.

differential gain The difference in output amplitude (expressed in percent or dB) of a small high frequency sinewave signal at two stated levels of a low frequency signal on which it is superimposed.

differential phase The difference in output phase of a small high frequency sinewave signal at two stated levels of a low frequency signal on which it is superimposed.

differential-mode interference An interference source that causes a change in potential of one side of a signal transmission path relative to the other side.

diffuse reflection The scattering effect that occurs when light, radio, or sound waves strike a rough surface.

diffusion The spreading or scattering of a wave, such as a radio wave.

diode A semiconductor or vacuum tube with two electrodes that passes electric current in one direction only. Diodes are used in rectifiers, gates, modulators, and detectors.

direct coupling A coupling method between stages that permits dc current to flow between the stages.

direct current An electrical signal in which the direction of current flow remains constant.

discharge The conversion of stored energy, as in a battery or capacitor, into an electric current.

discontinuity An abrupt nonuniform point of change in a transmission circuit that causes a disruption of normal operation.

discrete An individual circuit component.

discrete component A separately contained circuit element with its own external connections.

discriminator A device or circuit whose output amplitude and polarity vary according to how much the input signal varies from a standard or from another signal. A discriminator can be used to recover the modulating waveform in a frequency modulated signal.

dish An antenna system consisting of a parabolic shaped reflector with a signal feed element at the focal point. Dish antennas commonly are used for transmission and reception from microwave stations and communications satellites.

dispersion The wavelength dependence of a parameter.

display The representation of text and images on a cathode-ray tube, an array of light-emitting diodes, a liquid-crystal readout, or another similar device.

display device An output unit that provides a visual representation of data.

distortion The difference between the wave shape of an original signal and the signal after it has traversed a transmission circuit.

distortion-limited operation The condition prevailing when the shape of the signal, rather than the amplitude (or power), is the limiting factor in communication capability. This condition is reached when the system distorts the shape of the waveform beyond tolerable limits. For linear systems, *distortion-limited* operation is equivalent to *bandwidth-limited* operation.

disturbance The interference with normal conditions and communications by some external energy source.

disturbance current The unwanted current of any irregular phenomenon associated with transmission that tends to limit or interfere with the interchange of information.

disturbance power The unwanted power of any irregular phenomenon associated with transmission that tends to limit or interfere with the interchange of information.

disturbance voltage The unwanted voltage of any irregular phenomenon associated with transmission that tends to limit or interfere with the interchange of information.

diversity receiver A receiver using two antennas connected through circuitry that senses which antenna is receiving the stronger signal. Electronic gating permits the stronger source to be routed to the receiving system.

documentation A written description of a program. *Documentation* can be considered as any record that has permanence and can be read by humans or machines.

down-lead A lead-in wire from an antenna to a receiver.

downlink The portion of a communication link used for transmission of signals from a satellite or airborne platform to a surface terminal.

downstream A specified signal modification occurring after other given devices in a signal path.

downtime The time during which equipment is not capable of doing useful work because of malfunction. This does not include preventive maintenance time. In other words, *downtime* is measured from the occurrence of a malfunction to the correction of that malfunction.

drift A slow change in a nominally constant signal characteristic, such as frequency.

drift-space The area in a klystron tube in which electrons drift at their entering velocities and form electron *bunches*.

drive The input signal to a circuit, particularly to an amplifier.

driver An electronic circuit that supplies an isolated output to drive the input of another circuit.

drop-out value The value of current or voltage at which a relay will cease to be operated.

dropout The momentary loss of a signal.

dropping resistor A resistor designed to carry current that will make a required voltage available.

duplex separation The frequency spacing required in a communications system between the *forward* and *return* channels to maintain interference at an acceptably low level.

duplex signaling A configuration permitting signaling in both transmission directions simultaneously.

duty cycle The ratio of operating time to total elapsed time of a device that operates intermittently, expressed in percent.

dynamic A situation in which the operating parameters and/or requirements of a given system are continually changing.

dynamic range The maximum range or extremes in amplitude, from the lowest to the highest (noise floor to system clipping), that a system is capable of reproducing. The dynamic range is expressed in dB against a reference level.

dynamo A rotating machine, normally a dc generator.

dynamotor A rotating machine used to convert dc into ac.

E

earth A large conducting body with no electrical potential, also called *ground*.

earth capacitance The capacitance between a given circuit or component and a point at ground potential.

earth current A current that flows to earth/ground, especially one that follows from a fault in the system. *Earth current* may also refer to a current that flows in the earth, resulting from ionospheric disturbances, lightning, or faults on power lines.

earth fault A fault that occurs when a conductor is accidentally grounded/earthed, or when the resistance to earth of an insulator falls below a specified value.

earth ground A large conducting body that represents *zero level* in the scale of electrical potential. An *earth ground* is a connection made either accidentally or by design between a conductor and earth.

earth potential The potential taken to be the arbitrary zero in a scale of electric potential.

effective ground A connection to ground through a medium of sufficiently low impedance and adequate current-carrying capacity to prevent the buildup of voltages that might be hazardous to equipment or personnel.

effective resistance The increased resistance of a conductor to an alternating current resulting from the *skin effect*, relative to the direct-current resistance of the conductor. Higher frequencies tend to travel only on the outer skin of the conductor, whereas dc flows uniformly through the entire area.

efficiency The useful power output of an electrical device or circuit divided by the total power input, expressed in percent.

electric Any device or circuit that produces, operates on, transmits, or uses electricity.

electric charge An excess of either electrons or protons within a given space or material.

electric field strength The magnitude, measured in volts per meter, of the electric field in an electromagnetic wave.

electric flux The amount of electric charge, measured in coulombs, across a dielectric of specified area. *Electric flux* may also refer simply to electric lines of force.

electricity An energy force derived from the movement of negative and positive electric charges.

electrode An electrical terminal that emits, collects, or controls an electric current.

electrolysis A chemical change induced in a substance resulting from the passage of electric current through an electrolyte.

electrolyte A nonmetallic conductor of electricity in which current is carried by the physical movement of ions.

electromagnet An iron or steel core surrounded by a wire coil. The core becomes magnetized when current flows through the coil but loses its magnetism when the current flow is stopped.

electromagnetic compatibility The capability of electronic equipment or systems to operate in a specific electromagnetic environment, at designated levels of efficiency and within a defined margin of safety, without interfering with itself or other systems.

electromagnetic field The electric and magnetic fields associated with radio and light waves.

electromagnetic induction An electromotive force created with a conductor by the relative motion between the conductor and a nearby magnetic field.

electromagnetism The study of phenomena associated with varying magnetic fields, electromagnetic radiation, and moving electric charges.

electromotive force (EMF) An electrical potential, measured in volts, that can produce the movement of electrical charges.

electron A stable elementary particle with a negative charge that is mainly responsible for electrical conduction. Electrons move when under the influence of an electric field. This movement constitutes an *electric current.*

electron beam A stream of emitted electrons, usually in a vacuum.

electron gun A hot cathode that produces a finely focused stream of fast electrons, which are necessary for the operation of a vacuum tube, such as a cathode ray tube. The gun is made up of a hot cathode electron source, a control grid, accelerating anodes, and (usually) focusing electrodes.

electron lens A device used for focusing an electron beam in a cathode ray tube. Such focusing can be accomplished by either magnetic forces, in which external coils are used to create the proper magnetic field within the tube, or electrostatic forces, where metallic plates within the tube are charged electrically in such a way as to control the movement of electrons in the beam.

electron volt The energy acquired by an electron in passing through a potential difference of one volt in a vacuum.

electronic A description of devices (or systems) that are dependent on the flow of electrons in electron tubes, semiconductors, and other devices, and not solely on electron flow in ordinary wires, inductors, capacitors, and similar passive components.

Electronic Industries Association (EIA) A trade organization, based in Washington, DC, representing the manufacturers of electronic systems and parts, including communications systems. The association develops standards for electronic components and systems.

electronic switch A transistor, semiconductor diode, or a vacuum tube used as an on/off switch in an electrical circuit. Electronic switches can be controlled manually, by other circuits, or by computers.

electronics The field of science and engineering that deals with electron devices and their utilization.

electroplate The process of coating a given material with a deposit of metal by electrolytic action.

electrostatic The condition pertaining to electric charges that are at rest.

electrostatic field The space in which there is electric stress produced by static electric charges.

electrostatic induction The process of inducing static electric charges on a body by bringing it near other bodies that carry high electrostatic charges.

element A substance that consists of atoms of the same atomic number. Elements are the basic units in all chemical changes other than those in which *atomic changes,* such as fusion and fission, are involved.

EMI (electromagnetic interference) Undesirable electromagnetic waves that are radiated unintentionally from an electronic circuit or device into other circuits or devices, disrupting their operation.

emission (1—radiation) The radiation produced, or the production of radiation by a radio transmitting system. The emission is considered to be a *single emission* if the modulating signal and other characteristics are the same for every transmitter of the radio transmitting system and the spacing between antennas is not more than a few wavelengths. **(2—cathode)** The release of electrons from the cathode of a vacuum tube. **(3—parasitic)** A spurious radio frequency emission unintentionally generated at frequencies that are independent of the carrier frequency being amplified or modulated. **(4—secondary)** In an electron tube, emission of electrons by a plate or grid because of bombardment by *primary emission* electrons from the cathode of the tube. **(5—spurious)** An emission outside the radio frequency band authorized for a transmitter. **(6—thermonic)** An emission from a cathode resulting from high temperature.

emphasis The intentional alteration of the frequency-amplitude characteristics of a signal to reduce the adverse effects of noise in a communication system.

empirical A conclusion not based on pure theory, but on practical and experimental work.

emulation The use of one system to imitate the capabilities of another system.

enable To prepare a circuit for operation or to allow an item to function.

enabling signal A signal that permits the occurrence of a specified event.

encode The conversion of information from one form into another to obtain characteristics required by a transmission or storage system.

encoder A device that processes one or more input signals into a specified form for transmission and/or storage.

energized The condition when a circuit is switched on, or powered up.

energy spectral density A frequency-domain description of the energy in each of the frequency components of a pulse.

envelope The boundary of the family of curves obtained by varying a parameter of a wave.

envelope delay The difference in absolute delay between the fastest and slowest propagating frequencies within a specified bandwidth.

envelope delay distortion The maximum difference or deviation of the envelope-delay characteristic between any two specified frequencies.

envelope detection A demodulation process that senses the shape of the modulated RF envelope. A diode detector is one type of envelop detection device.

environmental An equipment specification category relating to temperature and humidity.

EQ (equalization) network A network connected to a circuit to correct or control its transmission frequency characteristics.

equalization (EQ) The reduction of frequency distortion and/or phase distortion of a circuit through the introduction of one or more networks to compensate for the difference in attenuation, time delay, or both, at the various frequencies in the transmission band.

equalize The process of inserting in a line a network with complementary transmission characteristics to those of the line, so that when the loss or delay in the line and that in the equalizer are combined, the overall loss or delay is approximately equal at all frequencies.

equalizer A network that corrects the transmission-frequency characteristics of a circuit to allow it to transmit selected frequencies in a uniform manner.

equatorial orbit The plane of a satellite orbit which coincides with that of the equator of the primary body.

equipment A general term for electrical apparatus and hardware, switching systems, and transmission components.

equipment failure The condition when a hardware fault stops the successful completion of a task.

equipment ground A protective ground consisting of a conducting path to ground of noncurrent carrying metal parts.

equivalent circuit A simplified network that emulates the characteristics of the real circuit it replaces. An equivalent circuit is typically used for mathematical analysis.

equivalent noise resistance A quantitative representation in resistance units of the spectral density of a noise voltage generator at a specified frequency.

error A collective term that includes all types of inconsistencies, transmission deviations, and control failures.

excitation The current that energizes field coils in a generator.

expandor A device with a nonlinear gain characteristic that acts to increase the gain more on larger input signals than it does on smaller input signals.

extremely high frequency (EHF) The band of microwave frequencies between the limits of 30 GHz and 300 GHz (wavelengths between 1 cm and 1 mm).

extremely low frequency The radio signals with operating frequencies below 300 Hz (wavelengths longer than 1000 km).

F

fail-safe operation A type of control architecture for a system that prevents improper functioning in the event of circuit or operator failure.

failure A detected cessation of ability to perform a specified function or functions within previously established limits. A *failure* is beyond adjustment by the operator by means of controls normally accessible during routine operation of the system. (This requires that measurable limits be established to define "satisfactory performance".)

failure effect The result of the malfunction or failure of a device or component.

failure in time (FIT) A unit value that indicates the reliability of a component or device. One failure in time corresponds to a failure rate of 10^{-9} per hour.

failure mode and effects analysis (FMEA) An iterative documented process performed to identify basic faults at the component level and determine their effects at higher levels of assembly.

failure rate The ratio of the number of actual failures to the number of times each item has been subjected to a set of specified stress conditions.

fall time The length of time during which a pulse decreases from 90 percent to 10 percent of its maximum amplitude.

farad The standard unit of capacitance equal to the value of a capacitor with a potential of one volt between its plates when the charge on one plate is one coulomb and there is an equal and opposite charge on the other plate. The farad is a large value and is more commonly expressed in *microfarads* or *picofarads*. The *farad* is named for the English chemist and physicist Michael Faraday (1791–1867).

fast frequency shift keying (FFSK) A system of digital modulation where the digits are represented by different frequencies that are related to the baud rate, and where transitions occur at the zero crossings.

fatigue The reduction in strength of a metal caused by the formation of crystals resulting from repeated flexing of the part in question.

fault A condition that causes a device, a component, or an element to fail to perform in a required manner. Examples include a short-circuit, broken wire, or intermittent connection.

fault to ground A fault caused by the failure of insulation and the consequent establishment of a direct path to ground from a part of the circuit that should not normally be grounded.

fault tree analysis (FTA) An iterative documented process of a systematic nature performed to identify basic faults, determine their causes and effects, and establish their probabilities of occurrence.

feature A distinctive characteristic or part of a system or piece of equipment, usually visible to end users and designed for their convenience.

Federal Communications Commission (FCC) The federal agency empowered by law to regulate all interstate radio and wireline communications services originating in the United States, including radio, television, facsimile, telegraph, data transmission, and telephone systems. The agency was established by the Communications Act of 1934.

feedback The return of a portion of the output of a device to the input. *Positive feedback* adds to the input, *negative feedback* subtracts from the input.

feedback amplifier An amplifier with the components required to feed a portion of the output back into the input to alter the characteristics of the output signal.

feedline A transmission line, typically coaxial cable, that connects a high frequency energy source to its load.

femto A prefix meaning *one quadrillionth* (10^{-15}).

ferrite A ceramic material made of powdered and compressed ferric oxide, plus other oxides (mainly cobalt, nickel, zinc, yttrium-iron, and manganese). These materials have low eddy current losses at high frequencies.

ferromagnetic material A material with low relative permeability and high coercive force so that it is difficult to magnetize and demagnetize. Hard ferromagnetic materials retain magnetism well, and are commonly used in permanent magnets.

fidelity The degree to which a system, or a portion of a system, accurately reproduces at its output the essential characteristics of the signal impressed upon its input.

field strength The strength of an electric, magnetic, or electromagnetic field.

filament A wire that becomes hot when current is passed through it, used either to emit light (for a light bulb) or to heat a cathode to enable it to emit electrons (for an electron tube).

film resistor A type of resistor made by depositing a thin layer of resistive material on an insulating core.

filter A network that passes desired frequencies but greatly attenuates other frequencies.

filtered noise White noise that has been passed through a filter. The power spectral density of filtered white noise has the same shape as the transfer function of the filter.

fitting A coupling or other mechanical device that joins one component with another.

fixed A system or device that is not changeable or movable.

flashover An arc or spark between two conductors.

flashover voltage The voltage between conductors at which flashover just occurs.

flat face tube The design of CRT tube with almost a flat face, giving improved legibility of text and reduced reflection of ambient light.

flat level A signal that has an equal amplitude response for all frequencies within a stated range.

flat loss A circuit, device, or channel that attenuates all frequencies of interest by the same amount, also called *flat slope*.

flat noise A noise whose power per unit of frequency is essentially independent of frequency over a specified frequency range.

flat response The performance parameter of a system in which the output signal amplitude of the system is a faithful reproduction of the input amplitude over some range of specified input frequencies.

floating A circuit or device that is not connected to any source of potential or to ground.

fluorescence The characteristic of a material to produce light when excited by an external energy source. Minimal or no heat results from the process.

flux The electric or magnetic lines of force resulting from an applied energy source.

flywheel effect The characteristic of an oscillator that enables it to sustain oscillations after removal of the control stimulus. This characteristic may be desirable, as in the case of a phase-locked loop employed in a synchronous system, or undesirable, as in the case of a voltage-controlled oscillator.

focusing A method of making beams of radiation converge on a target, such as the face of a CRT.

Fourier analysis A mathematical process for transforming values between the frequency domain and the time domain. This term also refers to the decomposition of a time-domain signal into its frequency components.

Fourier transform An integral that performs an actual transformation between the frequency domain and the time domain in Fourier analysis.

frame A segment of an analog or digital signal that has a repetitive characteristic, in that corresponding elements of successive *frames* represent the same things.

free electron An electron that is not attached to an atom and is, thus, mobile when an electromotive force is applied.

free running An oscillator that is not controlled by an external synchronizing signal.

free-running oscillator An oscillator that is not synchronized with an external timing source.

frequency The number of complete cycles of a periodic waveform that occur within a given length of time. Frequency is usually specified in cycles per second (*Hertz*). Frequency is the reciprocal of wavelength. The higher the frequency, the shorter the wavelength. In general, the higher the frequency of a signal, the more capacity it has to carry information, the smaller an antenna is required, and the more susceptible the signal is to absorption by the atmosphere and by physical structures. At microwave frequencies, radio signals take on a *line-of-sight* characteristic and require highly directional and focused antennas to be used successfully.

frequency accuracy The degree of conformity of a given signal to the specified value of a frequency.

frequency allocation The designation of radio-frequency bands for use by specific radio services.

frequency content The band of frequencies or specific frequency components contained in a signal.

frequency converter A circuit or device used to change a signal of one frequency into another of a different frequency.

frequency coordination The process of analyzing frequencies in use in various bands of the spectrum to achieve reliable performance for current and new services.

frequency counter An instrument or test set used to measure the frequency of a radio signal or any other alternating waveform.

frequency departure An unintentional deviation from the nominal frequency value.

frequency difference The algebraic difference between two frequencies. The two frequencies can be of identical or different nominal values.

frequency displacement The end-to-end shift in frequency that may result from independent frequency translation errors in a circuit.

frequency distortion The distortion of a multifrequency signal caused by unequal attenuation or amplification at the different frequencies of the signal. This term may also be referred to as *amplitude distortion*.

frequency domain A representation of signals as a function of frequency, rather than of time.

frequency modulation (FM) The modulation of a carrier signal so that its instantaneous frequency is proportional to the instantaneous value of the modulating wave.

frequency multiplier A circuit that provides as an output an exact multiple of the input frequency.

frequency offset A frequency shift that occurs when a signal is sent over an analog transmission facility in which the modulating and demodulating frequencies are not identical. A channel with frequency offset does not preserve the waveform of a transmitted signal.

frequency response The measure of system linearity in reproducing signals across a specified bandwidth. Frequency response is expressed as a frequency range with a specified amplitude tolerance in dB.

frequency response characteristic The variation in the transmission performance (gain or loss) of a system with respect to variations in frequency.

frequency reuse A technique used to expand the capacity of a given set of frequencies or channels by separating the signals either geographically or through the use of different polarization techniques. Frequency reuse is a common element of the *frequency coordination* process.

frequency selectivity The ability of equipment to separate or differentiate between signals at different frequencies.

frequency shift The difference between the frequency of a signal applied at the input of a circuit and the frequency of that signal at the output.

frequency shift keying (FSK) A commonly-used method of digital modulation in which a one and a zero (the two possible states) are each transmitted as separate frequencies.

frequency stability A measure of the variations of the frequency of an oscillator from its mean frequency over a specified period of time.

frequency standard An oscillator with an output frequency sufficiently stable and accurate that it is used as a reference.

frequency-division multiple access (FDMA) The provision of multiple access to a transmission facility, such as an earth satellite, by assigning each transmitter its own frequency band.

frequency-division multiplexing (FDM) The process of transmitting multiple analog signals by an orderly assignment of frequency slots, that is, by dividing transmission bandwidth into several narrow bands, each of which carries a single communication and is sent simultaneously with others over a common transmission path.

full duplex A communications system capable of transmission simultaneously in two directions.

full-wave rectifier A circuit configuration in which both positive and negative half-cycles of the incoming ac signal are rectified to produce a unidirectional (dc) current through the load.

functional block diagram A diagram illustrating the definition of a device, system, or problem on a logical and functional basis.

functional unit An entity of hardware and/or software capable of accomplishing a given purpose.

fundamental frequency The lowest frequency component of a complex signal.

fuse A protective device used to limit current flow in a circuit to a specified level. The fuse consists of a metallic link that melts and opens the circuit at a specified current level.

fuse wire A fine-gauge wire made of an alloy that overheats and melts at the relatively low temperatures produced when the wire carries overload currents. When used in a fuse, the wire is called a fuse (or fusible) link.

G

gain An increase or decrease in the level of an electrical signal. Gain is measured in terms of decibels or number-of-times of magnification. Strictly speaking, *gain* refers to an increase in level. Negative numbers, however, are commonly used to denote a decrease in level.

gain-bandwidth The gain times the frequency of measurement when a device is biased for maximum obtainable gain.

gain/frequency characteristic The gain-versus-frequency characteristic of a channel over the bandwidth provided, also referred to as *frequency response.*

gain/frequency distortion A circuit defect in which a change in frequency causes a change in signal amplitude.

galvanic A device that produces direct current by chemical action.

gang The mechanical connection of two or more circuit devices so that they can all be adjusted simultaneously.

gang capacitor A variable capacitor with more than one set of moving plates linked together.

gang tuning The simultaneous tuning of several different circuits by turning a single shaft on which ganged capacitors are mounted.

ganged One or more devices that are mechanically coupled, normally through the use of a shared shaft.

gas breakdown The ionization of a gas between two electrodes caused by the application of a voltage that exceeds a threshold value. The ionized path has a low impedance. Certain types of circuit and line protectors rely on gas breakdown to divert hazardous currents away from protected equipment.

gas tube A protection device in which a sufficient voltage across two electrodes causes a gas to ionize, creating a low impedance path for the discharge of dangerous voltages.

gas-discharge tube A gas-filled tube designed to carry current during gas breakdown. The gas-discharge tube is commonly used as a protective device, preventing high voltages from damaging sensitive equipment.

gauge A measure of wire diameter. In measuring wire gauge, the lower the number, the thicker the wire.

Gaussian distribution A statistical distribution, also called the *normal* distribution. The graph of a Gaussian distribution is a bell-shaped curve.

Gaussian noise Noise in which the distribution of amplitude follows a Gaussian model, that is, the noise is random but distributed about a reference voltage of zero.

Gaussian pulse A pulse that has the same form as its own Fourier transform.

generator A machine that converts mechanical energy into electrical energy, or one form of electrical energy into another form.

geosynchronous The attribute of a satellite in which the relative position of the satellite as viewed from the surface of a given planet is stationary. For earth, the geosynchronous position is 22,300 miles above the planet.

getter A metal used in vaporized form to remove residual gas from inside an electron tube during manufacture.

giga A prefix meaning one billion.

gigahertz (GHz) A measure of frequency equal to one billion cycles per second. Signals operating above 1 gigahertz are commonly known as *microwaves*, and begin to take on the characteristics of visible light.

glitch A general term used to describe a wide variety of momentary signal discontinuities.

graceful degradation An equipment failure mode in which the system suffers reduced capability, but does not fail altogether.

graticule A fixed pattern of reference markings used with oscilloscope CRTs to simplify measurements. The graticule may be etched on a transparent plate covering the front of the CRT or, for greater accuracy in readings, may be electrically generated within the CRT itself.

grid (1—general) A mesh electrode within an electron tube that controls the flow of electrons between the cathode and plate of the tube. **(2—bias)** The potential applied to a grid in an electron tube to control its center operating point. **(3—control)** The grid in an electron tube to which the input signal is usually applied. **(4—screen)** The grid in an electron tube, typically held at a steady potential, that screens the control grid from changes in anode potential. **(5—suppressor)** The grid in an electron tube near the anode (plate) that suppresses the emission of secondary electrons from the plate.

ground An electrical connection to earth or to a common conductor usually connected to earth.

ground clamp A clamp used to connect a ground wire to a ground rod or system.

ground loop An undesirable circulating ground current in a circuit grounded via multiple connections or at multiple points.

ground plane A conducting material at ground potential, physically close to other equipment, so that connections may be made readily to ground the equipment at the required points.

ground potential The point at zero electric potential.

ground return A conductor used as a path for one or more circuits back to the ground plane or central facility ground point.

ground rod A metal rod driven into the earth and connected into a mesh of interconnected rods so as to provide a low resistance link to ground.

ground window A single-point interface between the integrated ground plane of a building and an isolated ground plane.

ground wire A copper conductor used to extend a good low-resistance earth ground to protective devices in a facility.

grounded The connection of a piece of equipment to earth via a low resistance path.

grounding The act of connecting a device or circuit to ground or to a conductor that is grounded.

group delay A condition where the different frequency elements of a given signal suffer differing propagation delays through a circuit or a system. The delay at a lower frequency is different from the delay at a higher frequency, resulting in a time-related distortion of the signal at the receiving point.

group delay time The rate of change of the total phase shift of a waveform with angular frequency through a device or transmission facility.

group velocity The speed of a pulse on a transmission line.

guard band A narrow bandwidth between adjacent channels intended to reduce interference or crosstalk.

H

half-wave rectifier A circuit or device that changes only positive or negative half-cycle inputs of alternating current into direct current.

Hall effect The phenomenon by which a voltage develops between the edges of a current-carrying metal strip whose faces are perpendicular to an external magnetic field.

hard-wired Electrical devices connected through physical wiring.

harden The process of constructing military telecommunications facilities so as to protect them from damage by enemy action, especially *electromagnetic pulse* (EMP) radiation.

hardware Physical equipment, such as mechanical, magnetic, electrical, or electronic devices or components.

harmonic A periodic wave having a frequency that is an integral multiple of the fundamental frequency. For example, a wave with twice the frequency of the fundamental is called the *second harmonic*.

harmonic analyzer A test set capable of identifying the frequencies of the individual signals that make up a complex wave.

harmonic distortion The production of harmonics at the output of a circuit when a periodic wave is applied to its input. The level of the distortion is usually expressed as a percentage of the level of the input.

hazard A condition that could lead to danger for operating personnel.

headroom The difference, in decibels, between the typical operating signal level and a peak overload level.

heat loss The loss of useful electrical energy resulting from conversion into unwanted heat.

heat sink A device that conducts heat away from a heat-producing component so that it stays within a safe working temperature range.

heater In an electron tube, the filament that heats the cathode to enable it to emit electrons.

hecto A prefix meaning 100.

henry The standard unit of electrical inductance, equal to the self-inductance of a circuit or the mutual inductance of two circuits when there is an induced electromotive force of one volt and a current change of one ampere per second. The symbol for inductance is H, named for the American physicist Joseph Henry (1797–1878).

hertz (Hz) The unit of frequency that is equal to one cycle per second. Hertz is the reciprocal of the *period*, the interval after which the same portion of a periodic waveform recurs. Hertz was named for the German physicist Heinrich R. Hertz (1857–1894).

heterodyne The mixing of two signals in a nonlinear device in order to produce two additional signals at frequencies that are the sum and difference of the original frequencies.

heterodyne frequency The sum of, or the difference between, two frequencies, produced by combining the two signals together in a modulator or similar device.

heterodyne wavemeter A test set that uses the heterodyne principle to measure the frequencies of incoming signals.

high-frequency loss Loss of signal amplitude at higher frequencies through a given circuit or medium. For example, high frequency loss could be caused by passing a signal through a coaxial cable.

high Q An inductance or capacitance whose ratio of reactance to resistance is high.

high tension A high voltage circuit.

high-pass filter A network that passes signals of higher than a specified frequency but attenuates signals of all lower frequencies.

homochronous Signals whose corresponding significant instants have a constant but uncontrolled phase relationship with each other.

horn gap A lightning arrester utilizing a gap between two horns. When lightning causes a discharge between the horns, the heat produced lengthens the arc and breaks it.

horsepower The basic unit of mechanical power. One horsepower (hp) equals 550 foot-pounds per second or 746 watts.

hot A charged electrical circuit or device.

hot dip galvanized The process of galvanizing steel by dipping it into a bath of molten zinc.

hot standby System equipment that is fully powered but not in service. A *hot standby* can rapidly replace a primary system in the event of a failure.

hum Undesirable coupling of the 60 Hz power sine wave into other electrical signals and/or circuits.

HVAC An abbreviation for *heating, ventilation, and air conditioning* system.

hybrid system A communication system that accommodates both digital and analog signals.

hydrometer A testing device used to measure specific gravity, particularly the specific gravity of the dilute sulphuric acid in a lead-acid storage battery, to learn the state of charge of the battery.

hygrometer An instrument that measures the relative humidity of the atmosphere.

hygroscopic The ability of a substance to absorb moisture from the air.

hysteresis The property of an element evidenced by the dependence of the value of the output, for a given excursion of the input, upon the history of prior excursions and direction of the input. Originally, *hysteresis* was the name for magnetic phenomena only—the lagging of flux density behind the change in value of the magnetizing flux—but now, the term is also used to describe other inelastic behavior.

hysteresis loop The plot of magnetizing current against magnetic flux density (or of other similarly related pairs of parameters), which appears as a loop. The area within the loop is proportional to the power loss resulting from hysteresis.

hysteresis loss The loss in a magnetic core resulting from hysteresis.

I

I^2R **loss** The power lost as a result of the heating effect of current passing through resistance.

idling current The current drawn by a circuit, such as an amplifier, when no signal is present at its input.

image frequency A frequency on which a carrier signal, when heterodyned with the local oscillator in a superheterodyne receiver, will cause a sum or difference frequency that is the same as the intermediate frequency of the receiver. Thus, a signal on an *image frequency* will be demodulated along with the desired signal and will interfere with it.

impact ionization The ionization of an atom or molecule as a result of a high energy collision.

impedance The total passive opposition offered to the flow of an alternating current. *Impedance* consists of a combination of resistance, inductive reactance, and capacitive reactance. It is the vector sum of resistance and reactance ® $+ jX$) or the vector of magnitude Z at an angle θ.

impedance characteristic A graph of the impedance of a circuit showing how it varies with frequency.

impedance irregularity A discontinuity in an impedance characteristic caused, for example, by the use of different coaxial cable types.

impedance matching The adjustment of the impedances of adjoining circuit components to a common value so as to minimize reflected energy from the junction and to maximize energy transfer across it. Incorrect adjustment results in an *impedance mismatch.*

impedance matching transformer A transformer used between two circuits of different impedances with a turns ratio that provides for maximum power transfer and minimum loss by reflection.

impulse A short high energy surge of electrical current in a circuit or on a line.

impulse current A current that rises rapidly to a peak then decays to zero without oscillating.

impulse excitation The production of an oscillatory current in a circuit by impressing a voltage for a relatively short period compared with the duration of the current produced.

impulse noise A noise signal consisting of random occurrences of energy spikes, having random amplitude and bandwidth.

impulse response The amplitude-versus-time output of a transmission facility or device in response to an impulse.

impulse voltage A unidirectional voltage that rises rapidly to a peak and then falls to zero, without any appreciable oscillation.

in-phase The property of alternating current signals of the same frequency that achieve their peak positive, peak negative, and zero amplitude values simultaneously.

incidence angle The angle between the perpendicular to a surface and the direction of arrival of a signal.

increment A small change in the value of a quantity.

induce To produce an electrical or magnetic effect in one conductor by changing the condition or position of another conductor.

induced current The current that flows in a conductor because a voltage has been induced across two points in, or connected to, the conductor.

induced voltage A voltage developed in a conductor when the conductor passes through magnetic lines of force.

inductance The property of an inductor that opposes any change in a current that flows through it. The standard unit of inductance is the *Henry*.

induction The electrical and magnetic interaction process by which a changing current in one circuit produces a voltage change not only in its own circuit (*self inductance*) but also in other circuits to which it is linked magnetically.

inductive A circuit element exhibiting inductive reactance.

inductive kick A voltage surge produced when a current flowing through an inductance is interrupted.

inductive load A load that possesses a net inductive reactance.

inductive reactance The reactance of a circuit resulting from the presence of inductance and the phenomenon of induction.

inductor A coil of wire, usually wound on a core of high permeability, that provides high inductance without necessarily exhibiting high resistance.

inert An inactive unit, or a unit that has no power requirements.

infinite line A transmission line that appears to be of infinite length. There are no reflections back from the far end because it is terminated in its characteristic impedance.

infra low frequency (ILF) The frequency band from 300 Hz to 3000 Hz.

inhibit A control signal that prevents a device or circuit from operating.

injection The application of a signal to an electronic device.

input The waveform fed into a circuit, or the terminals that receive the input waveform.

insertion gain The gain resulting from the insertion of a transducer in a transmission system, expressed as the ratio of the power delivered to that part of the system following the transducer to the power delivered to that same part before insertion. If more than one component is involved in the input or output, the particular component used must be specified. This ratio is usually expressed in decibels. If the resulting number is negative, an *insertion loss* is indicated.

insertion loss The signal loss within a circuit, usually expressed in decibels as the ratio of input power to output power.

insertion loss-vs.-frequency characteristic The amplitude transfer characteristic of a system or component as a function of frequency. The amplitude response may be stated as actual gain, loss, amplification, or attenuation, or as a ratio of any one of these quantities at a particular frequency, with respect to that at a specified reference frequency.

inspection lot A collection of units of product from which a sample is drawn and inspected to determine conformance with acceptability criteria.

instantaneous value The value of a varying waveform at a given instant of time. The value can be in volts, amperes, or phase angle.

Institute of Electrical and Electronics Engineers (IEEE) The organization of electrical and electronics scientists and engineers formed in 1963 by the merger of the Institute of Radio Engineers (IRE) and the American Institute of Electrical Engineers (AIEE).

instrument multiplier A measuring device that enables a high voltage to be measured using a meter with only a low voltage range.

instrument rating The range within which an instrument has been designed to operate without damage.

insulate The process of separating one conducting body from another conductor.

insulation The material that surrounds and insulates an electrical wire from other wires or circuits. *Insulation* may also refer to any material that does not ionize easily and thus presents a large impedance to the flow of electrical current.

insulator A material or device used to separate one conducting body from another.

intelligence signal A signal containing information.

intensity The strength of a given signal under specified conditions.

interconnect cable A short distance cable intended for use between equipment (generally less than 3 m in length)

interface A device or circuit used to interconnect two pieces of electronic equipment.

interface device A unit that joins two interconnecting systems.

interference emission An emission that results in an electrical signal being propagated into and interfering with the proper operation of electrical or electronic equipment.

interlock A protection device or system designed to remove all dangerous voltages from a machine or piece of equipment when access doors or panels are opened or removed.

intermediate frequency A frequency that results from combining a signal of interest with a signal generated within a radio receiver. In superheterodyne receivers, all incoming signals are converted to a single intermediate frequency for which the amplifiers and filters of the receiver have been optimized.

intermittent A noncontinuous recurring event, often used to denote a problem that is difficult to find because of its unpredictable nature.

intermodulation The production, in a nonlinear transducer element, of frequencies corresponding to the sums and differences of the fundamentals and harmonics of two or more frequencies that are transmitted through the transducer.

intermodulation distortion (IMD) The distortion that results from the mixing of two input signals in a nonlinear system. The resulting output contains new frequencies that represent the sum and difference of the input signals and the sums and differences of their harmonics. IMD is also called *intermodulation noise.*

intermodulation noise In a transmission path or device, the noise signal that is contingent upon modulation and demodulation, resulting from nonlinear characteristics in the path or device.

internal resistance The actual resistance of a source of electric power. The total electromotive force produced by a power source is not available for external use; some of the energy is used in driving current through the source itself.

International Standards Organization (ISO) An international body concerned with worldwide standardization for a broad range of industrial products, including telecommunications equipment. Members are represented by national standards organizations, such as ANSI (American National Standards Institute) in the United States. ISO was established in 1947 as a specialized agency of the United Nations.

International Telecommunications Union (ITU) A specialized agency of the United Nations established to maintain and extend international cooperation for the maintenance, development, and efficient use of telecommunications. The union does this through standards and recommended regulations, and through technical and telecommunications studies.

International Telecommunications Satellite Consortium (Intelsat) A nonprofit cooperative of member nations that owns and operates a satellite system for international and, in many instances, domestic communications.

interoperability The condition achieved among communications and electronics systems or equipment when information or services can be exchanged directly between them or their users, or both.

interpolate The process of estimating unknown values based on a knowledge of comparable data that falls on both sides of the point in question.

interrupting capacity The rating of a circuit breaker or fuse that specifies the maximum current the device is designed to interrupt at its rated voltage.

interval The points or numbers lying between two specified endpoints.

inverse voltage The effective value of voltage across a rectifying device, which conducts a current in one direction during one half cycle of the alternating input, during the half cycle when current is not flowing.

inversion The change in the polarity of a pulse, such as from positive to negative.

inverter A circuit or device that converts a direct current into an alternating current.

ionizing radiation The form of electromagnetic radiation that can turn an atom into an ion by knocking one or more of its electrons loose. Examples of ionizing radiation include X rays, gamma rays, and cosmic rays

IR drop A drop in voltage because of the flow of current (I) through a resistance (R), also called *resistance drop*.

IR loss The conversion of electrical power to heat caused by the flow of electrical current through a resistance.

isochronous A signal in which the time interval separating any two significant instants is theoretically equal to a specified unit interval or to an integral multiple of the unit interval.

isolated ground A ground circuit that is isolated from all equipment framework and any other grounds, except for a single-point external connection.

isolated ground plane A set of connected frames that are grounded through a single connection to a ground reference point. That point and all parts of the frames are insulated from any other ground system in a building.

isolated pulse A pulse uninfluenced by other pulses in the same signal.

isophasing amplifier A timing device that corrects for small timing errors.

isotropic A quantity exhibiting the same properties in all planes and directions.

J

jack A receptacle or connector that makes electrical contact with the mating contacts of a plug. In combination, the plug and jack provide a ready means for making connections in electrical circuits.

jacket An insulating layer of material surrounding a wire in a cable.

jitter Small, rapid variations in a waveform resulting from fluctuations in a supply voltage or other causes.

joule The standard unit of work that is equal to the work done by one newton of force when the point at which the force is applied is displaced a distance of one meter in the direction of the force. The *joule* is named for the English physicist James Prescott Joule (1818–1889).

Julian date A chronological date in which days of the year are numbered in sequence. For example, the first day is 001, the second is 002, and the last is 365 (or 366 in a leap year).

K

Kelvin (K) The standard unit of thermodynamic temperature. Zero degrees Kelvin represents *absolute zero*. Water freezes at 273 K and water boils at 373 K under standard pressure conditions.

kilo A prefix meaning one thousand.

kilohertz (kHz) A unit of measure of frequency equal to 1,000 Hz.

kilovar A unit equal to one thousand volt-amperes.

kilovolt (kV) A unit of measure of electrical voltage equal to 1,000 V.

kilowatt A unit equal to one thousand watts.

Kirchoff's Law At any point in a circuit, there is as much current flowing into the point as there is flowing away from it.

klystron (1—general) A family of electron tubes that function as microwave amplifiers and oscillators. Simplest in form are two-cavity klystrons in which an electron beam passes through a cavity that is excited by a microwave input, producing a velocity-modulated beam which passes through a second cavity a precise distance away that is coupled to a tuned circuit, thereby producing an amplified output of the original input signal frequency. If part of the output is fed back to the input, an oscillator can be the result. **(2—multi-cavity)** An amplifier device for UHF and microwave signals based on velocity modulation of an electron beam. The beam is directed through an input cavity, where the input RF signal polarity initializes a *bunching effect* on electrons in the beam. The bunching effect excites subsequent cavities, which increase the bunching through an energy flywheel concept. Finally, the beam passes to an output cavity that couples the amplified signal to the load (antenna system). The beam falls onto a collector element that forms the return path for the current and dissipates the heat resulting from electron beam bombardment. **(3—reflex)** A klystron with only one cavity. The action is the same as in a two-cavity klystron but the beam is reflected back into the cavity in which it was first excited, after being sent out to a reflector. The one cavity, therefore, acts both as the original exciter (or buncher) and as the collector from which the output is taken.

knee In a response curve, the region of maximum curvature.

ku band Radio frequencies in the range of 15.35 GHz to 17.25 GHz, typically used for satellite telecommunications.

L

ladder network A type of filter with components alternately across the line and in the line.

lag The difference in phase between a current and the voltage that produced it, expressed in electrical degrees.

lagging current A current that lags behind the alternating electromotive force that produced it. A circuit that produces a *lagging current* is one containing inductance alone, or whose effective impedance is inductive.

lagging load A load whose combined inductive reactance exceeds its capacitive reactance. When an alternating voltage is applied, the current lags behind the voltage.

laminate A material consisting of layers of the same or different materials bonded together and built up to the required thickness.

latitude An angular measurement of a point on the earth above or below the equator. The equator represents 0°, the north pole +90°, and the south pole –90°.

layout A proposed or actual arrangement or allocation of equipment.

LC circuit An electrical circuit with both inductance (*L*) and capacitance (*C*) that is resonant at a particular frequency.

LC ratio The ratio of inductance to capacitance in a given circuit.

lead An electrical wire, usually insulated.

leading edge The initial portion of a pulse or wave in which voltage or current rise rapidly from zero to a final value.

leading load A reactive load in which the reactance of capacitance is greater than that of inductance. Current through such a load *leads* the applied voltage causing the current.

leakage The loss of energy resulting from the flow of electricity past an insulating material, the escape of electromagnetic radiation beyond its shielding, or the extension of magnetic lines of force beyond their intended working area.

leakage resistance The resistance of a path through which leakage current flows.

level The strength or intensity of a given signal.

level alignment The adjustment of transmission levels of single links and links in tandem to prevent overloading of transmission subsystems.

life cycle The predicted useful life of a class of equipment, operating under normal (specified) working conditions.

life safety system A system designed to protect life and property, such as emergency lighting, fire alarms, smoke exhaust and ventilating fans, and site security.

life test A test in which random samples of a product are checked to see how long they can continue to perform their functions satisfactorily. A form of *stress testing* is used, including temperature, current, voltage, and/or vibration effects, cycled at many times the rate that would apply in normal usage.

limiter An electronic device in which some characteristic of the output is automatically prevented from exceeding a predetermined value.

limiter circuit A circuit of nonlinear elements that restricts the electrical excursion of a variable in accordance with some specified criteria.

limiting A process by which some characteristic at the output of a device is prevented from exceeding a predetermined value.

line loss The total end-to-end loss in decibels in a transmission line.

line-up The process of adjusting transmission parameters to bring a circuit to its specified values.

linear A circuit, device, or channel whose output is directly proportional to its input.

linear distortion A distortion mechanism that is independent of signal amplitude.

linearity A constant relationship, over a designated range, between the input and output characteristics of a circuit or device.

lines of force A group of imaginary lines indicating the direction of the electric or magnetic field at all points along it.

lissajous pattern The looping patterns generated by a CRT spot when the horizontal (X) and vertical (Y) deflection signals are sinusoids. The lissajous pattern is useful for evaluating the delay or phase of two sinusoids of the same frequency.

live A device or system connected to a source of electric potential.

load The work required of an electrical or mechanical system.

load factor The ratio of the average load over a designated period of time to the peak load occurring during the same period.

load line A straight line drawn across a grouping of plate current/plate voltage characteristic curves showing the relationship between grid voltage and plate current for a particular plate load resistance of an electron tube.

logarithm The power to which a base must be raised to produce a given number. Common logarithms are to base 10.

logarithmic scale A meter scale with displacement proportional to the logarithm of the quantity represented.

long persistence The quality of a cathode ray tube that has phosphorescent compounds on its screen (in addition to fluorescent compounds) so that the image continues to glow after the original electron beam has ceased to create it by producing the usual fluorescence effect. Long persistence is often used in radar screens or where photographic evidence is needed of a display. Most such applications, however, have been superseded through the use of digital storage techniques.

longitude The angular measurement of a point on the surface of the earth in relation to the meridian of Greenwich (London). The earth is divided into 360° of longitude, beginning at the Greenwich mean. As one travels west around the globe, the longitude increases.

longitudinal current A current that travels in the same direction on both wires of a pair. The return current either flows in another pair or via a ground return path.

loss The power dissipated in a circuit, usually expressed in decibels, that performs no useful work.

loss deviation The change of actual loss in a circuit or system from a designed value.

loss variation The change in actual measured loss over time.

lossy The condition when the line loss per unit length is significantly greater than some defined normal parameter.

lossy cable A coaxial cable constructed to have high transmission loss so it can be used as an artificial load or as an attenuator.

lot size A specific quantity of similar material or a collection of similar units from a common source; in inspection work, the quantity offered for inspection and acceptance at any one time. The **lot size** may be a collection of raw material, parts, subassemblies inspected during production, or a consignment of finished products to be sent out for service.

low tension A low voltage circuit.

low-pass filter A filter network that passes all frequencies below a specified frequency with little or no loss, but that significantly attenuates higher frequencies.

lug A tag or projecting terminal onto which a wire may be connected by wrapping, soldering, or crimping.

lumped constant A resistance, inductance, or capacitance connected at a point, and not distributed uniformly throughout the length of a route or circuit.

M

mA An abbreviation for *milliamperes* (0.001 A).

magnet A device that produces a magnetic field and can attract iron, and attract or repel other magnets.

magnetic field An energy field that exists around magnetic materials and current-carrying conductors. Magnetic fields combine with electric fields in light and radio waves.

magnetic flux The field produced in the area surrounding a magnet or electric current. The standard unit of flux is the *Weber*.

magnetic flux density A vector quantity measured by a standard unit called the *Tesla*. The *magnetic flux density* is the number of magnetic lines of force per unit area, at right angles to the lines.

magnetic leakage The magnetic flux that does not follow a useful path.

magnetic pole A point that appears from the outside to be the center of magnetic attraction or repulsion at or near one end of a magnet.

magnetic storm A violent local variation in the earth's magnetic field, usually the result of sunspot activity.

magnetism A property of iron and some other materials by which external magnetic fields are maintained, other magnets being thereby attracted or repelled.

magnetization The exposure of a magnetic material to a magnetizing current, field, or force.

magnetizing force The force producing magnetization.

magnetomotive force The force that tends to produce lines of force in a magnetic circuit. The *magnetomotive force* bears the same relationship to a magnetic circuit that voltage does to an electrical circuit.

magnetron A high-power, ultra high frequency electron tube oscillator that employs the interaction of a strong electric field between an anode and cathode with the field of a strong permanent magnet to cause oscillatory electron flow through multiple internal cavity resonators. The magnetron may operate in a continuous or pulsed mode.

maintainability The probability that a failure will be repaired within a specified time after the failure occurs.

maintenance Any activity intended to keep a functional unit in satisfactory working condition. The term includes the tests, measurements, replacements, adjustments, and repairs necessary to keep a device or system operating properly.

malfunction An equipment failure or a fault.

manometer A test device for measuring gas pressure.

margin The difference between the value of an operating parameter and the value that would result in unsatisfactory operation. Typical *margin* parameters include signal level, signal-to-noise ratio, distortion, crosstalk coupling, and/or undesired emission level.

Markov model A statistical model of the behavior of a complex system over time in which the probabilities of the occurrence of various future states depend only on the present state of the system, and not on the path by which the present state was achieved. This term was named for the Russian mathematician Andrei Andreevich Markov (1856-1922).

master clock An accurate timing device that generates a synchronous signal to control other clocks or equipment.

master oscillator A stable oscillator that provides a standard frequency signal for other hardware and/or systems.

matched termination A termination that absorbs all the incident power and so produces no reflected waves or mismatch loss.

matching The connection of channels, circuits, or devices in a manner that results in minimal reflected energy.

matrix A logical network configured in a rectangular array of intersections of input/output signals.

Maxwell's equations Four differential equations that relate electric and magnetic fields to electromagnetic waves. The equations are a basis of electrical and electronic engineering.

mean An arithmetic average in which values are added and divided by the number of such values.

mean time between failures (MTBF) For a particular interval, the total functioning life of a population of an item divided by the total number of failures within the population during the measurement interval.

mean time to failure (MTTF) The measured operating time of a single piece of equipment divided by the total number of failures during the measured period of time. This measurement is normally made during that period between early life and wear-out failures.

mean time to repair (MTTR) The total corrective maintenance time on a component or system divided by the total number of corrective maintenance actions during a given period of time.

measurement A procedure for determining the amount of a quantity.

median A value in a series that has as many readings or values above it as below.

medium An electronic pathway or mechanism for passing information from one point to another.

mega A prefix meaning one million.

megahertz (MHz) A quantity equal to one million Hertz (cycles per second).

megohm A quantity equal to one million ohms.

metric system A decimal system of measurement based on the meter, the kilogram, and the second.

micro A prefix meaning one millionth.

micron A unit of length equal to one millionth of a meter (1/25,000 of an inch).

microphonic(s) Unintended noise introduced into an electronic system by mechanical vibration of electrical components.

microsecond One millionth of a second (0.000001 s).

microvolt A quantity equal to one-millionth of a volt.

milli A prefix meaning one thousandth.

milliammeter A test instrument for measuring electrical current, often part of a *multimeter*.

millihenry A quantity equal to one-thousandth of a henry.

milliwatt A quantity equal to one thousandth of a watt.

minimum discernible signal The smallest input that will produce a discernible change in the output of a circuit or device.

mixer A circuit used to combine two or more signals to produce a third signal that is a function of the input waveforms.

mixing ratio The ratio of the mass of water vapor to the mass of dry air in a given volume of air. The *mixing ratio* affects radio propagation.

mode An electromagnetic field distribution that satisfies theoretical requirements for propagation in a waveguide or oscillation in a cavity.

modified refractive index The sum of the refractive index of the air at a given height above sea level, and the ratio of this height to the radius of the earth.

modular An equipment design in which major elements are readily separable, and which the user may replace, reducing the mean-time-to-repair.

modulation The process whereby the amplitude, frequency, or phase of a single-frequency wave (the *carrier*) is varied in step with the instantaneous value of, or samples of, a complex wave (the *modulating wave*).

modulator A device that enables the intelligence in an information-carrying modulating wave to be conveyed by a signal at a higher frequency. A *modulator* modifies a carrier wave by amplitude, phase, and/or frequency as a function of a control signal that carries intelligence. Signals are *modulated* in this way to permit more efficient and/or reliable transmission over any of several media.

module An assembly replaceable as an entity, often as an interchangeable plug-in item. A *module* is not normally capable of being disassembled.

monostable A device that is stable in one state only. An input pulse causes the device to change state, but it reverts immediately to its stable state.

motor A machine that converts electrical energy into mechanical energy.

motor effect The repulsion force exerted between adjacent conductors carrying currents in opposite directions.

moving coil Any device that utilizes a coil of wire in a magnetic field in such a way that the coil is made to move by varying the applied current, or itself produces a varying voltage because of its movement.

ms An abbreviation for *millisecond* (0.001 s).

multimeter A test instrument fitted with several ranges for measuring voltage, resistance, and current, and equipped with an analog meter or digital display readout. The *multimeter* is also known as a *volt-ohm-milliammeter*, or *VOM*.

multiplex (MUX) The use of a common channel to convey two or more channels. This is done either by splitting of the common channel frequency band into narrower bands, each of which is used to constitute a distinct channel (*frequency division multiplex*), or by allotting this common channel to multiple users in turn to constitute different intermittent channels (*time division multiplex*).

multiplexer A device or circuit that combines several signals onto a single signal.

multiplexing A technique that uses a single transmission path to carry multiple channels. In *time division multiplexing* (TDM), path time is shared. For *frequency division multiplexing* (FDM) or *wavelength division multiplexing* (WDM), signals are divided into individual channels sent along the same path but at different frequencies.

multiplication Signal mixing that occurs within a multiplier circuit.

multiplier A circuit in which one or more input signals are mixed under the direction of one or more control signals. The resulting output is a composite of the input signals, the characteristics of which are determined by the scaling specified for the circuit.

mutual induction The property of the magnetic flux around a conductor that induces a voltage in a nearby conductor. The voltage generated in the secondary conductor in turn induces a voltage in the primary conductor. The inductance of two conductors so coupled is referred to as *mutual inductance*.

mV An abbreviation for *millivolt* (0.001 V).

mW An abbreviation for *milliwatt* (0.001 W).

N

nano A prefix meaning one billionth.

nanometer 1×10^{-9} meter.

nanosecond (ns) One billionth of a second (1×10^{-9} s).

narrowband A communications channel of restricted bandwidth, often resulting in degradation of the transmitted signal.

narrowband emission An emission having a spectrum exhibiting one or more sharp peaks that are narrow in width compared to the nominal bandwidth of the measuring instrument, and are far enough apart in frequency to be resolvable by the instrument.

National Electrical Code (NEC) A document providing rules for the installation of electric wiring and equipment in public and private buildings, published by the National Fire Protection Association. The NEC has been adopted as law by many states and municipalities in the U.S.

National Institute of Standards and Technology (NIST) A nonregulatory agency of the Department of Commerce that serves as a national reference and measurement laboratory for the physical and engineering sciences. Formerly called the *National Bureau of Standards*, the agency was renamed in 1988 and given the additional responsibility of aiding U.S. companies in adopting new technologies to increase their international competitiveness.

negative In a conductor or semiconductor material, an excess of electrons or a deficiency of positive charge.

negative feedback The return of a portion of the output signal from a circuit to the input but 180° out of phase. This type of feedback decreases signal amplitude but stabilizes the amplifier and reduces distortion and noise.

negative impedance An impedance characterized by a decrease in voltage drop across a device as the current through the device is increased, or a decrease in current through the device as the voltage across it is increased.

neutral A device or object having no electrical charge.

neutral conductor A conductor in a power distribution system connected to a point in the system that is designed to be at neutral potential. In a balanced system, the neutral conductor carries no current.

neutral ground An intentional ground applied to the neutral conductor or neutral point of a circuit, transformer, machine, apparatus, or system.

newton The standard unit of force. One *newton* is the force that, when applied to a body having a mass of 1 kg, gives it an acceleration of 1 m/s^2.

nitrogen A gas widely used to pressurize radio frequency transmission lines. If a small puncture occurs in the cable sheath, the nitrogen keeps moisture out so that service is not adversely affected.

node The points at which the current is at minimum in a transmission system in which standing waves are present.

noise Any random disturbance or unwanted signal in a communication system that tends to obscure the clarity or usefulness of a signal in relation to its intended use.

noise factor (NF) The ratio of the noise power measured at the output of a receiver to the noise power that would be present at the output if the thermal noise resulting from the resistive component of the source impedance were the only source of noise in the system.

noise figure A measure of the noise in dB generated at the input of an amplifier, compared with the noise generated by an impedance-method resistor at a specified temperature.

noise filter A network that attenuates noise frequencies.

noise generator A generator of wideband random noise.

noise power ratio (NPR) The ratio, expressed in decibels, of signal power to intermodulation product power plus residual noise power, measured at the baseband level.

noise suppressor A filter or digital signal processing circuit in a receiver or transmitter that automatically reduces or eliminates noise.

noise temperature The temperature, expressed in Kelvin, at which a resistor will develop a particular noise voltage. The noise temperature of a radio receiver is the value by which the temperature of the resistive component of the source impedance should be increased—if it were the only source of noise in the system—to cause the noise power at the output of the receiver to be the same as in the real system.

nominal The most common value for a component or parameter that falls between the maximum and minimum limits of a tolerance range.

nominal value A specified or intended value independent of any uncertainty in its realization.

nomogram A chart showing three or more scales across which a straight edge may be held in order to read off a graphical solution to a three-variable equation.

nonionizing radiation Electromagnetic radiation that does not turn an atom into an ion. Examples of nonionizing radiation include visible light and radio waves.

nonconductor A material that does not conduct energy, such as electricity, heat, or sound.

noncritical technical load That part of the technical power load for a facility not required for minimum acceptable operation.

noninductive A device or circuit without significant inductance.

nonlinearity A distortion in which the output of a circuit or system does not rise or fall in direct proportion to the input.

nontechnical load The part of the total operational load of a facility used for such purposes as general lighting, air conditioning, and ventilating equipment during normal operation.

normal A line perpendicular to another line or to a surface.

normal-mode noise Unwanted signals in the form of voltages appearing in line-to-line and line-to-neutral signals.

normalized frequency The ratio between the actual frequency and its nominal value.

normalized frequency departure The frequency departure divided by the nominal frequency value.

normalized frequency difference The algebraic difference between two normalized frequencies.

normalized frequency drift The frequency drift divided by the nominal frequency value.

normally closed Switch contacts that are closed in their nonoperated state, or relay contacts that are closed when the relay is de-energized.

normally open Switch contacts that are open in their nonoperated state, or relay contacts that are open when the relay is de-energized.

north pole The pole of a magnet that seeks the north magnetic pole of the earth.

notch filter A circuit designed to attenuate a specific frequency band; also known as a *band stop filter*.

notched noise A noise signal in which a narrow band of frequencies has been removed.

ns An abbreviation for *nanosecond*.

null A zero or minimum amount or position.

O

octave Any frequency band in which the highest frequency is twice the lowest frequency.

off-line A condition wherein devices or subsystems are not connected into, do not form a part of, and are not subject to the same controls as an operational system.

offset An intentional difference between the realized value and the nominal value.

ohm The unit of electric resistance through which one ampere of current will flow when there is a difference of one volt. The quantity is named for the German physicist Georg Simon Ohm (1787-1854).

Ohm's law A law that sets forth the relationship between voltage (E), current (I), and resistance (R). The law states that $E = I \times R$, $I = \dfrac{E}{R}$, and $R = \dfrac{E}{I}$. *Ohm's Law* is named for the German physicist Georg Simon Ohm (1787–1854).

ohmic loss The power dissipation in a line or circuit caused by electrical resistance.

ohmmeter A test instrument used for measuring resistance, often part of a *multimeter*.

ohms-per-volt A measure of the sensitivity of a voltmeter.

on-line A device or system that is energized and operational, and ready to perform useful work.

open An interruption in the flow of electrical current, as caused by a broken wire or connection.

open-circuit A defined loop or path that closes on itself and contains an infinite impedance.

open-circuit impedance The input impedance of a circuit when its output terminals are open, that is, not terminated.

open-circuit voltage The voltage measured at the terminals of a circuit when there is no load and, hence, no current flowing.

operating lifetime The period of time during which the principal parameters of a component or system remain within a prescribed range.

optimize The process of adjusting for the best output or maximum response from a circuit or system.

orbit The path, relative to a specified frame of reference, described by the center of mass of a satellite or other object in space, subjected solely to natural forces (mainly gravitational attraction).

order of diversity The number of independently fading propagation paths or frequencies, or both, used in a diversity reception system.

original equipment manufacturer (OEM) A manufacturer of equipment that is used in systems assembled and sold by others.

oscillation A variation with time of the magnitude of a quantity with respect to a specified reference when the magnitude is alternately greater than and smaller than the reference.

oscillator A nonrotating device for producing alternating current, the output frequency of which is determined by the characteristics of the circuit.

oscilloscope A test instrument that uses a display, usually a cathode-ray tube, to show the instantaneous values and waveforms of a signal that varies with time or some other parameter.

out-of-band energy Energy emitted by a transmission system that falls outside the frequency spectrum of the intended transmission.

outage duration The average elapsed time between the start and the end of an outage period.

outage probability The probability that an outage state will occur within a specified time period. In the absence of specific known causes of outages, the *outage probability* is the sum of all outage durations divided by the time period of measurement.

outage threshold A defined value for a supported performance parameter that establishes the minimum operational service performance level for that parameter.

output impedance The impedance presented at the output terminals of a circuit, device, or channel.

output stage The final driving circuit in a piece of electronic equipment.

ovenized crystal oscillator (OXO) A crystal oscillator enclosed within a temperature regulated heater (oven) to maintain a stable frequency despite external temperature variations.

overcoupling A degree of coupling greater than the *critical coupling* between two resonant circuits. *Overcoupling* results in a wide bandwidth circuit with two peaks in the response curve.

overload In a transmission system, a power greater than the amount the system was designed to carry. In a power system, an overload could cause excessive heating. In a communications system, distortion of a signal could result.

overshoot The first maximum excursion of a pulse beyond the 100% level. Overshoot is the portion of the pulse that exceeds its defined level temporarily before settling to the correct level. Overshoot amplitude is expressed as a percentage of the defined level.

P

pentode An electron tube with five electrodes, the cathode, control grid, screen grid, suppressor grid, and plate.

photocathode An electrode in an electron tube that will emit electrons when bombarded by photons of light.

picture tube A cathode-ray tube used to produce an image by variation of the intensity of a scanning beam on a phosphor screen.

pin A terminal on the base of a component, such as an electron tube.

plasma (1—arc) An ionized gas in an arc-discharge tube that provides a conducting path for the discharge. **(2—solar)** The ionized gas at extremely high temperature found in the sun.

plate (1—electron tube) The anode of an electron tube. **(2—battery)** An electrode in a storage battery. **(3—capacitor)** One of the surfaces in a capacitor. **(4—chassis)** A mounting surface to which equipment may be fastened.

propagation time delay The time required for a signal to travel from one point to another.

protector A device or circuit that prevents damage to lines or equipment by conducting dangerously high voltages or currents to ground. Protector types include spark gaps, semiconductors, varistors, and gas tubes.

proximity effect A nonuniform current distribution in a conductor, caused by current flow in a nearby conductor.

pseudonoise In a spread-spectrum system, a seemingly random series of pulses whose frequency spectrum resembles that of continuous noise.

pseudorandom A sequence of signals that appears to be completely random but have, in fact, been carefully drawn up and repeat after a significant time interval.

pseudorandom noise A noise signal that satisfies one or more of the standard tests for statistical randomness. Although it seems to lack any definite pattern, there is a sequence of pulses that repeats after a long time interval.

pseudorandom number sequence A sequence of numbers that satisfies one or more of the standard tests for statistical randomness. Although it seems to lack any definite pattern, there is a sequence that repeats after a long time interval.

pulsating direct current A current changing in value at regular or irregular intervals but which has the same direction at all times.

pulse One of the elements of a repetitive signal characterized by the rise and decay in time of its magnitude. A *pulse* is usually short in relation to the time span of interest.

pulse decay time The time required for the trailing edge of a pulse to decrease from 90 percent to 10 percent of its peak amplitude.

pulse duration The time interval between the points on the leading and trailing edges of a pulse at which the instantaneous value bears a specified relation to the peak pulse amplitude.

pulse duration modulation (PDM) The modulation of a pulse carrier by varying the width of the pulses according to the instantaneous values of the voltage samples of the modulating signal (also called *pulse width modulation*).

pulse edge The leading or trailing edge of a pulse, defined as the 50 percent point of the pulse rise or fall time.

pulse fall time The interval of time required for the edge of a pulse to fall from 90 percent to 10 percent of its peak amplitude.

pulse interval The time between the start of one pulse and the start of the next.

pulse length The duration of a pulse (also called *pulse width*).

pulse level The voltage amplitude of a pulse.

pulse period The time between the start of one pulse and the start of the next.

pulse ratio The ratio of the length of any pulse to the total pulse period.

pulse repetition period The time interval from the beginning of one pulse to the beginning of the next pulse.

pulse repetition rate The number of times each second that pulses are transmitted.

pulse rise time The time required for the leading edge of a pulse to rise from 10 percent to 90 percent of its peak amplitude.

pulse train A series of pulses having similar characteristics.

pulse width The measured interval between the 50 percent amplitude points of the leading and trailing edges of a pulse.

puncture A breakdown of insulation or of a dielectric, such as in a cable sheath or in the insulant around a conductor.

pW An abbreviation for picowatt, a unit of power equal to 10^{-12} W (–90 dBm).

Q

Q (quality factor) A figure of merit that defines how close a coil comes to functioning as a pure inductor. *High Q* describes an inductor with little energy loss resulting from resistance. *Q* is found by dividing the inductive reactance of a device by its resistance.

quadrature A state of alternating current signals separated by one quarter of a cycle (90°).

quadrature amplitude modulation (QAM) A process that allows two different signals to modulate a single carrier frequency. The two signals of interest amplitude modulate two samples of the carrier that are of the same frequency, but differ in phase by 90°. The two resultant signals can be added and transmitted. Both signals may be recovered at a decoder when they are demodulated 90° apart.

quadrature component The component of a voltage or current at an angle of 90° to a reference signal, resulting from inductive or capacitive reactance.

quadrature phase shift keying (QPSK) A type of phase shift keying using four phase states.

quality The absence of objectionable distortion.

quality assurance (QA) All those activities, including surveillance, inspection, control, and documentation, aimed at ensuring that a given product will meet its performance specifications.

quality control (QC) A function whereby management exercises control over the quality of raw material or intermediate products in order to prevent the production of defective devices or systems.

quantum noise Any noise attributable to the discrete nature of electromagnetic radiation. Examples include shot noise, photon noise, and recombination noise.

quantum-limited operation An operation wherein the minimum detectable signal is limited by quantum noise.

quartz A crystalline mineral that when electrically excited vibrates with a stable period. Quartz is typically used as the frequency-determining element in oscillators and filters.

quasi-peak detector A detector that delivers an output voltage that is some fraction of the peak value of the regularly repeated pulses applied to it. The fraction increases toward unity as the pulse repetition rate increases.

quick-break fuse A fuse in which the fusible link is under tension, providing for rapid operation.

quiescent An inactive device, signal, or system.

quiescent current The current that flows in a device in the absence of an applied signal.

R

rack An equipment rack, usually measuring 19 in (48.26 cm) wide at the front mounting rails.

rack unit (RU) A unit of measure of vertical space in an equipment enclosure. One rack unit is equal to 1.75 in (4.45 cm).

radiate The process of emitting electromagnetic energy.

radiation The emission and propagation of electromagnetic energy in the form of waves. *Radiation* is also called *radiant energy*.

radiation scattering The diversion of thermal, electromagnetic, or nuclear radiation from its original path as a result of interactions or collisions with atoms, molecules, or large particles in the atmosphere or other media between the source of radiation and a point some distance away. As a result of scattering, radiation (especially gamma rays and neutrons) will be received at such a point from many directions, rather than only from the direction of the source.

radio The transmission of signals over a distance by means of electromagnetic waves in the approximate frequency range of 150 kHz to 300 GHz. The term may also be used to describe the equipment used to transmit or receive electromagnetic waves.

radio detection The detection of the presence of an object by radio location without precise determination of its position.

radio frequency interference (RFI) The intrusion of unwanted signals or electromagnetic noise into various types of equipment resulting from radio frequency transmission equipment or other devices using radio frequencies.

radio frequency spectrum Those frequency bands in the electromagnetic spectrum that range from several hundred thousand cycles per second (*very low frequency*) to several billion cycles per second (*microwave frequencies*).

radio recognition In military communications, the determination by radio means of the "friendly" or "unfriendly" character of an aircraft or ship.

random noise Electromagnetic signals that originate in transient electrical disturbances and have random time and amplitude patterns. Random noise is generally undesirable; however, it may also be generated for testing purposes.

rated output power The power available from an amplifier or other device under specified conditions of operation.

RC constant The time constant of a resistor-capacitor circuit. The *RC constant* is the time in seconds required for current in an RC circuit to rise to 63 percent of its final steady value or fall to 37 percent of its original steady value, obtained by multiplying the resistance value in ohms by the capacitance value in farads.

RC network A circuit that contains resistors and capacitors, normally connected in series.

reactance The part of the impedance of a network resulting from inductance or capacitance. The *reactance* of a component varies with the frequency of the applied signal.

reactive power The power circulating in an ac circuit. It is delivered to the circuit during part of the cycle and is returned during the other half of the cycle. The *reactive power* is obtained by multiplying the voltage, current, and the sine of the phase angle between them.

reactor A component with inductive reactance.

received signal level (RSL) The value of a specified bandwidth of signals at the receiver input terminals relative to an established reference.

receiver Any device for receiving electrical signals and converting them to audible sound, visible light, data, or some combination of these elements.

receptacle An electrical socket designed to receive a mating plug.

reception The act of receiving, listening to, or watching information-carrying signals.

rectification The conversion of alternating current into direct current.

rectifier A device for converting alternating current into direct current. A *rectifier* normally includes filters so that the output is, within specified limits, smooth and free of ac components.

rectify The process of converting alternating current into direct current.

redundancy A system design that provides a back-up for key circuits or components in the event of a failure. Redundancy improves the overall reliability of a system.

redundant A configuration when two complete systems are available at one time. If the online system fails, the backup will take over with no loss of service.

reference voltage A voltage used for control or comparison purposes.

reflectance The ratio of reflected power to incident power.

reflection An abrupt change, resulting from an impedance mismatch, in the direction of propagation of an electromagnetic wave. For light, at the interface of two dissimilar materials, the incident wave is returned to its medium of origin.

reflection coefficient The ratio between the amplitude of a reflected wave and the amplitude of the incident wave. For large smooth surfaces, the reflection coefficient may be near unity.

reflection gain The increase in signal strength that results when a reflected wave combines, in phase, with an incident wave.

reflection loss The apparent loss of signal strength caused by an impedance mismatch in a transmission line or circuit. The loss results from the reflection of part of the signal back toward the source from the point of the impedance discontinuity. The greater the mismatch, the greater the loss.

reflectometer A device that measures energy traveling in each direction in a waveguide, used in determining the standing wave ratio.

refraction The bending of a sound, radio, or light wave as it passes obliquely from a medium of one density to a medium of another density that varies its speed.

regulation The process of adjusting the level of some quantity, such as circuit gain, by means of an electronic system that monitors an output and feeds back a controlling signal to constantly maintain a desired level.

regulator A device that maintains its output voltage at a constant level.

relative envelope delay The difference in envelope delay at various frequencies when compared with a reference frequency that is chosen as having zero delay.

relative humidity The ratio of the quantity of water vapor in the atmosphere to the quantity that would cause saturation at the ambient temperature.

relative transmission level The ratio of the signal power in a transmission system to the signal power at some point chosen as a reference. The ratio is usually determined by applying a standard test signal at the input to the system and measuring the gain or loss at the location of interest.

relay A device by which current flowing in one circuit causes contacts to operate that control the flow of current in another circuit.

relay armature The movable part of an electromechanical relay, usually coupled to spring sets on which contacts are mounted.

relay bypass A device that, in the event of a loss of power or other failure, routes a critical signal around the equipment that has failed.

release time The time required for a pulse to drop from steady-state level to zero, also referred to as the *decay time*.

reliability The ability of a system or subsystem to perform within the prescribed parameters of quality of service. *Reliability* is often expressed as the probability that a system or subsystem will perform its intended function for a specified interval under stated conditions.

reliability growth The action taken to move a hardware item toward its reliability potential, during development or subsequent manufacturing or operation.

reliability predictions The compiled failure rates for parts, components, subassemblies, assemblies, and systems. These generic failure rates are used as basic data to predict the reliability of a given device or system.

remote control A system used to control a device from a distance.

remote station A station or terminal that is physically remote from a main station or center but can gain access through a communication channel.

repeater The equipment between two circuits that receives a signal degraded by normal factors during transmission and amplifies the signal to its original level for re-transmission.

repetition rate The rate at which regularly recurring pulses are repeated.

reply A transmitted message that is a direct response to an original message.

repulsion The mechanical force that tends to separate like magnetic poles, like electric charges, or conductors carrying currents in opposite directions.

reset The act of restoring a device to its default or original state.

residual flux The magnetic flux that remains after a magnetomotive force has been removed.

residual magnetism The magnetism or flux that remains in a core after current ceases to flow in the coil producing the magnetomotive force.

residual voltage The vector sum of the voltages in all the phase wires of an unbalanced polyphase power system.

resistance The opposition of a material to the flow of electrical current. Resistance is equal to the voltage drop through a given material divided by the current flow through it. The standard unit of resistance is the *ohm*, named for the German physicist Georg Simon Ohm (1787–1854).

resistance drop The fall in potential (volts) between two points, the product of the current and resistance.

resistance-grounded A circuit or system grounded for safety through a resistance, which limits the value of the current flowing through the circuit in the event of a fault.

resistive load A load in which the voltage is in phase with the current.

resistivity The resistance per unit volume or per unit area.

resistor A device the primary function of which is to introduce a specified resistance into an electrical circuit.

resonance A tuned condition conducive to oscillation, when the reactance resulting from capacitance in a circuit is equal in value to the reactance resulting from inductance.

resonant frequency The frequency at which the inductive reactance and capacitive reactance of a circuit are equal.

resonator A resonant cavity.

return A return path for current, sometimes through ground.

reversal A change in magnetic polarity, in the direction of current flow.

reverse current A small current that flows through a diode when the voltage across it is such that normal forward current does not flow.

reverse voltage A voltage in the reverse direction from that normally applied.

rheostat A two-terminal variable resistor, usually constructed with a sliding or rotating shaft that can be used to vary the resistance value of the device.

ripple An ac voltage superimposed on the output of a dc power supply, usually resulting from imperfect filtering.

rise time The time required for a pulse to rise from 10 percent to 90 percent of its peak value.

roll-off A gradual attenuation of gain-frequency response at either or both ends of a transmission pass band.

root-mean-square (RMS) The square root of the average value of the squares of all the instantaneous values of current or voltage during one half-cycle of an alternating current. For an alternating current, the RMS voltage or current is equal to the amount of direct current or voltage that would produce the same heating effect in a purely resistive circuit. For a sinewave, the root-mean-square value is equal to 0.707 times the peak value. RMS is also called the *effective value*.

rotor The rotating part of an electric generator or motor.

RU An abbreviation for *rack unit*.

S

scan One sweep of the target area in a camera tube, or of the screen in a picture tube.

screen grid A grid in an electron tube that improves performance of the device by shielding the control grid from the plate.

self-bias The provision of bias in an electron tube through a voltage drop in the cathode circuit.

shot noise The noise developed in a vacuum tube or photoconductor resulting from the random number and velocity of emitted charge carriers.

slope The rate of change, with respect to frequency, of transmission line attenuation over a given frequency spectrum.

slope equalizer A device or circuit used to achieve a specified slope in a transmission line.

smoothing circuit A filter designed to reduce the amount of ripple in a circuit, usually a dc power supply.

snubber An electronic circuit used to suppress high frequency noise.

solar wind Charged particles from the sun that continuously bombard the surface of the earth.

solid A single wire conductor, as contrasted with a stranded, braided, or rope-type wire.

solid-state The use of semiconductors rather than electron tubes in a circuit or system.

source The part of a system from which signals or messages are considered to originate.

source terminated A circuit whose output is terminated for correct impedance matching with standard cable.

spare A system that is available but not presently in use.

spark gap A gap between two electrodes designed to produce a spark under given conditions.

specific gravity The ratio of the weight of a volume, liquid, or solid to the weight of the same volume of water at a specified temperature.

spectrum A continuous band of frequencies within which waves have some common characteristics.

spectrum analyzer A test instrument that presents a graphic display of signals over a selected frequency bandwidth. A cathode-ray tube is often used for the display.

spectrum designation of frequency A method of referring to a range of communication frequencies. In American practice, the designation is a two or three letter acronym for the name. The ranges are: below 300 Hz, ELF (extremely low frequency); 300 Hz–3000 Hz, ILF (infra low frequency); 3 kHz–30 kHz, VLF (very low frequency); 30 kHz–300 kHz, LF (low frequency); 300 kHz–3000 kHz, MF (medium frequency); 3 MHz–30 MHz, HF (high frequency); 30 MHz–300 MHz, VHF (very high frequency); 300 MHz–3000 MHz, UHF (ultra high frequency); 3 GHz–30 GHz, SHF (super high frequency); 30 GHz–300 GHz, EHF (extremely high frequency); 300 GHz–3000 GHz, THF (tremendously high frequency).

spherical antenna A type of satellite receiving antenna that permits more than one satellite to be accessed at any given time. A spherical antenna has a broader angle of acceptance than a parabolic antenna.

spike A high amplitude, short duration pulse superimposed on an otherwise regular waveform.

split-phase A device that derives a second phase from a single phase power supply by passing it through a capacitive or inductive reactor.

splitter A circuit or device that accepts one input signal and distributes it to several outputs.

splitting ratio The ratio of the power emerging from the output ports of a coupler.

sporadic An event occurring at random and infrequent intervals.

spread spectrum A communications technique in which the frequency components of a narrowband signal are spread over a wide band. The resulting signal resembles white noise. The technique is used to achieve signal security and privacy, and to enable the use of a common band by many users.

spurious signal Any portion of a given signal that is not part of the fundamental waveform. Spurious signals include transients, noise, and hum.

square wave A square or rectangular-shaped periodic wave that alternately assumes two fixed values for equal lengths of time, the transition being negligible in comparison with the duration of each fixed value.

square wave testing The use of a square wave containing many odd harmonics of the fundamental frequency as an input signal to a device. Visual examination of the output signal on an oscilloscope indicates the amount of distortion introduced.

stability The ability of a device or circuit to remain stable in frequency, power level, and/or other specified parameters.

standard The specific signal configuration, reference pulses, voltage levels, and other parameters that describe the input/output requirements for a particular type of equipment.

standard time and frequency signal A time-controlled radio signal broadcast at scheduled intervals on a number of different frequencies by government-operated radio stations to provide a method for calibrating instruments.

standing wave ratio (SWR) The ratio of the maximum to the minimum value of a component of a wave in a transmission line or waveguide, such as the maximum voltage to the minimum voltage.

static charge An electric charge on the surface of an object, particularly a dielectric.

station One of the input or output points in a communications system.

stator The stationary part of a rotating electric machine.

status The present condition of a device.

statute mile A unit of distance equal to 1,609 km or 5,280 ft.

steady-state A condition in which circuit values remain essentially constant, occurring after all initial transients or fluctuating conditions have passed.

steady-state condition A condition occurring after all initial transient or fluctuating conditions have damped out in which currents, voltages, or fields remain essentially constant or oscillate uniformly without changes in characteristics such as amplitude, frequency, or wave shape.

steep wavefront A rapid rise in voltage of a given signal, indicating the presence of high frequency odd harmonics of a fundamental wave frequency.

step up (or down) The process of increasing (or decreasing) the voltage of an electrical signal, as in a step-up (or step-down) transformer.

straight-line capacitance A capacitance employing a variable capacitor with plates so shaped that capacitance varies directly with the angle of rotation.

stray capacitance An unintended—and usually undesired—capacitance between wires and components in a circuit or system.

stray current A current through a path other than the intended one.

stress The force per unit of cross-sectional area on a given object or structure

subassembly A functional unit of a system.

subcarrier (SC) A carrier applied as modulation on another carrier, or on an intermediate subcarrier.

subharmonic A frequency equal to the fundamental frequency of a given signal divided by a whole number.

submodule A small circuit board or device that mounts on a larger module or device.

subrefraction A refraction for which the refractivity gradient is greater than standard.

subsystem A functional unit of a system.

superheterodyne receiver A radio receiver in which all signals are first converted to a common frequency for which the intermediate stages of the receiver have been optimized, both for tuning and filtering. Signals are converted by mixing them with the output of a local oscillator whose output is varied in accordance with the frequency of the received signals so as to maintain the desired *intermediate frequency*.

suppressor grid The fifth grid of a pentode electron tube, which provides screening between plate and screen grid.

surface leakage A leakage current from line to ground over the face of an insulator supporting an open wire route.

surface refractivity The refractive index, calculated from observations of pressure, temperature, and humidity at the surface of the earth.

surge A rapid rise in current or voltage, usually followed by a fall back to the normal value.

susceptance The reciprocal of reactance, and the imaginary component of admittance, expressed in siemens.

sweep The process of varying the frequency of a signal over a specified bandwidth.

sweep generator A test oscillator, the frequency of which is constantly varied over a specified bandwidth.

switching The process of making and breaking (connecting and disconnecting) two or more electrical circuits.

synchronization The process of adjusting the corresponding significant instants of signals—for example, the zero-crossings—to make them synchronous. The term *synchronization* is often abbreviated as *sync*.

synchronize The process of causing two systems to operate at the same speed.

synchronous In step or in phase, as applied to two or more devices; a system in which all events occur in a predetermined timed sequence.

synchronous detection A demodulation process in which the original signal is recovered by multiplying the modulated signal by the output of a synchronous oscillator locked to the carrier.

synchronous system A system in which the transmitter and receiver are operating in a fixed time relationship.

system standards The minimum required electrical performance characteristics of a specific collection of hardware and/or software.

systems analysis An analysis of a given activity to determine precisely what must be accomplished and how it is to be done.

T

tetrode A four element electron tube consisting of a cathode, control grid, screen grid, and plate.

thyratron A gas-filled electron tube in which plate current flows when the grid voltage reaches a predetermined level. At that point, the grid has no further control over the current, which continues to flow until it is interrupted or reversed.

tolerance The permissible variation from a standard.

torque A moment of force acting on a body and tending to produce rotation about an axis.

total harmonic distortion (THD) The ratio of the sum of the amplitudes of all signals harmonically related to the fundamental versus the amplitude of the fundamental signal. THD is expressed in percent.

trace The pattern on an oscilloscope screen when displaying a signal.

tracking The locking of tuned stages in a radio receiver so that all stages are changed appropriately as the receiver tuning is changed.

trade-off The process of weighing conflicting requirements and reaching a compromise decision in the design of a component or a subsystem.

transceiver Any circuit or device that receives and transmits signals.

transconductance The mutual conductance of an electron tube expressed as the change in plate current divided by the change in control grid voltage that produced it.

transducer A device that converts energy from one form to another.

transfer characteristics The intrinsic parameters of a system, subsystem, or unit of equipment which, when applied to the input of the system, subsystem, or unit of equipment, will fully describe its output.

transformer A device consisting of two or more windings wrapped around a single core or linked by a common magnetic circuit.

transformer ratio The ratio of the number of turns in the secondary winding of a transformer to the number of turns in the primary winding, also known as the *turns ratio*.

transient A sudden variance of current or voltage from a steady-state value. A transient normally results from changes in load or effects related to switching action.

transient disturbance A voltage pulse of high energy and short duration impressed upon the ac waveform. The overvoltage pulse can be one to 100 times the normal ac potential (or more) and can last up to 15 ms. Rise times measure in the nanosecond range.

transient response The time response of a system under test to a stated input stimulus.

transition A sequence of actions that occurs when a process changes from one state to another in response to an input.

transmission The transfer of electrical power, signals, or an intelligence from one location to another by wire, fiber optic, or radio means.

transmission facility A transmission medium and all the associated equipment required to transmit information.

transmission loss The ratio, in decibels, of the power of a signal at a point along a transmission path to the power of the same signal at a more distant point along the same path. This value is often used as a measure of the quality of the transmission medium for conveying signals. Changes in power level are normally expressed in decibels by calculating ten times the logarithm (base 10) of the ratio of the two powers.

transmission mode One of the field patterns in a waveguide in a plane transverse to the direction of propagation.

transmission system The set of equipment that provides single or multichannel communications facilities capable of carrying audio, video, or data signals.

transmitter The device or circuit that launches a signal into a passive medium, such as the atmosphere.

transparency The property of a communications system that enables it to carry a signal without altering or otherwise affecting the electrical characteristics of the signal.

tray The metal cabinet that holds circuit boards.

tremendously high frequency (THF) The frequency band from 300 GHz to 3000 GHz.

triangular wave An oscillation, the values of which rise and fall linearly, and immediately change upon reaching their peak maximum and minimum. A graphical representation of a triangular wave resembles a triangle.

trim The process of making fine adjustments to a circuit or a circuit element.

trimmer A small mechanically-adjustable component connected in parallel or series with a major component so that the net value of the two can be finely adjusted for tuning purposes.

triode A three-element electron tube, consisting of a cathode, control grid, and plate.

triple beat A third-order beat whose three beating carriers all have different frequencies, but are spaced at equal frequency separations.

troposphere The layer of the earth's atmosphere, between the surface and the stratosphere, in which about 80 percent of the total mass of atmospheric air is concentrated and in which temperature normally decreases with altitude.

trouble A failure or fault affecting the service provided by a system.

troubleshoot The process of investigating, localizing and (if possible) correcting a fault.

tube (1—electron) An evacuated or gas-filled tube enclosed in a glass or metal case in which the electrodes are maintained at different voltages, giving rise to a controlled flow of electrons from the cathode to the anode. **(2—cathode ray, CRT)** An electron beam tube used for the display of changing electrical phenomena, generally similar to a television picture tube. **(3—cold-cathode)** An electron tube whose cathode emits electrons without the need of a heating filament. **(4—gas)** A gas-filled electron tube in which the gas plays an essential role in operation of the device. **(5—mercury-vapor)** A tube filled with mercury vapor at low pressure, used as a rectifying device. **(6—metal)** An electron tube enclosed in a metal case. **(7—traveling wave, TWT)** A wide band microwave amplifier in which a stream of electrons interacts with a guided electromagnetic wave moving substantially in synchronism with the electron stream, resulting in a net transfer of energy from the electron stream to the wave. **(8—velocity-modulated)** An electron tube in which the velocity of the electron stream is continually changing, as in a klystron.

tune The process of adjusting the frequency of a device or circuit, such as for resonance or for maximum response to an input signal.

tuned trap A series resonant network bridged across a circuit that eliminates ("traps") the frequency of the resonant network.

tuner The radio frequency and intermediate frequency parts of a radio receiver that produce a low level output signal.

tuning The process of adjusting a given frequency; in particular, to adjust for resonance or for maximum response to a particular incoming signal.

turns ratio In a transformer, the ratio of the number of turns on the secondary to the number of turns on the primary.

tweaking The process of adjusting an electronic circuit to optimize its performance.

twin-line A feeder cable with two parallel, insulated conductors.

two-phase A source of alternating current circuit with two sinusoidal voltages that are 90° apart.

U

ultra high frequency (UHF) The frequency range from 300 MHz to 3000 MHz.

ultraviolet radiation Electromagnetic radiation in a frequency range between visible light and high-frequency X-rays.

unattended A device or system designed to operate without a human attendant.

unattended operation A system that permits a station to receive and transmit messages without the presence of an attendant or operator.

unavailability A measure of the degree to which a system, subsystem, or piece of equipment is not operable and not in a committable state at the start of a mission, when the mission is called for at a random point in time.

unbalanced circuit A two-wire circuit with legs that differ from one another in resistance, capacity to earth or to other conductors, leakage, or inductance.

unbalanced line A transmission line in which the magnitudes of the voltages on the two conductors are not equal with respect to ground. A coaxial cable is an example of an unbalanced line.

unbalanced modulator A modulator whose output includes the carrier signal.

unbalanced output An output with one leg at ground potential.

unbalanced wire circuit A circuit whose two sides are inherently electrically unlike.

uncertainty An expression of the magnitude of a possible deviation of a measured value from the true value. Frequently, it is possible to distinguish two components: the *systematic uncertainty* and the *random uncertainty*. The random uncertainty is expressed by the standard deviation or by a multiple of the standard deviation. The systematic uncertainty is generally estimated on the basis of the parameter characteristics.

undamped wave A signal with constant amplitude.

underbunching A condition in a traveling wave tube wherein the tube is not operating at its optimum bunching rate.

Underwriters Laboratories, Inc. A laboratory established by the National Board of Fire Underwriters which tests equipment, materials, and systems that may affect insurance risks, with special attention to fire dangers and other hazards to life.

ungrounded A circuit or line not connected to ground.

unicoupler A device used to couple a balanced circuit to an unbalanced circuit.

unidirectional A signal or current flowing in one direction only.

uniform transmission line A transmission line with electrical characteristics that are identical, per unit length, over its entire length.

unit An assembly of equipment and associated wiring that together forms a complete system or independent subsystem.

unity coupling In a theoretically perfect transformer, complete electromagnetic coupling between the primary and secondary windings with no loss of power.

unity gain An amplifier or active circuit in which the output amplitude is the same as the input amplitude.

unity power factor A power factor of 1.0, which means that the load is—in effect—a pure resistance, with ac voltage and current completely in phase.

unterminated A device or system that is not terminated.

up-converter A frequency translation device in which the frequency of the output signal is greater than that of the input signal. Such devices are commonly found in microwave radio and satellite systems.

uplink A transmission system for sending radio signals from the ground to a satellite or aircraft.

upstream A device or system placed ahead of other devices or systems in a signal path.

useful life The period during which a low, constant failure rate can be expected for a given device or system. The *useful life* is the portion of a product life cycle between break-in and wear out.

user A person, organization, or group that employs the services of a system for the transfer of information or other purposes.

V

VA An abbreviation for *volt-amperes*, volts times amperes.

vacuum relay A relay whose contacts are enclosed in an evacuated space, usually to provide reliable long-term operation.

vacuum switch A switch whose contacts are enclosed in an evacuated container so that spark formation is discouraged.

vacuum tube An electron tube. The most common vacuum tubes include the diode, triode, tetrode, and pentode.

validity check A test designed to ensure that the quality of transmission is maintained over a given system.

varactor A semiconductor that behaves like a capacitor under the influence of an external control voltage.

varactor diode A semiconductor device whose capacitance is a function of the applied voltage. A varactor diode, also called a *variable reactance diode* or simply a *varactor*, is often used to tune the operating frequency of a radio circuit.

variable frequency oscillator (VFO) An oscillator whose frequency can be set to any required value in a given range of frequencies.

variable impedance A capacitor, inductor, or resistor that is adjustable in value.

variable-gain amplifier An amplifier whose gain can be controlled by an external signal source.

variable-reluctance A transducer in which the input (usually a mechanical movement) varies the magnetic reluctance of a device.

variation monitor A device used for sensing a deviation in voltage, current, or frequency, which is capable of providing an alarm and/or initiating transfer to another power source when programmed limits of voltage, frequency, current, or time are exceeded.

varicap A diode used as a variable capacitor.

VCXO (voltage controlled crystal oscillator) A device whose output frequency is determined by an input control voltage.

vector A quantity having both magnitude and direction.

vector diagram A diagram using vectors to indicate the relationship between voltage and current in a circuit.

vector sum The sum of two vectors which, when they are at right angles to each other, equal the length of the hypotenuse of the right triangle so formed. In the general case, the vector sum of the two vectors equals the diagonal of the parallelogram formed on the two vectors.

velocity of light The speed of propagation of electromagnetic waves in a vacuum, equal to 299,792,458 m/s, or approximately 186,000 mi/s. For rough calculations, the figure of 300,000 km/s is used.

velocity of propagation The velocity of signal transmission. In free space, electromagnetic waves travel at the speed of light. In a cable, the velocity is substantially lower.

vernier A device that enables precision reading of a measuring set or gauge, or the setting of a dial with precision.

very low frequency (VLF) A radio frequency in the band 3 kHz to 30 kHz.

vestigial sideband A form of transmission in which one sideband is significantly attenuated. The carrier and the other sideband are transmitted without attenuation.

vibration testing A testing procedure whereby subsystems are mounted on a test base that vibrates, thereby revealing any faults resulting from badly soldered joints or other poor mechanical design features.

volt The standard unit of electromotive force, equal to the potential difference between two points on a conductor that is carrying a constant current of one ampere when the power dissipated between the two points is equal to one watt. One *volt* is equivalent to the potential difference across a resistance of one ohm when one ampere is flowing through it. The volt is named for the Italian physicist Alessandro Volta (1745–1827).

volt-ampere (VA) The apparent power in an ac circuit (volts times amperes).

volt-ohm-milliammeter (VOM) A general purpose multirange test meter used to measure voltage, resistance, and current.

voltage The potential difference between two points.

voltage drop A decrease in electrical potential resulting from current flow through a resistance.

voltage gradient The continuous drop in electrical potential, per unit length, along a uniform conductor or thickness of a uniform dielectric.

voltage level The ratio of the voltage at a given point to the voltage at an arbitrary reference point.

voltage reference circuit A stable voltage reference source.

voltage regulation The deviation from a nominal voltage, expressed as a percentage of the nominal voltage.

voltage regulator A circuit used for controlling and maintaining a voltage at a constant level.

voltage stabilizer A device that produces a constant or substantially constant output voltage despite variations in input voltage or output load current.

voltage to ground The voltage between any given portion of a piece of equipment and the ground potential.

voltmeter An instrument used to measure differences in electrical potential.

vox A voice-operated relay circuit that permits the equivalent of push-to-talk operation of a transmitter by the operator.

VSAT (very small aperture terminal) A satellite Ku-band earth station intended for fixed or portable use. The antenna diameter of a VSAT is on the order of 1.5 m or less.

W

watt The unit of power equal to the work done at one joule per second, or the rate of work measured as a current of one ampere under an electric potential of one volt. Designated by the symbol W, the watt is named after the Scottish inventor James Watt (1736-1819).

watt meter A meter indicating in watts the rate of consumption of electrical energy.

watt-hour The work performed by one watt over a one hour period.

wave A disturbance that is a function of time or space, or both, and is propagated in a medium or through space.

wave number The reciprocal of wavelength; the number of wave lengths per unit distance in the direction of propagation of a wave.

waveband A band of wavelengths defined for some given purpose.

waveform The characteristic shape of a periodic wave, determined by the frequencies present and their amplitudes and relative phases.

wavefront A continuous surface that is a locus of points having the same phase at a given instant. A *wavefront* is a surface at right angles to rays that proceed from the wave source. The surface passes through those parts of the wave that are in the same phase and travel in the same direction. For parallel rays the wavefront is a plane; for rays that radiate from a point, the wavefront is spherical.

waveguide Generally, a rectangular or circular pipe that constrains the propagation of an acoustic or electromagnetic wave along a path between two locations. The dimensions of a waveguide determine the frequencies for optimum transmission.

wavelength For a sinusoidal wave, the distance between points of corresponding phase of two consecutive cycles.

weber The unit of magnetic flux equal to the flux that, when linked to a circuit of one turn, produces an electromotive force of one volt as the flux is reduced at a uniform rate to zero in one second. The *weber* is named for the German physicist Wilhelm Eduard Weber (1804–1891).

weighted The condition when a correction factor is applied to a measurement.

weighting The adjustment of a measured value to account for conditions that would otherwise be different or appropriate during a measurement.

weighting network A circuit, used with a test instrument, that has a specified amplitude-versus-frequency characteristic.

wideband The passing or processing of a wide range of frequencies. The meaning varies with the context.

Wien bridge An ac bridge used to measure capacitance or inductance.

winding A coil of wire used to form an inductor.

wire A single metallic conductor, usually solid-drawn and circular in cross section.

working range The permitted range of values of an analog signal over which transmitting or other processing equipment can operate.

working voltage The rated voltage that may safely be applied continuously to a given circuit or device.

X

x-band A microwave frequency band from 5.2 GHz to 10.9 GHz.

x-cut A method of cutting a quartz plate for an oscillator, with the x-axis of the crystal perpendicular to the faces of the plate.

X ray An electromagnetic radiation of approximately 100 nm to 0.1 nm, capable of penetrating nonmetallic materials.

Y

y-cut A method of cutting a quartz plate for an oscillator, with the y-axis of the crystal perpendicular to the faces of the plate.

yield strength The magnitude of mechanical stress at which a material will begin to deform. Beyond the *yield strength* point, extension is no longer proportional to stress and rupture is possible.

yoke A material that interconnects magnetic cores. *Yoke* can also refer to the deflection windings of a CRT.

yttrium-iron garnet (YIG) A crystalline material used in microwave devices.

10

Reference Data and Tables

10.1 Standard Units

Name	Symbol	Quantity
ampere	A	electric current
ampere per meter	A/m	magnetic field strength
ampere per square meter	A/m^2	current density
becquerel	Bg	activity (of a radionuclide)
candela	cd	luminous intensity
coulomb	C	electric charge
coulomb per kilogram	C/kg	exposure (x and gamma rays)
coulomb per sq. meter	C/m^2	electric flux density
cubic meter	m^3	volume
cubic meter per kilogram	m^3/kg	specific volume
degree Celsius	°C	Celsius temperature
farad	F	capacitance
farad per meter	F/m	permittivity
henry	H	inductance
henry per meter	H/m	permeability
hertz	Hz	frequency
joule	J	energy, work, quantity of heat
joule per cubic meter	J/m^3	energy density
joule per kelvin	J/K	heat capacity
joule per kilogram K	$J/(kg•K)$	specific heat capacity
joule per mole	J/mol	molar energy
kelvin	K	thermodynamic temperature
kilogram	kg	mass
kilogram per cubic meter	kg/m^3	density, mass density
lumen	lm	luminous flux
lux	lx	luminance
meter	m	length
meter per second	m/s	speed, velocity
meter per second sq.	m/s^2	acceleration
mole	mol	amount of substance

newton	N	force
newton per meter	N/m	surface tension
ohm	Ω	electrical resistance
pascal	Pa	pressure, stress
pascal second	Pa•s	dynamic viscosity
radian	rad	plane angle
radian per second	rad/s	angular velocity
radian per second squared	rad/s^2	angular acceleration
second	s	time
siemens	S	electrical conductance
square meter	m^2	area
steradian	sr	solid angle
tesla	T	magnetic flux density
volt	V	electrical potential
volt per meter	V/m	electric field strength
watt	W	power, radiant flux
watt per meter kelvin	W/(m•K)	thermal conductivity
watt per square meter	W/m^2	heat (power) flux density
weber	Wb	magnetic flux

10.2 Standard Prefixes

Multiple	Prefix	Symbol
10^{18}	exa	E
10^{15}	peta	P
10^{12}	tera	T
10^{9}	giga	G
10^{6}	mega	M
10^{3}	kilo	k
10^{2}	hecto	h
10	deka	da
10^{-1}	deci	d
10^{-2}	centi	c
10^{-3}	milli	m
10^{-6}	micro	μ
10^{-9}	nano	n
10^{-12}	pico	p
10^{-15}	femto	f
10^{-18}	atto	a

10.3 Common Standard Units

Unit	Symbol
centimeter	cm
cubic centimeter	cm^3
cubic meter per second	m^3/s
gigahertz	GHz
gram	g
kilohertz	kHz
kilohm	$k\Omega$
kilojoule	kJ
kilometer	km
kilovolt	kV
kilovoltampere	kVA
kilowatt	kW
megahertz	MHz
megavolt	MV
megawatt	MW
megohm	$M\Omega$
microampere	μA
microfarad	μF
microgram	μg
microhenry	μH
microsecond	μs
microwatt	μW
milliampere	mA
milligram	mg
millihenry	mH
millimeter	mm
millisecond	ms
millivolt	mV
milliwatt	mW
nanoampere	nA
nanofarad	nF
nanometer	nm
nanosecond	ns
nanowatt	nW
picoampere	pA
picofarad	pF
picosecond	ps
picowatt	pW

10.4 Conversion Reference Data

A

To Convert	Into	Multiply By
abcoulomb	statcoulombs	2.998×10^{10}
acre	sq. chain (Gunters)	10
acre	rods	160
acre	square links (Gunters)	1×10^5
acre	Hectare or	
	sq. hectometer	0.4047
acre-feet	cubic feet	43,560.0
acre-feet	gallons	3.259×10^5
acres	sq. feet	43,560.0
acres	sq. meters	4,047
acres	sq. miles	1.562×10^{-3}
acres	sq. yards	4,840
ampere-hours	coulombs	3,600.0
ampere-hours	faradays	0.03731
amperes/sq. cm	amps/sq. in	6.452
amperes/sq. cm	amps/sq. meter	10^4
amperes/sq. in	amps/sq. cm	0.1550
amperes/sq. in	amps/sq. meter	1,550.0
amperes/sq. meter	amps/sq. cm	10^{-4}
amperes/sq. meter	amps/sq. in	6.452×10^{-4}
ampere-turns	gilberts	1.257
ampere-turns/cm	amp-turns/in	2.540
ampere-turns/cm	amp-turns/meter	100.0
ampere-turns/cm	gilberts/cm	1.257
ampere-turns/in	amp-turns/cm	0.3937
ampere-turns/in	amp-turns/m	39.37
ampere-turns/in	gilberts/cm	0.4950
ampere-turns/meter	amp-turns/cm	0.01
ampere-turns/meter	amp-turns/in	0.0254
ampere-turns/meter	gilberts/cm	0.01257
Angstrom unit	inch	3937×10^{-9}
Angstrom unit	meter	1×10^{-10}
Angstrom unit	micron or (Mu)	1×10^{-4}
are	acre (U.S.)	0.02471
ares	sq. yards	119.60
ares	acres	0.02471
ares	sq. meters	100.0
astronomical unit	kilometers	1.495×10^8
atmospheres	ton/sq. in	0.007348
atmospheres	cm of mercury	76.0
atmospheres	ft of water (at 4°C)	33.90
atmospheres	in of mercury (at 0°C)	29.92
atmospheres	kg/sq. cm	1.0333
atmospheres	kg/sq. m	10,332
atmospheres	pounds/sq. in	14.70
atmospheres	tons/sq. ft	1.058

B

To Convert	Into	Multiply By
barrels (U.S., dry)	cubic inches	7056
barrels (U.S., dry)	quarts (dry)	105.0
barrels (U.S., liquid)	gallons	31.5
barrels (oil)	gallons (oil)	42.0
bars	atmospheres	0.9869
bars	dynes/sq. cm	10^4
bars	kg/sq. m	1.020×10^4
bars	pounds/sq. ft	2,089
bars	pounds/sq. in	14.50
baryl	dyne/sq. cm	1.000
bolt (U.S. cloth)	meters	36.576
Btu	liter-atmosphere	10.409
Btu	ergs	1.0550×10^{10}
Btu	foot-lb	778.3
Btu	gram-calories	252.0
Btu	horsepower-hr	3.931×10^{-4}
Btu	joules	1,054.8
Btu	kilogram-calories	0.2520
Btu	kilogram-meters	107.5
Btu	kilowatt-hr	2.928×10^{-4}
Btu/hr	foot-pounds/s	0.2162
Btu/hr	gram-calories/s	0.0700
Btu/hr	horsepower-hr	3.929×10^{-4}
Btu/hr	watts	0.2931
Btu/min	foot-lbs/s	12.96
Btu/min	horsepower	0.02356
Btu/min	kilowatts	0.01757
Btu/min	watts	17.57
Btu/sq. ft/min	watts/sq. in	0.1221
bucket (br. dry)	cubic cm	1.818×10^4
bushels	cubic ft	1.2445
bushels	cubic in	2,150.4
bushels	cubic m	0.03524
bushels	liters	35.24
bushels	pecks	4.0
bushels	pints (dry)	64.0
bushels	quarts (dry)	32.0

C

To Convert	Into	Multiply By
calories, gram (mean)	Btu (mean)	3.9685×10^{-3}
candle/sq. cm	Lamberts	3.142
candle/sq. in	Lamberts	0.4870
centares (centiares)	sq. meters	1.0
Centigrade	Fahrenheit	$(C° \times 9/5) + 32$
centigrams	grams	0.01
centiliter	ounce fluid (U.S.)	0.3382
centiliter	cubic inch	0.6103

centiliter	drams	2.705
centiliter	liters	0.01
centimeter	feet	3.281×10^{-2}
centimeter	inches	0.3937
centimeter	kilometers	10^{-5}
centimeter	meters	0.01
centimeter	miles	6.214×10^{-6}
centimeter	millimeters	10.0
centimeter	mils	393.7
centimeter	yards	1.094×10^{-2}
centimeter-dynes	cm-grams	1.020×10^{-3}
centimeter-dynes	meter-kg	1.020×10^{-8}
centimeter-dynes	pound-ft	7.376×10^{-8}
centimeter-grams	cm-dynes	980.7
centimeter-grams	meter-kg	10^{-5}
centimeter-grams	pound-ft	7.233×10^{-5}
centimeters of mercury	atmospheres	0.01316
centimeters of mercury	feet of water	0.4461
centimeters of mercury	kg/sq. meter	136.0
centimeters of mercury	pounds/sq. ft	27.85
centimeters of mercury	pounds/sq. in	0.1934
centimeters/sec	feet/min	1.9686
centimeters/sec	feet/sec	0.03281
centimeters/sec	kilometers/hr	0.036
centimeters/sec	knots	0.1943
centimeters/sec	meters/min	0.6
centimeters/sec	miles/hr	0.02237
centimeters/sec	miles/min	3.728×10^{-4}
centimeters/sec/sec	feet/sec/sec	0.03281
centimeters/sec/sec	km/hr/sec	0.036
centimeters/sec/sec	meters/sec/sec	0.01
centimeters/sec/sec	miles/hr/sec	0.02237
chain	inches	792.00
chain	meters	20.12
chains (surveyor's or Gunter's)	yards	22.00
circular mils	sq. cm	5.067×10^{-6}
circular mils	sq. mils	0.7854
circular mils	sq. inches	7.854×10^{-7}
circumference	Radians	6.283
cord feet	cubic feet	16
cords	cord feet	8
coulomb	statcoulombs	2.998×10^{9}
coulombs	faradays	1.036×10^{-5}
coulombs/sq. cm	coulombs/sq. in	64.52
coulombs/sq. cm	coulombs/sq. meter	10^{4}
coulombs/sq. in	coulombs/sq. cm	0.1550
coulombs/sq. in	coulombs/sq. meter	1,550
coulombs/sq. meter	coulombs/sq. cm	10^{-4}
coulombs/sq. meter	coulombs/sq. in	6.452×10^{-4}

cubic centimeters	cubic feet	3.531×10^{-5}
cubic centimeters	cubic inches	0.06102
cubic centimeters	cubic meters	10^{-6}
cubic centimeters	cubic yards	1.308×10^{-6}
cubic centimeters	gallons (U.S. liq.)	2.642×10^{-4}
cubic centimeters	liters	0.001
cubic centimeters	pints (U.S. liq.)	2.113×10^{-3}
cubic centimeters	quarts (U.S. liq.)	1.057×10^{-3}
cubic feet	bushels (dry)	0.8036
cubic feet	cubic cm	28,320.0
cubic feet	cubic inches	1,728.0
cubic feet	cubic meters	0.02832
cubic feet	cubic yards	0.03704
cubic feet	gallons (U.S. liq.)	7.48052
cubic feet	liters	28.32
cubic feet	pints (U.S. liq.)	59.84
cubic feet	quarts (U.S. liq.)	29.92
cubic feet/min	cubic cm/sec	472.0
cubic feet/min	gallons/sec	0.1247
cubic feet/min	liters/sec	0.4720
cubic feet/min	pounds of water/min	62.43
cubic feet/sec	million gal/day	0.646317
cubic feet/sec	gallons/min	448.831
cubic inches	cubic cm	16.39
cubic inches	cubic feet	5.787×10^{-4}
cubic inches	cubic meters	1.639×10^{-5}
cubic inches	cubic yards	2.143×10^{-5}
cubic inches	gallons	4.329×10^{-3}
cubic inches	liters	0.01639
cubic inches	mil-feet	1.061×10^{5}
cubic inches	pints (U.S. liq.)	0.03463
cubic inches	quarts (U.S. liq.)	0.01732
cubic meters	bushels (dry)	28.38
cubic meters	cubic cm	10^{6}
cubic meters	cubic feet	35.31
cubic meters	cubic inches	61,023.0
cubic meters	cubic yards	1.308
cubic meters	gallons (U.S. liq.)	264.2
cubic meters	liters	1,000.0
cubic meters	pints (U.S. liq.)	2,113.0
cubic meters	quarts (U.S. liq.)	1,057.
cubic yards	cubic cm	7.646×10^{5}
cubic yards	cubic feet	27.0
cubic yards	cubic inches	46,656.0
cubic yards	cubic meters	0.7646
cubic yards	gallons (U.S. liq.)	202.0
cubic yards	liters	764.6
cubic yards	pints (U.S. liq.)	1,615.9
cubic yards	quarts (U.S. liq.)	807.9
cubic yards/min	cubic ft/sec	0.45

cubic yards/min	gallons/sec	3.367
cubic yards/min	liters/sec	12.74

D

To Convert	Into	Multiply By
Dalton	gram	1.650×10^{-24}
days	seconds	86,400.0
decigrams	grams	0.1
deciliters	liters	0.1
decimeters	meters	0.1
degrees (angle)	quadrants	0.01111
degrees (angle)	radians	0.01745
degrees (angle)	seconds	3,600.0
degrees/sec	radians/sec	0.01745
degrees/sec	revolutions/min	0.1667
degrees/sec	revolutions/sec	2.778×10^{-3}
dekagrams	grams	10.0
dekaliters	liters	10.0
dekameters	meters	10.0
drams (apothecaries or troy)	ounces (avoirdupois)	0.1371429
drams (apothecaries or troy)	ounces (troy)	0.125
drams (U.S., fluid or apothecaries)	cubic cm	3.6967
drams	grams	1.7718
drams	grains	27.3437
drams	ounces	0.0625
dyne/cm	erg/sq. millimeter	0.01
dyne/sq. cm	atmospheres	9.869×10^{-7}
dyne/sq. cm	inch of mercury at 0°C	2.953×10^{-5}
dyne/sq. cm	inch of water at 4°C	4.015×10^{-4}
dynes	grams	1.020×10^{-3}
dynes	joules/cm	10^{-7}
dynes	joules/meter (newtons)	10^{-5}
dynes	kilograms	1.020×10^{-6}
dynes	poundals	7.233×10^{-5}
dynes	pounds	2.248×10^{-6}
dynes/sq. cm	bars	10^{-6}

E

To Convert	Into	Multiply By
ell	cm	114.30
ell	inches	45
em, pica	inch	0.167
em, pica	cm	0.4233
erg/sec	Dyne-cm/sec	1.000
ergs	Btu	9.480×10^{-11}
ergs	dyne-centimeters	1.0
ergs	foot-pounds	7.367×10^{-8}

ergs	gram-calories	0.2389×10^{-7}
ergs	gram-cm	1.020×10^{-3}
ergs	horsepower-hr	3.7250×10^{-14}
ergs	joules	10^{-7}
ergs	kg-calories	2.389×10^{-11}
ergs	kg-meters	1.020×10^{-8}
ergs	kilowatt-hr	0.2778×10^{-13}
ergs	watt-hours	0.2778×10^{-10}
ergs/sec	Btu/min	$5,688 \times 10^{-9}$
ergs/sec	ft-lb/min	4.427×10^{-6}
ergs/sec	ft-lb/sec	7.3756×10^{-8}
ergs/sec	horsepower	1.341×10^{-10}
ergs/sec	kg-calories/min	1.433×10^{-9}
ergs/sec	kilowatts	10^{-10}

F

To Convert	Into	Multiply By
farad	microfarads	10^6
Faraday/sec	ampere (absolute)	9.6500×10^4
faradays	ampere-hours	26.80
faradays	coulombs	9.649×10^4
fathom	meter	1.828804
fathoms	feet	6.0
feet	centimeters	30.48
feet	kilometers	3.048×10^{-4}
feet	meters	0.3048
feet	miles (naut.)	1.645×10^{-4}
feet	miles (stat.)	1.894×10^{-4}
feet	millimeters	304.8
feet	mils	1.2×10^4
feet of water	atmospheres	0.02950
feet of water	in of mercury	0.8826
feet of water	kg/sq. cm	0.03048
feet of water	kg/sq. meter	304.8
feet of water	pounds/sq. ft	62.43
feet of water	pounds/sq. in	0.4335
feet/min	cm/sec	0.5080
feet/min	feet/sec	0.01667
feet/min	km/hr	0.01829
feet/min	meters/min	0.3048
feet/min	miles/hr	0.01136
feet/sec	cm/sec	30.48
feet/sec	km/hr	1.097
feet/sec	knots	0.5921
feet/sec	meters/min	18.29
feet/sec	miles/hr	0.6818
feet/sec	miles/min	0.01136
feet/sec/sec	cm/sec/sec	30.48
feet/sec/sec	km/hr/sec	1.097
feet/sec/sec	meters/sec/sec	0.3048

feet/sec/sec	miles/hr/sec	0.6818
feet/100 feet	per centigrade	1.0
foot-candle	lumen/sq. meter	10.764
foot-pounds	Btu	1.286×10^{-3}
foot-pounds	ergs	1.356×10^{7}
foot-pounds	gram-calories	0.3238
foot-pounds	hp-hr	5.050×10^{-7}
foot-pounds	joules	1.356
foot-pounds	kg-calories	3.24×10^{-4}
foot-pounds	kg-meters	0.1383
foot-pounds	kilowatt-hr	3.766×10^{-7}
foot-pounds/min	Btu/min	1.286×10^{-3}
foot-pounds/min	foot-pounds/sec	0.01667
foot-pounds/min	horsepower	3.030×10^{-5}
foot-pounds/min	kg-calories/min	3.24×10^{-4}
foot-pounds/min	kilowatts	2.260×10^{-5}
foot-pounds/sec	Btu/hr	4.6263
foot-pounds/sec	Btu/min	0.07717
foot-pounds/sec	horsepower	1.818×10^{-3}
foot-pounds/sec	kg-calories/min	0.01945
foot-pounds/sec	kilowatts	1.356×10^{-3}
Furlongs	miles (U.S.)	0.125
furlongs	rods	40.0
furlongs	feet	660.0

G

To Convert	**Into**	**Multiply By**
gallons	cubic cm	3,785.0
gallons	cubic feet	0.1337
gallons	cubic inches	231.0
gallons	cubic meters	3.785×10^{-3}
gallons	cubic yards	4.951×10^{-3}
gallons	liters	3.785
gallons (liq. Br. Imp.)	gallons (U.S. liq.)	1.20095
gallons (U.S.)	gallons (Imp.)	0.83267
gallons of water	pounds of water	8.3453
gallons/min	cubic ft/sec	2.228×10^{-3}
gallons/min	liters/sec	0.06308
gallons/min	cubic ft/hr	8.0208
gausses	lines/sq. in	6.452
gausses	webers/sq. cm	10^{-8}
gausses	webers/sq. in	6.452×10^{-8}
gausses	webers/sq. meter	10^{-4}
gilberts	ampere-turns	0.7958
gilberts/cm	amp-turns/cm	0.7958
gilberts/cm	amp-turns/in	2.021
gilberts/cm	amp-turns/meter	79.58
gills	liters	0.1183
gills	pints (liq.)	0.25
gills (British)	cubic cm	142.07

grade	radian	0.01571
grains	drams (avoirdupois)	0.03657143
grains (troy)	grains (avdp.)	1.0
grains (troy)	grams	0.06480
grains (troy)	ounces (avdp.)	2.0833×10^{-3}
grains (troy)	pennyweight (troy)	0.04167
grains/Imp. gal	parts/million	14.286
grains/U.S. gal	parts/million	17.118
grains/U.S. gal	pounds/million gal	142.86
gram-calories	Btu	3.9683×10^{-3}
gram-calories	ergs	4.1868×10^{7}
gram-calories	foot-pounds	3.0880
gram-calories	horsepower-hr	1.5596×10^{-6}
gram-calories	kilowatt-hr	1.1630×10^{-6}
gram-calories	watt-hr	1.1630×10^{-3}
gram-calories/sec	Btu/hr	14.286
gram-centimeters	Btu	9.297×10^{-8}
gram-centimeters	ergs	980.7
gram-centimeters	joules	9.807×10^{-5}
gram-centimeters	kg-calories	2.343×10^{-8}
gram-centimeters	kg-meters	10^{-5}
grams	dynes	980.7
grams	grains	15.43
grams	joules/cm	9.807×10^{-5}
grams	joules/meter (newtons)	9.807×10^{-3}
grams	kilograms	0.001
grams	milligrams	1,000
grams	ounces (avdp.)	0.03527
grams	ounces (troy)	0.03215
grams	poundals	0.07093
grams	pounds	2.205×10^{-3}
grams/cm	pounds/inch	5.600×10^{-3}
grams/cubic cm	pounds/cubic ft	62.43
grams/cubic cm	pounds/cubic in	0.03613
grams/cubic cm	pounds/mil-foot	3.405×10^{-7}
grams/liter	grains/gal	58.417
grams/liter	pounds/1,000 gal	8.345
grams/liter	pounds/cubic ft	0.062427
grams/liter	parts/million	1,000.0
grams/sq. cm	pounds/sq. ft	2.0481

H

To Convert	Into	Multiply By
hand	cm	10.16
hectares	acres	2.471
hectares	sq. feet	1.076×10^{5}
hectograms	grams	100.0
hectoliters	liters	100.0
hectometers	meters	100.0
hectowatts	watts	100.0

henries	millihenries	1,000.0
horsepower	Btu/min	42.44
horsepower	foot-lb/min	33,000
horsepower	foot-lb/sec	550.0
horsepower	kg-calories/min	10.68
horsepower	kilowatts	0.7457
horsepower	watts	745.7
horsepower (boiler)	Btu/hr	33.479
horsepower (boiler)	kilowatts	9.803
horsepower (metric) (542.5 ft lb./sec)	horsepower (550 ft lb./sec)	0.9863
horsepower (550 ft lb./sec)	horsepower (metric) (542.5 ft lb./sec)	1.014
horsepower-hr	Btu	2,547
horsepower-hr	ergs	2.6845×10^{13}
horsepower-hr	foot-lb	1.98×10^{6}
horsepower-hr	gram-calories	641,190
horsepower-hr	joules	2.684×10^{6}
horsepower-hr	kg-calories	641.1
horsepower-hr	kg-meters	2.737×10^{5}
horsepower-hr	kilowatt-hr	0.7457
hours	days	4.167×10^{-2}
hours	weeks	5.952×10^{-3}
hundredweights (long)	pounds	112
hundredweights (long)	tons (long)	0.05
hundredweights (short)	ounces (avoirdupois)	1,600
hundredweights (short)	pounds	100
hundredweights (short)	tons (metric)	0.0453592
hundredweights (short)	tons (long)	0.0446429

I

To Convert	**Into**	**Multiply By**
inches	centimeters	2.540
inches	meters	2.540×10^{-2}
inches	miles	1.578×10^{-5}
inches	millimeters	25.40
inches	mils	1,000.0
inches	yards	2.778×10^{-2}
inches of mercury	atmospheres	0.03342
inches of mercury	feet of water	1.133
inches of mercury	kg/sq. cm	0.03453
inches of mercury	kg/sq. meter	345.3
inches of mercury	pounds/sq. ft	70.73
inches of mercury	pounds/sq. in	0.4912
inches of water (at 4°C)	atmospheres	2.458×10^{-3}
inches of water (at 4°C)	inches of mercury	0.07355
inches of water (at 4°C)	kg/sq. cm	2.540×10^{-3}
inches of water (at 4°C)	ounces/sq. in	0.5781
inches of water (at 4°C)	pounds/sq. ft	5.204
inches of water (at 4°C)	pounds/sq. in	0.03613

international ampere	ampere (absolute)	0.9998
international Volt	volts (absolute)	1.0003
international volt	joules (absolute)	1.593×10^{-19}
international volt	joules	9.654×10^4

J

To Convert	Into	Multiply By
joules	Btu	9.480×10^{-4}
joules	ergs	10^7
joules	foot-pounds	0.7376
joules	kg-calories	2.389×10^{-4}
joules	kg-meters	0.1020
joules	watt-hr	2.778×10^{-4}
joules/cm	grams	1.020×10^4
joules/cm	dynes	10^7
joules/cm	joules/meter (newtons)	100.0
joules/cm	poundals	723.3
joules/cm	pounds	22.48

K

To Convert	Into	Multiply By
kilogram-calories	Btu	3.968
kilogram-calories	foot-pounds	3,088
kilogram-calories	hp-hr	1.560×10^{-3}
kilogram-calories	joules	4,186
kilogram-calories	kg-meters	426.9
kilogram-calories	kilojoules	4.186
kilogram-calories	kilowatt-hr	1.163×10^{-3}
kilogram meters	Btu	9.294×10^{-3}
kilogram meters	ergs	9.804×10^7
kilogram meters	foot-pounds	7.233
kilogram meters	joules	9.804
kilogram meters	kg-calories	2.342×10^{-3}
kilogram meters	kilowatt-hr	2.723×10^{-6}
kilograms	dynes	980,665
kilograms	grams	1,000.0
kilograms	joules/cm	0.09807
kilograms	joules/meter (newtons)	9.807
kilograms	poundals	70.93
kilograms	pounds	2.205
kilograms	tons (long)	9.842×10^{-4}
kilograms	tons (short)	1.102×10^{-3}
kilograms/cubic meter	grams/cubic cm	0.001
kilograms/cubic meter	pounds/cubic ft	0.06243
kilograms/cubic meter	pounds/cubic in	3.613×10^{-5}
kilograms/cubic meter	pounds/mil-foot	3.405×10^{-10}
kilograms/meter	pounds/ft	0.6720
kilograms/sq. cm	dynes	980,665
kilograms/sq. cm	atmospheres	0.9678
kilograms/sq. cm	feet of water	32.81

kilograms/sq. cm	inches of mercury	28.96
kilograms/sq. cm	pounds/sq. ft	2,048
kilograms/sq. cm	pounds/sq. in	14.22
kilograms/sq. meter	atmospheres	9.678×10^{-5}
kilograms/sq. meter	bars	98.07×10^{-6}
kilograms/sq. meter	feet of water	3.281×10^{-3}
kilograms/sq. meter	inches of mercury	2.896×10^{-3}
kilograms/sq. meter	pounds/sq. ft	0.2048
kilograms/sq. meter	pounds/sq. in	1.422×10^{-3}
kilograms/sq. mm	kg/sq. meter	10^{6}
kilolines	maxwells	1,000.0
kiloliters	liters	1,000.0
kilometers	centimeters	10^{5}
kilometers	feet	3,281
kilometers	inches	3.937×10^{4}
kilometers	meters	1,000.0
kilometers	miles	0.6214
kilometers	millimeters	10^{4}
kilometers	yards	1,094
kilometers/hr	cm/sec	27.78
kilometers/hr	feet/min	54.68
kilometers/hr	feet/sec	0.9113
kilometers/hr	knots	0.5396
kilometers/hr	meters/min	16.67
kilometers/hr	miles/hr	0.6214
kilometers/hr/sec	cm/sec/sec	27.78
kilometers/hr/sec	feet/sec/sec	0.9113
kilometers/hr/sec	meters/sec/sec	0.2778
kilometers/hr/sec	miles/hr/sec	0.6214
kilowatt-hr	Btu	3,413
kilowatt-hr	ergs	3.600×10^{13}
kilowatt-hr	foot-lb	2.655×10^{6}
kilowatt-hr	gram-calories	859,850
kilowatt-hr	horsepower-hr	1.341
kilowatt-hr	joules	3.6×10^{6}
kilowatt-hr	kg-calories	860.5
kilowatt-hr	kg-meters	3.671×10^{5}
kilowatt-hr	pounds of water raised from 62° to 212°F	22.75
kilowatts	Btu/min	56.92
kilowatts	foot-lb/min	4.426×10^{4}
kilowatts	foot-lb/sec	737.6
kilowatts	horsepower	1.341
kilowatts	kg-calories/min	14.34
kilowatts	watts	1,000.0
knots	feet/hr	6,080
knots	kilometers/hr	1.8532
knots	nautical miles/hr	1.0
knots	statute miles/hr	1.151
knots	yards/hr	2,027

knots	feet/sec	1.689

L

To Convert	Into	Multiply By
league	miles (approx.)	3.0
light year	miles	5.9×10^{12}
light year	kilometers	9.4637×10^{12}
lines/sq. cm	gausses	1.0
lines/sq. in	gausses	0.1550
lines/sq. in	webers/sq. cm	1.550×10^{-9}
lines/sq. in	webers/sq. in	10^{-8}
lines/sq. in	webers/sq. meter	1.550×10^{-5}
links (engineer's)	inches	12.0
links (surveyor's)	inches	7.92
liters	bushels (U.S. dry)	0.02838
liters	cubic cm	1,000.0
liters	cubic feet	0.03531
liters	cubic inches	61.02
liters	cubic meters	0.001
liters	cubic yards	1.308×10^{-3}
liters	gallons (U.S. liq.)	0.2642
liters	pints (U.S. liq.)	2.113
liters	quarts (U.S. liq.)	1.057
liters/min	cubic ft/sec	5.886×10^{-4}
liters/min	gal/sec	4.403×10^{-3}
lumen	spherical candle power	0.07958
lumen	watt	0.001496
lumens/sq. ft	foot-candles	1.0
lumens/sq. ft	lumen/sq. meter	10.76
lux	foot-candles	0.0929

M

To Convert	Into	Multiply By
maxwells	kilolines	0.001
maxwells	webers	10^{-8}
megalines	maxwells	10^{6}
megohms	microhms	10^{12}
megohms	ohms	10^{6}
meter-kilograms	cm-dynes	9.807×10^{7}
meter-kilograms	cm-grams	10^{5}
meter-kilograms	pound-feet	7.233
meters	centimeters	100.0
meters	feet	3.281
meters	inches	39.37
meters	kilometers	0.001
meters	miles (naut.)	5.396×10^{-4}
meters	miles (stat.)	6.214×10^{-4}
meters	millimeters	1,000.0
meters	yards	1.094
meters	varas	1.179

meters/min	cm/sec	1,667
meters/min	feet/min	3.281
meters/min	feet/sec	0.05468
meters/min	km/hr	0.06
meters/min	knots	0.03238
meters/min	miles/hr	0.03728
meters/sec	feet/min	196.8
meters/sec	feet/sec	3.281
meters/sec	kilometers/hr	3.6
meters/sec	kilometers/min	0.06
meters/sec	miles/hr	2.237
meters/sec	miles/min	0.03728
meters/sec/sec	cm/sec/sec	100.0
meters/sec/sec	ft/sec/sec	3.281
meters/sec/sec	km/hr/sec	3.6
meters/sec/sec	miles/hr/sec	2.237
microfarad	farads	10^{-6}
micrograms	grams	10^{-6}
microhms	megohms	10^{-12}
microhms	ohms	10^{-6}
microliters	liters	10^{-6}
microns	meters	1×10^{-6}
miles (naut.)	feet	6,080.27
miles (naut.)	kilometers	1.853
miles (naut.)	meters	1,853
miles (naut.)	miles (statute)	1.1516
miles (naut.)	yards	2,027
miles (statute)	centimeters	1.609×10^{5}
miles (statute)	feet	5,280
miles (statute)	inches	6.336×10^{4}
miles (statute)	kilometers	1.609
miles (statute)	meters	1,609
miles (statute)	miles (naut.)	0.8684
miles (statute)	yards	1,760
miles/hr	cm/sec	44.70
miles/hr	feet/min	88
miles/hr	feet/sec	1.467
miles/hr	km/hr	1.609
miles/hr	km/min	0.02682
miles/hr	knots	0.8684
miles/hr	meters/min	26.82
miles/hr	miles/min	0.1667
miles/hr/sec	cm/sec/sec	44.70
miles/hr/sec	feet/sec/sec	1.467
miles/hr/sec	km/hr/sec	1.609
miles/hr/sec	meters/sec/sec	0.4470
miles/min	cm/sec	2,682
miles/min	feet/sec	88
miles/min	km/min	1.609
miles/min	knots/min	0.8684

miles/min	miles/hr	60
mil-feet	cubic inches	9.425×10^{-6}
milliers	kilograms	1,000
milligrams	grains	0.01543236
milligrams	grams	0.001
milligrams/liter	parts/million	1.0
millihenries	henries	0.001
milliliters	liters	0.001
millimeters	centimeters	0.1
millimeters	feet	3.281×10^{-3}
millimeters	inches	0.03937
millimeters	kilometers	10^{-6}
millimeters	meters	0.001
millimeters	miles	6.214×10^{-7}
millimeters	mils	39.37
millimeters	yards	1.094×10^{-3}
millimicrons	meters	1×10^{-9}
million gal/day	cubic ft/sec	1.54723
mils	centimeters	2.540×10^{-3}
mils	feet	8.333×10^{-5}
mils	inches	0.001
mils	kilometers	2.540×10^{-8}
mils	yards	2.778×10^{-5}
miner's inches	cubic ft/min	1.5
minims (British)	cubic cm	0.059192
minims (U.S., fluid)	cubic cm	0.061612
minutes (angles)	degrees	0.01667
minutes (angles)	quadrants	1.852×10^{-4}
minutes (angles)	radians	2.909×10^{-4}
minutes (angles)	seconds	60.0
myriagrams	kilograms	10.0
myriameters	kilometers	10.0
myriawatts	kilowatts	10.0

N

To Convert	Into	Multiply By
nepers	decibels	8.686
Newton	dynes	1 x 105

O

To Convert	Into	Multiply By
ohm (international)	ohm (absolute)	1.0005
ohms	megohms	10^{-6}
ohms	microhms	10^{6}
ounces	drams	16.0
ounces	grains	437.5
ounces	grams	28.349527
ounces	pounds	0.0625
ounces	ounces (troy)	0.9115
ounces	tons (long)	2.790×10^{-5}

ounces	tons (metric)	2.835×10^{-5}
ounces (fluid)	cubic inches	1.805
ounces (fluid)	liters	0.02957
ounces (troy)	grains	480.0
ounces (troy)	grams	31.103481
ounces (troy)	ounces (avdp.)	1.09714
ounces (troy)	pennyweights (troy)	20.0
ounces (troy)	pounds (troy)	0.08333
ounces/sq. inch	dynes/sq. cm	4,309
ounces/sq. in	pounds/sq. in	0.0625

P

To Convert	Into	Multiply By
parsec	miles	19×10^{12}
parsec	kilometers	3.084×10^{13}
parts/million	grains/U.S. gal	0.0584
parts/million	grains/Imp. gal	0.07016
parts/million	pounds/million gal	8.345
pecks (British)	cubic inches	554.6
pecks (British)	liters	9.091901
pecks (U.S.)	bushels	0.25
pecks (U.S.)	cubic inches	537.605
pecks (U.S.)	liters	8.809582
pecks (U.S.)	quarts (dry)	8
pennyweights (troy)	grains	24.0
pennyweights (troy)	ounces (troy)	0.05
pennyweights (troy)	grams	1.55517
pennyweights (troy)	pounds (troy)	4.1667×10^{-3}
pints (dry)	cubic inches	33.60
pints (liq.)	cubic cm	473.2
pints (liq.)	cubic feet	0.01671
pints (liq.)	cubic inches	28.87
pints (liq.)	cubic meters	4.732×10^{-4}
pints (liq.)	cubic yards	6.189×10^{-4}
pints (liq.)	gallons	0.125
pints (liq.)	liters	0.4732
pints (liq.)	quarts (liq.)	0.5
Planck's quantum	erg - second	6.624×10^{-27}
poise	gram/cm sec	1.00
poundals	dynes	13,826
poundals	grams	14.10
poundals	joules/cm	1.383×10^{-3}
poundals	joules/meter (newtons)	0.1383
poundals	kilograms	0.01410
poundals	pounds	0.03108
pound-feet	cm-dynes	1.356×10^{7}
pound-feet	cm-grams	13,825
pound-feet	meter-kg	0.1383
pounds	drams	256
pounds	dynes	44.4823×10^{4}

pounds	grains	7,000
pounds	grams	453.5924
pounds	joules/cm	0.04448
pounds	joules/meter (newtons)	4.448
pounds	kilograms	0.4536
pounds	ounces	16.0
pounds	ounces (troy)	14.5833
pounds	poundals	32.17
pounds	pounds (troy)	1.21528
pounds	tons (short)	0.0005
pounds (avoirdupois)	ounces (troy)	14.5833
pounds (troy)	grains	5,760
pounds (troy)	grams	373.24177
pounds (troy)	ounces (avdp.)	13.1657
pounds (troy)	ounces (troy)	12.0
pounds (troy)	pennyweights (troy)	240.0
pounds (troy)	pounds (avdp.)	0.822857
pounds (troy)	tons (long)	3.6735×10^{-4}
pounds (troy)	tons (metric)	3.7324×10^{-4}
pounds (troy)	tons (short)	4.1143×10^{-4}
pounds of water	cubic ft	0.01602
pounds of water	cubic inches	27.68
pounds of water	gallons	0.1198
pounds of water/min	cubic ft/sec	2.670×10^{-4}
pounds/cubic ft	grams/cubic cm	0.01602
pounds/cubic ft	kg/cubic meter	16.02
pounds/cubic ft	pounds/cubic in	5.787×10^{-4}
pounds/cubic ft	pounds/mil-foot	5.456×10^{-9}
pounds/cubic in	gm/cubic cm	27.68
pounds/cubic in	kg/cubic meter	2.768×10^{4}
pounds/cubic in	pounds/cubic ft	1,728
pounds/cubic in	pounds/mil-foot	9.425×10^{-6}
pounds/ft	kg/meter	1.488
pounds/in	gm/cm	178.6
pounds/mil-foot	gm/cubic cm	2.306×10^{6}
pounds/sq. ft	atmospheres	4.725×10^{-4}
pounds/sq. ft	feet of water	0.01602
pounds/sq. ft	inches of mercury	0.01414
pounds/sq. ft	kg/sq. meter	4.882
pounds/sq. ft	pounds/sq. in	6.944×10^{-3}
pounds/sq. in	atmospheres	0.06804
pounds/sq. in	feet of water	2.307
pounds/sq. in	inches of mercury	2.036
pounds/sq. in	kg/sq. meter	703.1
pounds/sq. in	pounds/sq. ft	144.0

Q

To Convert	Into	Multiply By
quadrants (angle)	degrees	90.0
quadrants (angle)	minutes	5,400.0

quadrants (angle)	radians	1.571
quadrants (angle)	seconds	3.24×10^5
quarts (dry)	cubic inches	67.20
quarts (liq.)	cubic cm	946.4
quarts (liq.)	cubic feet	0.03342
quarts (liq.)	cubic inches	57.75
quarts (liq.)	cubic meters	9.464×10^{-4}
quarts (liq.)	cubic yards	1.238×10^{-3}
quarts (liq.)	gallons	0.25
quarts (liq.)	liters	0.9463

R

To Convert	Into	Multiply By
radians	degrees	57.30
radians	minutes	3,438
radians	quadrants	0.6366
radians	seconds	2.063×10^5
radians/sec	degrees/sec	57.30
radians/sec	revolutions/min	9.549
radians/sec	revolutions/sec	0.1592
radians/sec/sec	revolutions/min/min	573.0
radians/sec/sec	revolutions/min/sec	9.549
radians/sec/sec	revolutions/sec/sec	0.1592
revolutions	degrees	360.0
revolutions	quadrants	4.0
revolutions	radians	6.283
revolutions/min	degrees/sec	6.0
revolutions/min	radians/sec	0.1047
revolutions/min	revolutions/sec	0.01667
revolutions/min/min	radians/sec/sec	1.745×10^{-3}
revolutions/min/min	revolutions/min/sec	0.01667
revolutions/min/min	revolutions/sec/sec	2.778×10^{-4}
revolutions/sec	degrees/sec	360.0
revolutions/sec	radians/sec	6.283
revolutions/sec	revolutions/min	60.0
revolutions/sec/sec	radians/sec/sec	6.283
revolutions/sec/sec	revolutions/min/min	3,600.0
revolutions/sec/sec	revolutions/min/sec	60.0
rod	chain (Gunters)	0.25
rod	meters	5.029
rods	feet	16.5
rods (surveyors' meas.)	yards	5.5

S

To Convert	Into	Multiply By
scruples	grains	20
seconds (angle)	degrees	2.778×10^{-4}
seconds (angle)	minutes	0.01667
seconds (angle)	quadrants	3.087×10^{-6}
seconds (angle)	radians	4.848×10^{-6}

slug	kilogram	14.59
slug	pounds	32.17
sphere	steradians	12.57
square centimeters	circular mils	1.973×10^{5}
square centimeters	sq. feet	1.076×10^{-3}
square centimeters	sq. inches	0.1550
square centimeters	sq. meters	0.0001
square centimeters	sq. miles	3.861×10^{-11}
square centimeters	sq. millimeters	100.0
square centimeters	sq. yards	1.196×10^{-4}
square feet	acres	2.296×10^{-5}
square feet	circular mils	1.833×10^{8}
square feet	sq. cm	929.0
square feet	sq. inches	144.0
square feet	sq. meters	0.09290
square feet	sq. miles	3.587×10^{-8}
square feet	sq. millimeters	9.290×10^{4}
square feet	sq. yards	0.1111
square inches	circular mils	1.273×10^{6}
square inches	sq. cm	6.452
square inches	sq. feet	6.944×10^{-3}
square inches	sq. millimeters	645.2
square inches	sq. mils	10^{6}
square inches	sq. yards	7.716×10^{-4}
square kilometers	acres	247.1
square kilometers	sq. cm	10^{10}
square kilometers	sq. ft	10.76×10^{6}
square kilometers	sq. inches	1.550×10^{9}
square kilometers	sq. meters	10^{6}
square kilometers	sq. miles	0.3861
square kilometers	sq. yards	1.196×10^{6}
square meters	acres	2.471×10^{-4}
square meters	sq. cm	10^{4}
square meters	sq. feet	10.76
square meters	sq. inches	1,550
square meters	sq. miles	3.861×10^{-7}
square meters	sq. millimeters	10^{6}
square meters	sq. yards	1.196
square miles	acres	640.0
square miles	sq. feet	27.88×10^{6}
square miles	sq. km	2.590
square miles	sq. meters	2.590×10^{6}
square miles	sq. yards	3.098×10^{6}
square millimeters	circular mils	1,973
square millimeters	sq. cm	0.01
square millimeters	sq. feet	1.076×10^{-5}
square millimeters	sq. inches	1.550×10^{-3}
square mils	circular mils	1.273
square mils	sq. cm	6.452×10^{-6}
square mils	sq. inches	10^{-6}

square yards	acres	2.066×10^{-4}
square yards	sq. cm	8,361
square yards	sq. feet	9.0
square yards	sq. inches	1,296
square yards	sq. meters	0.8361
square yards	sq. miles	3.228×10^{-7}
square yards	sq. millimeters	8.361×10^{5}

T

To Convert	Into	Multiply By
temperature (°C)+273	absolute temperature (°C)	1.0
temperature (°C)+17.78	temperature (°F)	1.8
temperature (°F)+460	absolute temperature (°F)	1.0
temperature (°F)−32	temperature (°C)	5/9
tons (long)	kilograms	1,016
tons (long)	pounds	2,240
tons (long)	tons (short)	1.120
tons (metric)	kilograms	1,000
tons (metric)	pounds	2,205
tons (short)	kilograms	907.1848
tons (short)	ounces	32,000
tons (short)	ounces (troy)	29,166.66
tons (short)	pounds	2,000
tons (short)	pounds (troy)	2,430.56
tons (short)	tons (long)	0.89287
tons (short)	tons (metric)	0.9078
tons (short)/sq. ft	kg/sq. meter	9,765
tons (short)/sq. ft	pounds/sq. in	2,000
tons of water/24 hr	pounds of water/hr	83.333
tons of water/24 hr	gallons/min	0.16643
tons of water/24 hr	cubic ft/hr	1.3349

V

To Convert	Into	Multiply By
volt (absolute)	statvolts	0.003336
volt/inch	volt/cm	0.39370

W

To Convert	Into	Multiply By
watt-hours	Btu	3.413
watt-hours	ergs	3.60×10^{10}
watt-hours	foot-pounds	2,656
watt-hours	gram-calories	859.85
watt-hours	horsepower-hr	1.341×10^{-3}
watt-hours	kilogram-calories	0.8605
watt-hours	kilogram-meters	367.2
watt-hours	kilowatt-hr	0.001
watt (international)	watt (absolute)	1.0002
watts	Btu/hr	3.4129
watts	Btu/min	0.05688

watts	ergs/sec	107
watts	foot-lb/min	44.27
watts	foot-lb/sec	0.7378
watts	horsepower	1.341×10^{-3}
watts	horsepower (metric)	1.360×10^{-3}
watts	kg-calories/min	0.01433
watts	kilowatts	0.001
watts (Abs.)	Btu (mean)/min	0.056884
watts (Abs.)	joules/sec	1
webers	maxwells	10^8
webers	kilolines	10^5
webers/sq. in	gausses	1.550×10^7
webers/sq. in	lines/sq. in	10^8
webers/sq. in	webers/sq. cm	0.1550
webers/sq. in	webers/sq. meter	1,550
webers/sq. meter	gausses	10^4
webers/sq. meter	lines/sq. in	6.452×10^4
webers/sq. meter	webers/sq. cm	10^{-4}
webers/sq. meter	webers/sq. in	6.452×10^{-4}

Y

To Convert	Into	Multiply By
yards	centimeters	91.44
yards	kilometers	9.144×10^{-4}
yards	meters	0.9144
yards	miles (naut.)	4.934×10^{-4}
yards	miles (stat.)	5.682×10^{-4}
yards	millimeters	914.4

10.5 Reference Tables

Table 10.1 Power Conversion Factors (decibels to watts)

dBm	dBw	Watts		Multiple Prefix
+150	+120	1,000,000,000,000	10^{12}	1 Terawatt
+140	+110	100,000,000,000	10^{11}	100 Gigawatts
+130	+100	10,000,000,000	10^{10}	10 Gigawatts
+120	+90	1,000,000,000	10^{9}	1 Gigawatt
+110	+80	100,000,000	10^{8}	100 Megawatts
+100	+70	10,000,000	10^{7}	10 Megawatts
+90	+60	1,000,000	10^{6}	1 Megawatt
+80	+50	100,000	10^{5}	100 Kilowatts
+70	+40	10,000	10^{4}	10 Kilowatts
+60	+30	1,000	10^{3}	1 Kilowatt
+50	+20	100	10^{2}	1 Hectrowatt
+40	+10	10	10	1 Decawatt
+30	0	1	1	1 Watt
+20	−10	0.1	10^{-1}	1 Deciwatt
+10	−20	0.01	10^{-2}	1 Centiwatt
0	−30	0.001	10^{-3}	1 Milliwatt
−10	−40	0.0001	10^{-4}	100 Microwatts
−20	−50	0.00001	10^{-5}	10 Microwatts
−30	−60	0.000,001	10^{-6}	1 Microwatt
−40	−70	0.0,000,001	10^{-7}	100 Nanowatts
−50	−80	0.00,000,001	10^{-8}	10 Nanowatts
−60	−90	0.000,000,001	10^{-9}	1 Nanowatt
−70	−100	0.0,000,000,001	10^{-10}	100 Picowatts
−80	−110	0.00,000,000,001	10^{-11}	10 Picowatts
−90	−120	0,000,000,000,001	10^{-12}	1 Picowatt

Table 10.2 Relationships of Voltage Standing Wave Ratio and Key Operating Parameters

SWR	Reflection Coefficient	Return Loss	Power Ratio	Percent Reflected
1.01:1	0.0050	46.1 dB	0.00002	0.002
1.02:1	0.0099	40.1 dB	0.00010	0.010
1.04:1	0.0196	34.2 dB	0.00038	0.038
1.06:1	0.0291	30.7 dB	0.00085	0.085
1.08:1	0.0385	28.3 dB	0.00148	0.148
1.10:1	0.0476	26.4 dB	0.00227	0.227
1.20:1	0.0909	20.8 dB	0.00826	0.826
1.30:1	0.1304	17.7 dB	0.01701	1.7
1.40:1	0.1667	15.6 dB	0.02778	2.8
1.50:1	0.2000	14.0 dB	0.04000	4.0
1.60:1	0.2308	12.7 dB	0.05325	5.3
1.70:1	0.2593	11.7 dB	0.06722	6.7
1.80:1	0.2857	10.9 dB	0.08163	8.2
1.90:1	0.3103	10.2 dB	0.09631	9.6
2.00:1	0.3333	9.5 dB	0.11111	11.1
2.20:1	0.3750	8.5 dB	0.14063	14.1
2.40:1	0.4118	7.7 dB	0.16955	17.0
2.60:1	0.4444	7.0 dB	0.19753	19.8
2.80:1	0.4737	6.5 dB	0.22438	22.4
3.00:1	0.5000	6.0 dB	0.25000	25.0
3.50:1	0.5556	5.1 dB	0.30864	30.9
4.00:1	0.6000	4.4 dB	0.36000	36.0
4.50:1	0.6364	3.9 dB	0.40496	40.5
5.00:1	0.6667	3.5 dB	0.44444	44.4
6.00:1	0.7143	2.9 dB	0.51020	51.0
7.00:1	0.7500	2.5 dB	0.56250	56.3
8.00:1	0.7778	2.2 dB	0.60494	60.5
9.00:1	0.8000	1.9 dB	0.64000	64.0
10.00:1	0.8182	1.7 dB	0.66942	66.9
15.00:1	0.8750	1.2 dB	0.76563	76.6
20.00:1	0.9048	0.9 dB	0.81859	81.9
30.00:1	0.9355	0.6 dB	0.87513	97.5
40.00:1	0.9512	0.4 dB	0.90482	90.5
50.00:1	0.9608	0.3 dB	0.92311	92.3

Table 10.3 Specifications of Standard Copper Wire Sizes

Wire Size AWG	Diam. in mils	Circular mil Area	Turns per Linear Inch[1] Enam.	SCE	DCC	Ohms per 100 ft[2]	Current Carrying Capacity[3]	Diam. in mm
1	289.3	83810	-	-	-	0.1239	119.6	7.348
2	257.6	05370	-	-	-	0.1563	94.8	6.544
3	229.4	62640	-	-	-	0.1970	75.2	5.827
4	204.3	41740	-	-	-	0.2485	59.6	5.189
5	181.9	33100	-	-	-	0.3133	47.3	4.621
6	162.0	26250	-	-	-	0.3951	37.5	4.115
7	144.3	20820	-	-	-	0.4982	29.7	3.665
8	128.5	16510	7.6	-	7.1	0.6282	23.6	3.264
9	114.4	13090	8.6	-	7.8	0.7921	18.7	2.906
10	101.9	10380	9.6	9.1	8.9	0.9989	14.8	2.588
11	90.7	8234	10.7	-	9.8	1.26	11.8	2.305
12	80.8	6530	12.0	11.3	10.9	1.588	9.33	2.063
13	72.0	5178	13.5	-	12.8	2.003	7.40	1.828
14	64.1	4107	15.0	14.0	13.8	2.525	5.87	1.628
15	57.1	3257	16.8	-	14.7	3.184	4.65	1.450
16	50.8	2583	18.9	17.3	16.4	4.016	3.69	1.291
17	45.3	2048	21.2	-	18.1	5.064	2.93	1.150
18	40.3	1624	23.6	21.2	19.8	6.386	2.32	1.024
19	35.9	1288	26.4	-	21.8	8.051	1.84	0.912
20	32.0	1022	29.4	25.8	23.8	10.15	1.46	0.812
21	28.5	810	33.1	-	26.0	12.8	1.16	0.723
22	25.3	642	37.0	31.3	30.0	16.14	0.918	0.644
23	22.6	510	41.3	-	37.6	20.36	0.728	0.573
24	20.1	404	46.3	37.6	35.6	25.67	0.577	0.511
25	17.9	320	51.7	-	38.6	32.37	0.458	0.455
26	15.9	254	58.0	46.1	41.8	40.81	0.363	0.406
27	14.2	202	64.9	-	45.0	51.47	0.288	0.361
28	12.6	160	72.7	54.6	48.5	64.9	0.228	0.321
29	11.3	127	81.6	-	51.8	81.83	0.181	0.286
30	10.0	101	90.5	64.1	55.5	103.2	0.144	0.255
31	8.9	50	101	-	59.2	130.1	0.114	0.227
32	8.0	63	113	74.1	61.6	164.1	0.090	0.202
33	7.1	50	127	-	66.3	206.9	0.072	0.180
34	6.3	40	143	86.2	70.0	260.9	0.057	0.160
35	5.6	32	158	-	73.5	329.0	0.045	0.143
36	5.0	25	175	103.1	T7.0	414.8	0.036	0.127
37	4.5	20	198	-	80.3	523.1	0.028	0.113
38	4.0	16	224	116.3	83.6	659.6	0.022	0.101
39	3.5	12	248	-	86.6	831.8	0.018	0.090

1. Based on 25.4 mm.
2. Ohms per 1,000 ft measured at 20°C.
3. Current-carrying capacity at 700 cm/amp.

Table 10.4 Celcius-to-Fahrenheit Conversion Table

°Celsius	°Fahrenheit	°Celsius	°Fahrenheit
−50	−58	125	257
−45	−49	130	266
−40	−40	135	275
−35	−31	140	284
−30	−22	145	293
−25	−13	150	302
−20	4	155	311
−15	5	160	320
−10	14	165	329
−5	23	170	338
0	32	175	347
5	41	180	356
10	50	185	365
15	59	190	374
20	68	195	383
25	77	200	392
30	86	205	401
35	95	210	410
40	104	215	419
45	113	220	428
50	122	225	437
55	131	230	446
60	140	235	455
65	149	240	464
70	158	245	473
75	167	250	482
80	176	255	491
85	185	260	500
90	194	265	509
95	203	270	518
100	212	275	527
105	221	280	536
110	230	285	545
115	239	290	554
120	248	295	563

Table 10.5 Inch-to-Millimeter Conversion Table

Inch	0	1/8	1/4	3/8	1/2	5/8	3/4	7/8	Inch
0	0.0	3.18	6.35	9.52	12.70	15.88	19.05	22.22	0
1	25.40	28.58	31.75	34.92	38.10	41.28	44.45	47.62	1
2	50.80	53.98	57.15	60.32	63.50	66.68	69.85	73.02	2
3	76.20	79.38	82.55	85.72	88.90	92.08	95.25	98.42	3
4	101.6	104.8	108.0	111.1	114.3	117.5	120.6	123.8	4
5	127.0	130.2	133.4	136.5	139.7	142.9	146.0	149.2	5
6	152.4	155.6	158.8	161.9	165.1	168.3	171.4	174.6	6
7	177.8	181.0	184.2	187.3	190.5	193.7	196.8	200.0	7
8	203.2	206.4	209.6	212.7	215.9	219.1	222.2	225.4	8
9	228.6	231.8	235.0	238.1	241.3	244.5	247.6	250.8	9
10	254.0	257.2	260.4	263.5	266.7	269.9	273.0	276.2	10
11	279	283	286	289	292	295	298	302	11
12	305	308	311	314	317	321	324	327	12
13	330	333	337	340	343	346	349	352	13
14	356	359	362	365	368	371	375	378	14
15	381	384	387	391	394	397	400	403	15
16	406	410	413	416	419	422	425	429	16
17	432	435	438	441	445	448	451	454	17
18	457	460	464	467	470	473	476	479	18
19	483	486	489	492	495	498	502	505	19
20	508	511	514	518	521	524	527	530	20

Table 10.6 Conversion of Millimeters to Decimal Inches

mm	Inches	mm	Inches	mm	Inches
1	0.039370	46	1.811020	91	3.582670
2	0.078740	47	1.850390	92	3.622040
3	0.118110	48	1.889760	93	3.661410
4	0.157480	49	1.929130	94	3.700780
5	0.196850	50	1.968500	95	3.740150
6	0.236220	51	2.007870	96	3.779520
7	0.275590	52	2.047240	97	3.818890
8	0.314960	53	2.086610	98	3.858260
9	0.354330	54	2.125980	99	3.897630
10	0.393700	55	2.165350	100	3.937000
11	0.433070	56	2.204720	105	4.133848
12	0.472440	57	2.244090	110	4.330700
13	0.511810	58	2.283460	115	4.527550
14	0.551180	59	2.322830	120	4.724400
15	0.590550	60	2.362200	125	4.921250
16	0.629920	61	2.401570	210	8.267700
17	0.669290	62	2.440940	220	8.661400
18	0.708660	63	2.480310	230	9.055100
19	0.748030	64	2.519680	240	9.448800
20	0.787400	65	2.559050	250	9.842500
21	0.826770	66	2.598420	260	10.236200
22	0.866140	67	2.637790	270	10.629900
23	0.905510	68	2.677160	280	11.032600
24	0.944880	69	2.716530	290	11.417300
25	0.984250	70	2.755900	300	11.811000
26	1.023620	71	2.795270	310	12.204700
27	1.062990	72	2.834640	320	12.598400
28	1.102360	73	2.874010	330	12.992100
29	1.141730	74	2.913380	340	13.385800
30	1.181100	75	2.952750	350	13.779500
31	1.220470	76	2.992120	360	14.173200
32	1.259840	77	3.031490	370	14.566900
33	1.299210	78	3.070860	380	14.960600
34	1.338580	79	3.110230	390	15.354300
35	1.377949	80	3.149600	400	15.748000
36	1.417319	81	3.188970	500	19.685000
37	1.456689	82	3.228340	600	23.622000
38	1.496050	83	3.267710	700	27.559000
39	1.535430	84	3.307080	800	31.496000
40	1.574800	85	3.346450	900	35.433000
41	1.614170	86	3.385820	1000	39.370000
42	1.653540	87	3.425190	2000	78.740000
43	1.692910	88	3.464560	3000	118.110000
44	1.732280	89	3.503903	4000	157.480000
45	1.771650	90	3.543300	5000	196.850000

Table 10.7 Convertion of Common Fractions to Decimal and Millimeter Units

Common Fractions	Decimal Fractions	mm (approx.)	Common Fractions	Decimal Fractions	mm (appox.)
1/128	0.008	0.20	1/2	0.500	12.70
1/64	0.016	0.40	33/64	0.516	13.10
1/32	0.031	0.79	17/32	0.531	13.49
3/64	0.047	1.19	35/64	0.547	13.89
1/16	0.063	1.59	9/16	0.563	14.29
5/64	0.078	1.98	37/64	0.578	14.68
3/32	0.094	2.38	19/32	0.594	15.08
7/64	0.109	2.78	39/64	0.609	15.48
1/8	0.125	3.18	5/8	0.625	15.88
9/64	0.141	3.57	41/64	0.641	16.27
5/32	0.156	3.97	21/32	0.656	16.67
11/64	0.172	4.37	43/64	0.672	17.07
3/16	0.188	4.76	11/16	0.688	17.46
13/64	0.203	5.16	45/64	0.703	17.86
7/32	0.219	5.56	23/32	0.719	18.26
15/64	0.234	5.95	47/64	0.734	18.65
1/4	0.250	6.35	3/4	0.750	19.05
17/64	0.266	6.75	49/64	0.766	19.45
9/32	0.281	7.14	25/32	0.781	19.84
19/64	0.297	7.54	51/64	0.797	20.24
5/16	0.313	7.94	13/16	0.813	20.64
21/64	0.328	8.33	53/64	0.828	21.03
11/32	0.344	8.73	27/32	0.844	21.43
23/64	0.359	9.13	55/64	0.859	21.83
3/8	0.375	9.53	7/8	0.875	22.23
25/64	0.391	9.92	57/64	0.891	22.62
13/32	0.406	10.32	29/32	0.906	23.02
27/64	0.422	10.72	59/64	0.922	23.42
7/16	0.438	11.11	15/16	0.938	23.81
29/64	0.453	11.51	61/64	0.953	24.21
15/32	0.469	11.91	31/32	0.969	24.61
31/64	0.484	12.30	63/64	0.984	25.00

Table 10.8 Decimal Equivalent Size of Drill Numbers

Drill no.	Decimal Equiv.	Drill no.	Decimal Equiv.	Drill no.	Decimal Equiv.
80	0.0135	53	0.0595	26	0.1470
79	0.0145	52	0.0635	25	0.1495
78	0.0160	51	0.0670	24	0.1520
77	0.0180	50	0.0700	23	0.1540
76	0.0200	49	0.0730	22	0.1570
75	0.0210	48	0.0760	21	0.1590
74	0.0225	47	0.0785	20	0.1610
73	0.0240	46	0.0810	19	0.1660
72	0.0250	45	0.0820	18	0.1695
71	0.0260	44	0.0860	17	0.1730
70	0.0280	43	0.0890	16	0.1770
69	0.0292	42	0.0935	15	0.1800
68	0.0310	41	0.0960	14	0.1820
67	0.0320	40	0.0980	13	0.1850
66	0.0330	39	0.0995	12	0.1890
65	0.0350	38	0.1015	11	0.1910
64	0.0360	37	0.1040	10	0.1935
63	0.0370	36	0.1065	9	0.1960
62	0.0380	35	0.1100	8	0.1990
61	0.0390	34	0.1110	7	0.2010
60	0.0400	33	0.1130	6	0.2040
59	0.0410	32	0.1160	5	0.2055
58	0.0420	31	0.1200	4	0.2090
57	0.0430	30	0.1285	3	0.2130
56	0.0465	29	0.1360	2	0.2210
55	0.0520	28	0.1405	1	0.2280
54	0.0550	27	0.1440		

Table 10.9 Decimal Equivalent Size of Drill Letters

Letter Drill	Decimal Equiv.	Letter Drill	Decimal Equiv.	Letter Drill	Decimal Equiv.
A	0.234	J	0.277	S	0.348
B	0.238	K	0.281	T	0.358
C	0.242	L	0.290	U	0.368
D	0.246	M	0.295	V	0.377
E	0.250	N	0.302	W	0.386
F	0.257	O	0.316	X	0.397
G	0.261	P	0.323	Y	0.404
H	0.266	Q	0.332	Z	0.413
I	0.272	R	0.339		

Table 10.10 Conversion Ratios for Length

Known Quantity	Multiply by	Quantity to Find
inches (in)	2.54	centimeters (cm)
feet (ft)	30	centimeters (cm)
yards (yd)	0.9	meters (m)
miles (mi)	1.6	kilometers (km)
millimeters (mm)	0.04	inches (in)
centimeters (cm)	0.4	inches (in)
meters (m)	3.3	feet (ft)
meters (m)	1.1	yards (yd)
kilometers (km)	0.6	miles (mi)
centimeters (cm)	10	millimeters (mm)
decimeters (dm)	10	centimeters (cm)
decimeters (dm)	100	millimeters (mm)
meters (m)	10	decimeters (dm)
meters (m)	1000	millimeters (mm)
dekameters (dam)	10	meters (m)
hectometers (hm)	10	dekameters (dam)
hectometers (hm)	100	meters (m)
kilometers (km)	10	hectometers (hm)
kilometers (km)	1000	meters (m)

Table 10.11 Conversion Ratios for Area

Known Quantity	Multiply by	Quantity to Find
square inches (in^2)	6.5	square centimeters (cm^2)
square feet (ft^2)	0.09	square meters (m^2)
square yards (yd^2)	0.8	square meters (m^2)
square miles (mi^2)	2.6	square kilometers (km^2)
acres	0.4	hectares (ha)
square centimeters (cm^2)	0.16	square inches (in^2)
square meters (m^2)	1.2	square yards (yd^2)
square kilometers (km^2)	0.4	square miles (mi^2)
hectares (ha)	2.5	acres
square centimeters (cm^2)	100	square millimeters (mm^2)
square meters (m^2)	10,000	square centimeters (cm^2)
square meters (m^2)	1,000,000	square millimeters (mm^2)
ares (a)	100	square meters (m^2)
hectares (ha)	100	ares (a)
hectares (ha)	10,000	square meters (m^2)
square kilometers (km^2)	100	hectares (ha)
square kilometers (km^2)	1,000	square meters (m^2)

Table 10.12 Conversion Ratios for Mass

Known Quantity	Multiply by	Quantity to Find
ounces (oz)	28	grams (g)
pounds (lb)	0.45	kilograms (kg)
tons	0.9	tonnes (t)
grams (g)	0.035	ounces (oz)
kilograms (kg)	2.2	pounds (lb)
tonnes (t)	100	kilograms (kg)
tonnes (t)	1.1	tons
centigrams (cg)	10	milligrams (mg)
decigrams (dg)	10	centigrams (cg)
decigrams (dg)	100	milligrams (mg)
grams (g)	10	decigrams (dg)
grams (g)	1000	milligrams (mg)
dekagram (dag)	10	grams (g)
hectogram (hg)	10	dekagrams (dag)
hectogram (hg)	100	grams (g)
kilograms (kg)	10	hectograms (hg)
kilograms (kg)	1000	grams (g)

Table 10.13 Conversion Ratios for Volume

Known Quantity	Multiply by	Quantity to Find
milliliters (mL)	0.03	fluid ounces (fl oz)
liters (L)	2.1	pints (pt)
liters (L)	1.06	quarts (qt)
liters (L)	0.26	gallons (gal)
gallons (gal)	3.8	liters (L)
quarts (qt)	0.95	liters (L)
pints (pt)	0.47	liters (L)
cups (c)	0.24	liters (L)
fluid ounces (fl oz)	30	milliliters (mL)
teaspoons (tsp)	5	milliliters (mL)
tablespoons (tbsp)	15	milliliters (mL)
liters (L)	100	milliliters (mL)

Table 10.14 Conversion Ratios for Cubic Measure

Known Quantity	Multiply by	Quantity to Find
cubic meters (m^3)	35	cubic feet (ft^3)
cubic meters (m^3)	1.3	cubic yards (yd^3)
cubic yards (yd^3)	0.76	cubic meters (m^3)
cubic feet (ft^3)	0.028	cubic meters (m^3)
cubic centimeters (cm^3)	1000	cubic millimeters (mm^3)
cubic decimeters (dm^3)	1000	cubic centimeters (cm^3)
cubic decimeters (dm^3)	1,000,000	cubic millimeters (mm^3)
cubic meters (m^3)	1000	cubic decimeters (dm^3)
cubic meters (m^3)	1	steres
cubic feet (ft^3)	1728	cubic inches (in^3)
cubic feet (ft^3)	28.32	liters (L)
cubic inches (in^3)	16.39	cubic centimeters (cm^3)
cubic meters (m^3)	264	gallons (gal)
cubic yards (yd^3)	27	cubic feet (ft^3)
cubic yards (yd^3)	202	gallons (gal)
gallons (gal)	231	cubic inches (in^3)

Table 10.15 Conversion Ratios for Electrical Quantities

Known Quantity	Multiply by	Quantity to Find
Btu per minute	0.024	horsepower (hp)
Btu per minute	17.57	watts (W)
horsepower (hp)	33,000	foot-pounds per min (ft-lb/min)
horsepower (hp)	746	watts (W)
kilowatts (kW)	57	Btu per minute
kilowatts (kW)	1.34	horsepower (hp)

10.6 International Standards and Constants

Table 10.16 Names and Symbols for the SI Base Units (*From* [1]. *Used with permission.*)

Physical quantity	Name of SI unit	Symbol for SI unit
length	meter	m
mass	kilogram	kg
time	second	s
electric current	ampere	A
thermodynamic temperature	kelvin	K
amount of substance	mole	mol
luminous intensity	candela	cd

Table 10.17 Units in Use Together with the SI (These units are not part of the SI, but it is recognized that they will continue to be used in appropriate contexts. From [1]. Used with permission.)

Physical quantity	Name of unit	Symbol for unit	Value in SI units
time	minute	min	60 s
time	hour	h	3600 s
time	day	d	86 400 s
plane angle	degree	°	$(\pi/180)$ rad
plane angle	minute	′	$(\pi/10\ 800)$ rad
plane angle	second	″	$(\pi/648\ 000)$ rad
length	ångström[a]	Å	10^{-10} m
area	barn	b	10^{-28} m^2
volume	litre	l, L	$dm^3 = 10^{-3} m^3$
mass	tonne	t	$Mg = 10^3$ kg
pressure	bar[a]	bar	10^5 Pa $= 10^5$ N m^{-2}
energy	electronvolt[b]	eV ($= e \times V$)	$\approx 1.60218 \times 10^{-19}$ J
mass	unified atomic mass unit[b,c]	u ($= m_a(^{12}C)/12$)	$\approx 1.66054 \times 10^{-27}$ kg

[a]The ångström and the bar are approved by CIPM for temporary use with SI units, until CIPM makes a further recommendation. However, they should not be introduced where they are not used at present.

[b]The values of these units in terms of the corresponding SI units are not exact, since they depend on the values of the physical constants e (for the electronvolt) and N_A (for the unified atomic mass unit), which are determined by experiment.

[c]The unified atomic mass unit is also sometimes called the dalton, with symbol Da, although the name and symbol have not been approved by CGPM.

Table 10.18 Derived Units with Special Names and Symbols (*From* [1]. *Used with permission.*)

Physical quantity	Name of SI unit	Symbol for SI unit	Expression in terms of SI base units
frequency[a]	hertz	Hz	s^{-1}
force	newton	N	$m\,kg\,s^{-2}$
pressure, stress	pascal	Pa	$N\,m^{-2} = m^{-1}\,kg\,s^{-2}$
energy, work, heat	joule	J	$N\,m = m^2\,kg\,s^{-2}$
power, radiant flux	watt	W	$J\,s^{-1} = m^2\,kg\,s^{-3}$
electric charge	coulomb	C	$A\,s$
electric potential, electromotive force	volt	V	$J\,C^{-1} = m^2\,kg\,s^{-3}\,A^{-1}$
electric resistance	ohm	Ω	$V\,A^{-1} = m^2\,kg\,s^{-3}\,A^{-2}$
electric conductance	siemens	S	$Ω^{-1} = m^{-2}\,kg^{-1}\,s^3\,A^2$
electric capacitance	farad	F	$C\,V^{-1} = m^{-2}\,kg^{-1}\,s^4\,A^2$
magnetic flux density	tesla	T	$V\,s\,m^{-2} = kg\,s^{-2}\,A^{-1}$
magnetic flux	weber	Wb	$V\,s = m^2\,kg\,s^{-2}\,A^{-1}$
inductance	henry	H	$V\,A^{-1}\,s = m^2\,kg\,s^{-2}\,A^{-2}$
Celsius temperature[b]	degree Celsius	°C	K
luminous flux	lumen	lm	$cd\,sr$
illuminance	lux	lx	$cd\,sr\,m^{-2}$
activity (radioactive)	becquerel	Bq	s^{-1}
absorbed dose (of radiation)	gray	Gy	$J\,kg^{-1} = m^2\,s^{-2}$
dose equivalent (dose equivalent index)	sievert	Sv	$J\,kg^{-1} = m^2\,s^{-2}$
plane angle	radian	rad	$1 = m\,m^{-1}$
solid angle	steradian	sr	$1 = m^2\,m^{-2}$

[a]For radial (circular) frequency and for angular velocity the unit rad s^{-1}, or simply s^{-1}, should be used, and this may not be simplified to Hz. The unit Hz should be used only for frequency in the sense of cycles per second.

[b]The Celsius temperature θ is defined by the equation:

$$\theta/°C = T/K - 273.15$$

The SI unit of Celsius temperature interval is the degree Celsius, °C, which is equal to the kelvin, K. °C should be treated as a single symbol, with no space between the ° sign and the letter C. (The symbol °K and the symbol °, should no longer be used.)

Table 10.19 The Greek Alphabet (*From* [1]. *Used with permission.*)

Greek letter		Greek name	English equivalent	Greek letter			Greek name	English equivalent
A	α		Alpha	a	N	ν		Nu	n
B	β		Beta	b	Ξ	ξ		Xi	x
Γ	γ		Gamma	g	O	o		Omicron	ŏ
Δ	δ		Delta	d	Π	π		Pi	p
E	ϵ		Epsilon	ĕ	P	ρ		Rho	r
Z	ζ		Zeta	z	Σ	σ	ς	Sigma	s
H	η		Eta	ē	T	τ		Tau	t
Θ	θ	ϑ	Theta	th	Υ	υ		Upsilon	u
I	ι		Iota	i	Φ	ϕ	φ	Phi	ph
K	κ		Kappa	k	X	χ		Chi	ch
Λ	λ		Lambda	l	Ψ	ψ		Psi	ps
M	μ		Mu	m	Ω	ω		Omega	ō

Table 10.20 Constants (*From* [1]. *Used with permission.*)

π Constants

π	= 3.14159	26535	89793	23846	26433	83279	50288	41971	69399	37511
$1/\pi$	= 0.31830	98861	83790	67153	77675	26745	02872	40689	19291	48091
π^2	= 9.8690	44010	89358	61883	44909	99876	15113	53136	99407	24079
$\log_e \pi$	= 1.14472	98858	49400	17414	34273	51353	05871	16472	94812	91531
$\log_{10} \pi$	= 0.49714	98726	94133	85435	12682	88290	89887	36516	78324	38044
$\log_{10} \sqrt{2\pi}$	= 0.39908	99341	79057	52478	25035	91507	69595	02099	34102	92128

Constants Involving e

e	= 2.71828	18284	59045	23536	02874	71352	66249	77572	47093	69996
$1/e$	= 0.36787	94411	71442	32159	55237	70161	46086	74458	11131	03177
e^2	= 7.38905	60989	30650	22723	04274	06575	00781	31803	15570	55185
$M = \log_{10} e$	= 0.43429	44819	03251	82765	11289	18916	60508	22943	97005	80367
$1/M = \log_e 10$	= 2.30258	50929	94045	68401	79914	54684	36420	76011	01488	62877
$\log_{10} M$	= 9.63778	43113	00536	78912	29674	98645	−10			

Numerical Constants

$\sqrt{2}$	= 1.41421	35623	73095	04880	16887	24209	69807	85696	71875	37695
$\sqrt[3]{2}$	= 1.25992	10498	94873	16476	72106	07278	22835	05702	51464	70151
$\log_e 2$	= 0.69314	71805	59945	30941	72321	21458	17656	80755	00134	36026
$\log_{10} 2$	= 0.30102	99956	63981	19521	37388	94724	49302	67881	89881	46211
$\sqrt{3}$	= 1.73205	08075	68877	29352	74463	41505	87236	69428	05253	81039
$\sqrt[3]{3}$	= 1.44224	95703	07408	38232	16383	10780	10958	83918	69253	49935
$\log_e 3$	= 1.09861	22886	68109	69139	52452	36922	52570	46474	90557	82275
$\log_{10} 3$	= 0.47712	12547	19662	43729	50279	03255	11530	92001	28864	19070

Table 10.21 Symbols and Terminology for Physical and Chemical Quantities: Classical Mechanics (*From* [1]. *Used with permission.*)

Name	Symbol	Definition	SI unit
mass	m		kg
reduced mass	μ	$\mu = m_1 m_2 / (m_1 + m_2)$	kg
density, mass density	ρ	$\rho = m/V$	kg m^{-3}
relative density	d	$d = \rho/\rho^\theta$	1
surface density	ρ_A, ρ_S	$\rho_A = m/A$	kg m^{-2}
specific volume	v	$v = V/M = 1/\rho$	m^3 kg^{-1}
momentum	p	$p = mv$	kg ms^{-1}
angular momentum, action	L	$L = r \times p$	J s
moment of inertia	I, J	$I = \sum m_i r_i^2$	kg m^2
force	F	$F = dp/dt = ma$	N
torque, moment of a force	$T, (M)$	$T = r \times F$	N m
energy	E		J
potential energy	E_p, V, Φ	$E_p = -\int F \cdot ds$	J
kinetic energy	E_k, T, K	$E_k = (1/2)mv^2$	J
work	W, w	$W = \int F \cdot ds$	J
Hamilton function	H	$H(q, p)$ $= T(q, p) + V(q)$	J
Lagrange function	L	$L(q, \dot{q})$ $= T(q, \dot{q}) - V(q)$	J
pressure	p, P	$p = F/A$	Pa, N m^{-2}
surface tension	γ, σ	$\gamma = dW/dA$	N m^{-1}, J m^{-2}
weight	$G, (W, P)$	$G = mg$	N
gravitational constant	G	$F = Gm_1 m_2 / r^2$	N m^2 kg^{-2}
normal stress	σ	$\sigma = F/A$	Pa
shear stress	τ	$\tau = F/A$	Pa
linear strain, relative elongation	ε, e	$\varepsilon - \Delta l/l$	1
modulus of elasticity, Young's modulus	E	$E = \sigma/\varepsilon$	Pa
shear strain	γ	$\gamma = \Delta x/d$	1
shear modulus	G	$G = \tau/\gamma$	Pa
volume strain, bulk strain	θ	$\theta = \Delta V/V_0$	1
bulk modulus, compression modulus	K	$K = -V_0(dp/dV)$	Pa
viscosity, dynamic viscosity	η, μ	$\tau_{x,z} = \eta(dv_x/dz)$	Pa s
fluidity	ϕ	$\phi = 1/\eta$	m kg^{-1}s
kinematic viscosity	v	$v = \eta/\rho$	m^2 s^{-1}
friction coefficient	$\mu, (f)$	$F_{frict} = \mu F_{norm}$	1
power	P	$P = dW/dt$	W
sound energy flux	P, P_a	$P = dE/dt$	W
acoustic factors			
reflection factor	ρ	$\rho = P_r/P_0$	1
acoustic absorption factor	$\alpha_a, (\alpha)$	$\alpha_a = 1 - \rho$	1
transmission factor	τ	$\tau = P_{tr}/P_0$	1
dissipation factor	δ	$\delta = \alpha_a - \tau$	1

Table 10.22 Symbols and Terminology for Physical and Chemical Quantities: Electricity and Magnetism (*From* [1]. *Used with permission.*)

Name	Symbol	Definition	SI unit
quantity of electricity, electric charge	Q		C
charge density	ρ	$\rho = Q/V$	$\mathrm{C\,m^{-3}}$
surface charge density	σ	$\sigma = Q/A$	$\mathrm{C\,m^{-2}}$
electric potential	V, ϕ	$V = dW/dQ$	$\mathrm{V, J\,C^{-1}}$
electric potential difference	$U, \Delta V, \Delta\phi$	$U = V_2 - V_1$	V
electromotive force	E	$E = \int(F/Q) \cdot ds$	V
electric field strength	E	$E = F/Q = -\mathrm{grad}\,V$	$\mathrm{V\,m^{-1}}$
electric flux	Ψ	$\Psi = \int D \cdot dA$	C
electric displacement	D	$D = \varepsilon E$	$\mathrm{C\,m^{-2}}$
capacitance	C	$C = Q/U$	$\mathrm{F, C\,V^{-1}}$
permittivity	ε	$D = \varepsilon E$	$\mathrm{F\,m^{-1}}$
permittivity of vacuum	ε_0	$\varepsilon_0 = \mu_0^{-1} c_0^{-2}$	$\mathrm{F\,m^{-1}}$
relative permittivity	ε_r	$\varepsilon_r = \varepsilon/\varepsilon_0$	1
dielectric polarization (dipole moment per volume)	P	$P = D - \varepsilon_0 E$	$\mathrm{C\,m^{-2}}$
electric susceptibility	χ_e	$\chi_e = \varepsilon_r - 1$	1
electric dipole moment	p, μ	$p = Qr$	$\mathrm{C\,m}$
electric current	I	$I = dQ/dt$	A
electric current density	j, J	$I = \int j \cdot dA$	$\mathrm{A\,m^{-2}}$
magnetic flux density, magnetic induction	B	$F = Qv \times B$	T
magnetic flux	Φ	$\Phi = \int B \cdot dA$	Wb
magnetic field strength	H	$B = \mu H$	$\mathrm{A\,M^{-1}}$
permeability	μ	$B = \mu H$	$\mathrm{N\,A^{-2}, H\,m^{-1}}$
permeability of vacuum	μ_0		$\mathrm{H\,m^{-1}}$
relative permeability	μ_r	$\mu_r = \mu/\mu_0$	1
magnetization (magnetic dipole moment per volume)	M	$M = B/\mu_0 - H$	$\mathrm{A\,m^{-1}}$
magnetic susceptibility	$\chi, \kappa, (\chi_m)$	$\chi = \mu_r - 1$	1
molar magnetic susceptibility	χ_m	$\chi_m = V_m \chi$	$\mathrm{m^3\,mol^{-1}}$
magnetic dipole moment	m, μ	$E_p = -m \cdot B$	$\mathrm{A\,m^2, J\,T^{-1}}$
electrical resistance	R	$R = U/I$	Ω
conductance	G	$G = 1/R$	S
loss angle	δ	$\delta = (\pi/2) + \phi_I - \phi_U$	1, rad
reactance	X	$X = (U/I)\sin\delta$	Ω
impedance (complex impedance)	Z	$Z = R + iX$	Ω
admittance (complex admittance)	Y	$Y = 1/Z$	S
susceptance	B	$Y = G + iB$	S
resistivity	ρ	$\rho = E/j$	$\Omega\,\mathrm{m}$
conductivity	κ, γ, σ	$\kappa = 1/\rho$	$\mathrm{S\,m^{-1}}$
self-inductance	L	$E = -L(dI/dt)$	H
mutual inductance	M, L_{12}	$E_1 = L_{12}(dI_2/dt)$	H
magnetic vector potential	A	$B = \nabla \times A$	$\mathrm{Wb\,m^{-1}}$
Poynting vector	S	$S = E \times H$	$\mathrm{W\,m^{-2}}$

Table 10.23 Symbols and Terminology for Physical and Chemical Quantities: Electromagnetic Radiation (*From* [1]. *Used with permission.*)

Name	Symbol	Definition	SI unit
wavelength	λ		m
speed of light in vacuum	c_0		m s^{-1}
in a medium	c	$c = c_0/n$	m s^{-1}
wavenumber in vacuum	$\tilde{v}$	$\tilde{v} = v/c_0 = 1/n\lambda$	m^{-1}
wavenumber (in a medium)	σ	$\sigma = 1/\lambda$	m^{-1}
frequency	v	$v = c/\lambda$	Hz
circular frequency,	ω	$\omega = 2\pi v$	s^{-1}, rad s^{-1}
pulsatance			
refractive index	n	$n = c_0/c$	1
Planck constant	h		J s
Planck constant/2π	$\hbar$	$\hbar = h/2\pi$	J s
radiant energy	Q, W		J
radiant energy density	ρ, w	$\rho = Q/V$	J m^{-3}
spectral radiant energy density			
in terms of frequency	ρ_v, w_v	$\rho_v = d\rho/dv$	J m^{-3}Hz^{-1}
in terms of wavenumber	$\rho_{\tilde{v}}, w_{\tilde{v}}$	$\rho_{\tilde{v}} = d\rho/d\tilde{v}$	J m^{-2}
in terms of wavelength	ρ_λ, w_λ	$\rho_\lambda = d\rho/d\lambda$	J m^{-4}
Einstein transition probabilities			
spontaneous emission	A_{nm}	$dN_n/dt = -A_{nm}N_n$	s^{-1}
stimulated emission	B_{nm}	$dN_n/dt = -\rho_{\tilde{v}}(\tilde{v}_{nm}) \times B_{nm}N_n$	s kg^{-1}
stimulated absorption	B_{mn}	$dN_n/dt = \rho_{\tilde{v}}(\tilde{v}_{nm})B_{mn}N_m$	s kg^{-1}
radiant power,	Φ, P	$\Phi = dQ/dt$	W
radiant energy per time			
radiant intensity	I	$I = d\Phi/d\Omega$	W sr^{-1}
radiant exitance	M	$M = d\Phi/dA_{\text{source}}$	W m^{-2}
(emitted radiant flux)			
irradiance	$E, (I)$	$E = d\Phi/dA$	W m^{-2}
(radiant flux received)			
emittance	ε	$\varepsilon = M/M_{bb}$	1
Stefan-Boltzman constant	σ	$M_{bb} = \sigma T^4$	W m^{-2} K^{-4}
first radiation constant	c_1	$c_1 = 2\pi hc_0^2$	W m^2
second radiation constant	c_2	$c_2 = hc_0/k$	K m
transmittance,	τ, T	$\tau = \Phi_{tr}/\Phi_0$	1
transmission factor			
absorptance,	α	$\alpha = \Phi_{abs}/\Phi_0$	1
absorption factor			
reflectance,	ρ	$\rho = \Phi_{refl}/\Phi_0$	1
reflection factor			
(decadic) absorbance	A	$A = \lg(1 - \alpha_i)$	1
napierian absorbance	B	$B = \ln(1 - \alpha_i)$	1
absorption coefficient			
(linear) decadic	a, K	$a = A/l$	m^{-1}
(linear) napierian	α	$\alpha = B/l$	m^{-1}
molar (decadic)	ε	$\varepsilon = a/c = A/cl$	m^2 mol^{-1}
molar napierian	κ	$\kappa = \alpha/c = B/cl$	m^2 mol^{-1}
absorption index	k	$k = \alpha/4\pi\tilde{v}$	1
complex refractive index	$\hat{n}$	$\hat{n} = n + ik$	1
molar refraction	R, R_m	$R = \dfrac{n^2 - 1}{n^2 + 2}V_m$	m^3 mol^{-1}
angle of optical rotation	α		1, rad

Table 10.24 Symbols and Terminology for Physical and Chemical Quantities: Solid State (*From* [1]. *Used with permission.*)

Name	Symbol	Definition	SI unit
lattice vector	R, R_0		m
fundamental translation vectors for the crystal lattice	$a_1; a_2; a_3,$ $a; b; c$	$R = n_1 a_1 + n_2 a_2 + n_3 a_3$	m
(circular) reciprocal lattice vector	G	$G \cdot R = 2\pi m$	m^{-1}
(circular) fundamental translation vectors for the reciprocal lattice	$b_1; b_2; b_3,$ $a^*; b^*; c^*$	$a_i \cdot b_k = 2\pi \delta_{ik}$	m^{-1}
lattice plane spacing	d		m
Bragg angle	θ	$n\lambda = 2d \sin\theta$	1, rad
order of reflection	n		1
order parameters			
short range	σ		1
long range	s		1
Burgers vector	b		m
particle position vector	r, R_j		m
equilibrium position vector of an ion	R_0		m
equilibrium position vector of an ion	R_0		m
displacement vector of an ion	u	$u = R - R_0$	m
Debye–Waller factor	B, D		1
Debye circular wavenumber	q_D		m^{-1}
Debye circular frequency	ω_D		s^{-1}
Grüneisen parameter	γ, Γ	$\gamma = \alpha V / \kappa C_V$	1
Madelung constant	$\alpha, \mathcal{M}$	$E_{coul} = \dfrac{\alpha N_A z_+ z_- e^2}{4\pi \varepsilon_0 R_0}$	1
density of states	N_E	$N_E = dN(E)/dE$	$J^{-1}\, m^{-3}$
(spectral) density of vibrational modes	N_ω, g	$N_\omega = dN(\omega)/d\omega$	$s\, m^{-3}$
resistivity tensor	ρ_{ik}	$E = \rho \cdot j$	$\Omega\, m$
conductivity tensor	σ_{ik}	$\sigma = \rho^{-1}$	$S\, m^{-1}$
thermal conductivity tensor	λ_{ik}	$J_q = -\lambda \cdot \text{grad}\, T$	$W\, m^{-1}\, K^{-1}$
residual resistivity	ρ_R		$\Omega\, m$
relaxation time	τ	$\tau = l/v_F$	s
Lorenz coefficient	L	$L = \lambda/\sigma T$	$V^2\, K^{-2}$
Hall coefficient	A_H, R_H	$E = \rho \cdot j + R_H (B \times j)$	$m^3\, C^{-1}$
thermoelectric force	E		V
Peltier coefficient	Π		V
Thomson coefficient	$\mu, (\tau)$		$V\, K^{-1}$
work function	Φ	$\Phi = E_\infty - E_F$	J
number density, number concentration	$n, (p)$		m^{-3}
gap energy	E_g		J
donor ionization energy	E_d		J
acceptor ionization energy	E_a		J
Fermi energy	E_F, ε_F		J
circular wave vector, propagation vector	k, q	$k = 2\pi/\lambda$	m^{-1}
Bloch function	$u_k(r)$	$\psi(r) = u_k(r)\exp(i k \cdot r)$	$m^{-3/2}$
charge density of electrons	ρ	$\rho(r) = -e\psi^*(r)\psi(r)$	$C\, m^{-3}$
effective mass	m^*		kg
mobility	μ	$\mu = v_{drift}/E$	$m^2\, V^{-1}\, s^{-1}$
mobility ratio	b	$b = \mu_n/\mu_p$	1
diffusion coefficient	D	$dN/dt = -DA(dn/dx)$	$m^2\, s^{-1}$
diffusion length	L	$L = \sqrt{D\tau}$	m
characteristic (Weiss) temperature	ϕ, ϕ_W		K
Curie temperature	T_C		K
Néel temperature	T_N		K

10.7 Resistive Properties

Table 10.25 Electrical Resistivity of Various Substances in $10^{-8}\ \Omega \bullet$ m (*From* [1]. *Used with permission.*)

T/K	Aluminum	Barium	Beryllium	Calcium	Cesium	Chromium	Copper
1	0.000100	0.081	0.0332	0.045	0.0026		0.00200
10	0.000193	0.189	0.0332	0.047	0.243		0.00202
20	0.000755	0.94	0.0336	0.060	0.86		0.00280
40	0.0181	2.91	0.0367	0.175	1.99		0.0239
60	0.0959	4.86	0.067	0.40	3.07		0.0971
80	0.245	6.83	0.075	0.65	4.16		0.215
100	0.442	8.85	0.133	0.91	5.28	1.6	0.348
150	1.006	14.3	0.510	1.56	8.43	4.5	0.699
200	1.587	20.2	1.29	2.19	12.2	7.7	1.046
273	2.417	30.2	3.02	3.11	18.7	11.8	1.543
293	2.650	33.2	3.56	3.36	20.5	12.5	1.678
298	2.709	34.0	3.70	3.42	20.8	12.6	1.712
300	2.733	34.3	3.76	3.45	21.0	12.7	1.725
400	3.87	51.4	6.76	4.7		15.8	2.402
500	4.99	72.4	9.9	6.0		20.1	3.090
600	6.13	98.2	13.2	7.3		24.7	3.792
700	7.35	130	16.5	8.7		29.5	4.514
800	8.70	168	20.0	10.0		34.6	5.262
900	10.18	216	23.7	11.4		39.9	6.041

T/K	Gold	Hafnium	Iron	Lead	Lithium	Magnesium	Manganese
1	0.0220	1.00	0.0225		0.007	0.0062	7.02
10	0.0226	1.00	0.0238		0.008	0.0069	18.9
20	0.035	1.11	0.0287		0.012	0.0123	54
40	0.141	2.52	0.0758		0.074	0.074	116
60	0.308	4.53	0.271		0.345	0.261	131
80	0.481	6.75	0.693	4.9	1.00	0.557	132
100	0.650	9.12	1.28	6.4	1.73	0.91	132
150	1.061	15.0	3.15	9.9	3.72	1.84	136
200	1.462	21.0	5.20	13.6	5.71	2.75	139
273	2.051	30.4	8.57	19.2	8.53	4.05	143
293	2.214	33.1	9.61	20.8	9.28	4.39	144
298	2.255	33.7	9.87	21.1	9.47	4.48	144
300	2.271	34.0	9.98	21.3	9.55	4.51	144
400	3.107	48.1	16.1	29.6	13.4	6.19	147
500	3.97	63.1	23.7	38.3		7.86	149
600	4.87	78.5	32.9			9.52	151
700	5.82		44.0			11.2	152
800	6.81		57.1			12.8	
900	7.86					14.4	

T/K	Molybdenum	Nickel	Palladium	Platinum	Potassium	Rubidium	Silver
1	0.00070	0.0032	0.0200	0.002	0.0008	0.0131	0.00100
10	0.00089	0.0057	0.0242	0.0154	0.0160	0.109	0.00115
20	0.00261	0.0140	0.0563	0.0484	0.117	0.444	0.0042
40	0.0457	0.068	0.334	0.409	0.480	1.21	0.0539
60	0.206	0.242	0.938	1.107	0.90	1.94	0.162
80	0.482	0.545	1.75	1.922	1.34	2.65	0.289
100	0.858	0.96	2.62	2.755	1.79	3.36	0.418
150	1.99	2.21	4.80	4.76	2.99	5.27	0.726
200	3.13	3.67	6.88	6.77	4.26	7.49	1.029
273	4.85	6.16	9.78	9.6	6.49	11.5	1.467
293	5.34	6.93	10.54	10.5	7.20	12.8	1.587
298	5.47	7.12	10.73	10.7	7.39	13.1	1.617
300	5.52	7.20	10.80	10.8	7.47	13.3	1.629
400	8.02	11.8	14.48	14.6			2.241
500	10.6	17.7	17.94	18.3			2.87
600	13.1	25.5	21.2	21.9			3.53
700	15.8	32.1	24.2	25.4			4.21
800	18.4	35.5	27.1	28.7			4.91
900	21.2	38.6	29.4	32.0			5.64

T/K	Sodium	Strontium	Tantalum	Tungsten	Vanadium	Zinc	Zirconium
1	0.0009	0.80	0.10	0.000016		0.0100	0.250
10	0.0015	0.80	0.102	0.000137	0.0145	0.0112	0.253
20	0.016	0.92	0.146	0.00196	0.039	0.0387	0.357
40	0.172	1.70	0.751	0.0544	0.304	0.306	1.44
60	0.447	2.68	1.65	0.266	1.11	0.715	3.75
80	0.80	3.64	2.62	0.606	2.41	1.15	6.64
100	1.16	4.58	3.64	1.02	4.01	1.60	9.79
150	2.03	6.84	6.19	2.09	8.2	2.71	17.8
200	2.89	9.04	8.66	3.18	12.4	3.83	26.3
273	4.33	12.3	12.2	4.82	18.1	5.46	38.8
293	4.77	13.2	13.1	5.28	19.7	5.90	42.1
298	4.88	13.4	13.4	5.39	20.1	6.01	42.9
300	4.93	13.5	13.5	5.44	20.2	6.06	43.3
400		17.8	18.2	7.83	28.0	8.37	60.3
500		22.2	22.9	10.3	34.8	10.82	76.5
600		26.7	27.4	13.0	41.1	13.49	91.5
700		31.2	31.8	15.7	47.2		104.2
800		35.6	35.9	18.6	53.1		114.9
900			40.1	21.5	58.7		123.1

Table 10.26 Electrical Resistivity of Various Metallic elements at (approximately) Room Temperature (*From* [1]. *Used with permission.*)

Element	T/K	Electrical Resistivity $10^{-8}\ \Omega \cdot m$	Element	T/K	Electrical Resistivity $10^{-8}\ \Omega \cdot m$
Antimony	273	39	Polonium	273	40
Bismuth	273	107	Praseodymium	290–300	70.0
Cadmium	273	6.8	Promethium	290–300	75
Cerium	290–300	82.8	Protactinium	273	17.7
Cobalt	273	5.6	Rhenium	273	17.2
Dysprosium	290–300	92.6	Rhodium	273	4.3
Erbium	290–300	86.0	Ruthenium	273	7.1
Europium	290–300	90.0	Samarium	290–300	94.0
Gadolinium	290–300	131	Scandium	290–300	56.2
Gallium	273	13.6	Terbium	290–300	115
Holmium	290–300	81.4	Thallium	273	15
Indium	273	8.0	Thorium	273	14.7
Iridium	273	4.7	Thulium	290–300	67.6
Lanthanum	290–300	61.5	Tin	273	11.5
Lutetium	290–300	58.2	Titanium	273	39
Mercury	273	94.1	Uranium	273	28
Neodymium	290–300	64.3	Ytterbium	290–300	25.0
Niobium	273	15.2	Yttrium	290–300	59.6
Osmium	273	8.1			

Table 10.27 Electrical Resistivity of Selected Alloys in Units of 10^{-8} $\Omega \cdot$ m (From [1]. *Used with permission.*)

Wt % Al	273 K	293 K	300 K	350 K	400 K
	Alloy—Aluminum-Copper				
99[a]	2.51	2.74	2.82	3.38	3.95
95[a]	2.88	3.10	3.18	3.75	4.33
90[b]	3.36	3.59	3.67	4.25	4.86
85[b]	3.87	4.10	4.19	4.79	5.42
80[b]	4.33	4.58	4.67	5.31	5.99
70[b]	5.03	5.31	5.41	6.16	6.94
60[b]	5.56	5.88	5.99	6.77	7.63
50[b]	6.22	6.55	6.67	7.55	8.52
40[c]	7.57	7.96	8.10	9.12	10.2
30[c]	11.2	11.8	12.0	13.5	15.2
25[f]	16.3[aa]	17.2	17.6	19.8	22.2
15[h]	—	12.3	—	—	—
19[g]	1.8[aa]	11.0	11.1	11.7	12.3
5[e]	9.43	9.61	9.68	10.2	10.7
1[b]	4.46	4.60	4.65	5.00	5.37

Wt % Cu	273 K	293 K	300 K	350 K	400 K
	Alloy—Copper-Nickel				
99[c]	2.71	2.85	2.91	3.27	3.62
95[c]	7.60	7.71	7.82	8.22	8.62
90[c]	13.69	13.89	13.96	14.40	14.81
85[c]	19.63	19.83	19.90	2032	20.70
80[c]	25.46	25.66	25.72	26.12[aa]	26.44[aa]
70[i]	36.67	36.72	36.76	36.85	36.89
60[i]	45.43	45.38	45.35	45.20	45.01
50[i]	50.19	50.05	50.01	49.73	49.50
40[c]	47.42	47.73	47.82	48.28	48.49
30[i]	40.19	41.79	42.34	44.51	45.40
25[c]	33.46	35.11	35.69	39.67[aa]	42.81[aa]
15[c]	22.00	23.35	23.85	27.60	31.38
10[c]	16.65	17.82	18.26	21.51	25.19
5[c]	11.49	12.50	12.90	15.69	18.78
1[c]	7.23	8.08	8.37	10.63[aa]	13.18[aa]

Alloy—Aluminum-Magnesium

99[c]	2.96	3.18	3.26	3.82	4.39
95[c]	5.05	5.28	5.36	5.93	6.51
90[c]	7.52	7.76	7.85	8.43	9.02
85	—	—	—	—	—
80	—	—	—	—	—
70	—	—	—	—	—
60	—	—	—	—	—
50	—	—	—	—	—
40	—	—	—	—	—
30	—	—	—	—	—
25	—	—	—	—	—
15	—	—	—	—	—
10[b]	17.1	17.4	17.6	18.4	19.2
5[b]	13.1	13.4	13.5	14.3	15.2
1[a]	5.92	6.25	6.37	7.20	8.03

Alloy—Copper-Palladium

Wt % Cu					
99[c]	2.10	2.23	2.27	2.59	2.92
95[c]	4.21	4.35	4.40	4.74	5.08
90[c]	6.89	7.03	7.08	7.41	7.74
85[c]	9.48	9.61	9.66	10.01	10.36
80[c]	11.99	12.12	12.16	12.51[aa]	12.87
70[c]	16.87	17.01	17.06	17.41	17.78
60[c]	21.73	21.87	21.92	22.30	22.69
50[c]	27.62	27.79	27.86	28.25	28.64
40[c]	35.31	35.51	35.57	36.03	36.47
30[c]	46.50	46.66	46.71	47.11	47.47
25[c]	46.25	46.45	46.52	46.99[aa]	47.43[aa]
15[c]	36.52	36.99	37.16	38.28	39.35
10[c]	28.90	29.51	29.73	31.19[aa]	32.56[aa]
5[c]	20.00	20.75	21.02	22.84[aa]	24.54[aa]
1[c]	11.90	12.67	12.93[aa]	14.82[aa]	16.68[aa]

Alloy—Copper-Gold

Wt % Cu					
99[c]	1.73	1.86[aa]	1.91[aa]	2.24[aa]	2.58[aa]
95[c]	2.41	2.54[aa]	2.59[aa]	2.92[aa]	3.26[aa]
90[c]	3.29	4.42[aa]	3.46[aa]	3.79[aa]	4.12[aa]
85[c]	4.20	4.33	4.38[aa]	4.71[aa]	5.05[aa]
80[c]	5.15	5.28	5.32	5.65	5.99
70[c]	7.12	7.25	7.30	7.64	7.99
60[c]	9.18	9.13	9.36	9.70	10.05
50[c]	11.07	11.20	11.25	11.60	11.94
40[c]	12.70	12.85	12.90[aa]	13.27[aa]	13.65[aa]
30[c]	13.77	13.93	13.99[aa]	14.38[aa]	14.78[aa]
25[c]	13.93	14.09	14.14	14.54	14.94
15[c]	12.75	12.91	12.96[aa]	13.36[aa]	13.77
10[c]	10.70	10.86	10.91	11.31	11.72
5[c]	7.25	7.41[aa]	7.46	7.87	8.28
1[c]	3.40	3.57	3.62	4.03	4.45

Alloy—Copper-Zinc

Wt % Cu					
99[b]	1.84	1.97	2.02	2.36	2.71
95[b]	2.78	2.92	2.97	3.33	3.69
90[b]	3.66	3.81	3.86	4.25	4.63
85[b]	4.37	4.54	4.60	5.02	5.44
80[b]	5.01	5.19	5.26	5.71	6.17
70[b]	5.87	6.08	6.15	6.67	7.19
60	—	—	—	—	—
50	—	—	—	—	—
40	—	—	—	—	—
30	—	—	—	—	—
25	—	—	—	—	—
15	—	—	—	—	—
10	—	—	—	—	—
5	—	—	—	—	—
1	—	—	—	—	—

Alloy—Gold-Palladium

Wt % Au					
99[c]	2.69	2.86	2.91	3.32	3.73
95[c]	5.21	5.35	5.41	5.79	6.17
90[i]	8.01	8.17	8.22	8.56	8.93
85[b]	10.50[aa]	10.66	10.72[aa]	11.100[aa]	11.48[aa]
80[b]	12.75	12.93	12.99	13.45	13.93
70[c]	18.23	18.46	18.54	19.10	19.67
60[b]	26.70	26.94	27.02	27.63[aa]	28.23[aa]
50[a]	27.23	27.63	27.76	28.64[aa]	29.42[aa]
40[a]	24.65	25.23	25.42	26.74	27.95
30[b]	20.82	21.49	21.72	23.35	24.92
25[b]	18.86	19.53	19.77	21.51	23.19
15[a]	15.08	15.77	16.01	17.80	19.61
10[a]	13.25	13.95	14.20[aa]	16.00[aa]	17.81[aa]
5[a]	11.49[aa]	12.21	12.46[aa]	14.26[aa]	16.07[aa]
1[a]	10.07	10.85[aa]	11.12[aa]	12.99[aa]	14.80[aa]

Alloy—Gold-Silver

Wt % Au					
99[b]	2.58	2.75	2.80[aa]	3.22[aa]	3.63[aa]
95[a]	4.58	4.74	4.79	5.19	5.59
90[i]	6.57	6.73	6.78	7.19	7.58
85[j]	8.14	8.30	8.36[aa]	8.75	9.15
80[j]	9.34	9.50	9.55	9.94	10.33
70[j]	10.70	10.86	10.91	11.29	11.68[aa]
60[j]	10.92	11.07	11.12	11.50	11.87
50[j]	10.23	10.37	10.42	10.78	11.14
40[j]	8.92	9.06	9.11	9.46[aa]	9.81
30[a]	7.34	7.47	7.52	7.85	8.19
25[a]	6.46	6.59	6.63	6.96	7.30[aa]
15[a]	4.55	4.67	4.72	5.03	5.34
10[a]	3.54	3.66	3.71	4.00	4.31
5[i]	2.52	2.64[aa]	2.68[aa]	2.96[aa]	3.25[aa]
1[b]	1.69	1.80	1.84[aa]	2.12[aa]	2.42[aa]

Alloy—Iron-Nickel

Wt % Fe				
99[a]	10.9	12.0	12.4	—	18.7
95[c]	18.7	19.9	20.2	—	26.8
90[c]	24.2	25.5	25.9	—	33.2
85[c]	27.8	29.2	29.7	—	37.3
80[c]	30.1	31.6	32.2	—	40.0
70[b]	32.3	33.9	34.4	—	42.4
60[c]	53.8	57.1	58.2	—	73.9
50[d]	28.4	30.6	31.4	—	43.7
40[d]	19.6	21.6	22.5	—	34.0
30[c]	15.3	17.1	17.7	—	27.4
25[b]	14.3	15.9	16.4	—	25.1
15[c]	12.6	13.8	14.2	—	21.1
10[c]	11.4	12.5	12.9	—	18.9
5[c]	9.66	10.6	10.9	—	16.1[aa]
1[b]	7.17	7.94	8.12	—	12.8

[a] Uncertainty in resistivity is ±2%.
[b] Uncertainty in resistivity is ±3%.
[c] Uncertainty in resistivity is ±5%.
[d] Uncertainty in resistivity is ±7% below 300 K and ±5% at 300 and 400 K.
[e] Uncertainty in resistivity is ±7%.
[f] Uncertainty in resistivity is ±8%.
[g] Uncertainty in resistivity is ±10%.
[h] Uncertainty in resistivity is ±12%.
[i] Uncertainty in resistivity is ±4%.
[j] Uncertainty in resistivity is ±1%.
[k] Uncertainty in resistivity is ±3% up to 300 K and ± 4% above 300 K.
[m] Uncertainty in resistivity is ±2% up to 300 K and ± 4% above 300 K.
[a] Crystal usually a mixture of α-hcp and fcc lattice.
[aa] In temperature range where no experimental data are available.

Table 10.28 Resistivity of Selected Ceramics (*From* [1]. *Used with permission.*)

Ceramic	Resistivity, $\Omega \cdot cm$
Borides	
Chromium diboride (CrB_2)	21×10^{-6}
Hafnium diboride (HfB_2)	$10-12 \times 10^{-6}$ at room temp.
Tantalum diboride (TaB_2)	68×10^{-6}
Titanium diboride (TiB_2) (polycrystalline)	
85% dense	$26.5-28.4 \times 10^{-6}$ at room temp.
85% dense	9.0×10^{-6} at room temp.
100% dense, extrapolated values	$8.7-14.1 \times 10^{-6}$ at room temp.
	3.7×10^{-6} at liquid air temp.
Titanium diboride (TiB_2) (monocrystalline)	
Crystal length 5 cm, 39 deg. and 59 deg. orientation with respect to growth axis	$6.6 \pm 0.2 \times 10^{-6}$ at room temp.
Crystal length 1.5 cm, 16.5 deg. and 90 deg. orientation with respect to growth axis	$6.7 \pm 0.2 \times 10^{-6}$ at room temp.
Zirconium diboride (ZrB_2)	9.2×10^{-6} at 20°C
	1.8×10^{-6} at liquid air temp.
Carbides: boron carbide (B_4C)	$0.3-0.8$

10.8 Dielectrics and Semiconductors

Table 10.29 Dielectric Constants of Ceramics (*From* [1]. *Used with permission.*)

Material	Dielectric constant, 10^6 Hz	Dielectric strength V/mil	Volume resistivity $\Omega \cdot cm$ (23°C)	Loss factor[a]
Alumina	4.5–8.4	40–160	$10^{11}-10^{14}$	0.0002–0.01
Corderite	4.5–5.4	40–250	$10^{12}-10^{14}$	0.004–0.012
Forsterite	6.2	240	10^{14}	0.0004
Porcelain (dry process)	6.0–8.0	40–240	$10^{12}-10^{14}$	0.0003–0.02
Porcelain (wet process)	6.0–7.0	90–400	$10^{12}-10^{14}$	0.006–0.01
Porcelain, zircon	7.1–10.5	250–400	$10^{13}-10^{15}$	0.0002–0.008
Steatite	5.5–7.5	200–400	$10^{13}-10^{15}$	0.0002–0.004
Titanates (Ba, Sr, Ca, Mg, and Pb)	15–12.000	50–300	10^8-10^{15}	0.0001–0.02
Titanium dioxide	14–110	100–210	$10^{13}-10^{18}$	0.0002–0.005

[a] Power factor × dielectric constant equals loss factor.

Table 10.30 Dielectric Constants of Solids in the Temperature Range 17–22°C (*From* [1]. *Used with permission.*)

Material	Freq., Hz	Dielectric constant	Material	Freq., Hz	Dielectric Constant
Acetamide	4×10^8	4.0	Phenanthrene	4×10^8	2.80
Acetanilide	–	2.9	Phenol (10°C)	4×10^8	4.3
Acetic acid (2°C)	4×10^8	4.1	Phosphorus, red	10^8	4.1
Aluminum oleate	4×10^8	2.40	Phosphorus, yellow	10^8	3.6
Ammonium bromide	10^8	7.1	Potassium aluminum		
Ammonium chloride	10^8	7.0	sulfate	10^6	3.8
Antimony trichloride	10^8	5.34	Potassium carbonate		
Apatite ⊥ optic axis	3×10^8	9.50	(15°C)	10^8	5.6
Apatite ‖ optic axis	3×10^8	7.41	Potassium chlorate	6×10^7	5.1
Asphalt	$<3 \times 10^6$	2.68	Potassium chloride	10^4	5.03
Barium chloride (anhyd.)	6×10^7	11.4	Potassium chromate	6×10^7	7.3
Barium chloride ($2H_2O$)	6×10^7	9.4	Potassium iodide	6×10^7	5.6
Barium nitrate	6×10^7	5.9	Potassium nitrate	6×10^7	5.0
Barium sulfate(15°C)	10^8	11.4	Potassium sulfate	6×10^7	5.9
Beryl ⊥ optic axis	10^4	7.02	Quartz ⊥ optic axis	3×10^7	4.34
Beryl ‖ optic axis	10^4	6.08	Quartz ‖ optic axis	3×10^7	4.27
Calcite ⊥ optic axis	10^4	8.5	Resorcinol	4×10^8	3.2
Calcite ‖ optic axis	10^4	8.0	Ruby⊥ optic axis	10^4	13.27
Calcium carbonate	10^6	6.14	Ruby ‖ optic axis	10^4	11.28
Calcium fluoride	10^4	7.36	Rutile ⊥ optic axis	10^8	86
Calcium sulfate ($2H_2O$)	10^4	5.66	Rutile ‖ optic axis	10^8	170
Cassiterite ⊥ optic axis	10^{12}	23.4	Selenium	10^8	6.6
Cassiterite ‖ optic axis	10^{12}	24	Silver bromide	10^6	12.2
d-Cocaine	5×10^8	3.10	Silver chloride	10^6	11.2
Cupric oleate	4×10^8	2.80	Silver cyanide	10^6	5.6
Cupric oxide (15°C)	10^8	18.1	Smithsonite ⊥ optic axis	10^{12}	9.3
Cupric sulfate (anhyd.)	6×10^7	10.3			
Cupric sulfate ($5H_2O$)	6×10^7	7.8	Smithsonite ‖ optic axis	10^{10}	9.4
Diamond	10^8	5.5			
Diphenylymethane	4×10^8	2.7	Sodium carbonate (anhyd.)	6×10^7	8.4
Dolomite ⊥ optic axis	10^8	8.0			
Dolomite ‖ optic axis	10^8	6.8	Sodium carbonate ($10H_2O$)	6×10^7	5.3
Ferrous oxide (15°C)	10^8	14.2			
Iodine	10^8	4	Sodium chloride	10^4	6.12
Lead acetate	10^6	2.6	Sodium nitrate	–	5.2
Lead carbonate (15°C)	10^8	18.6	Sodium oleate	4×10^8	2.75
Lead chloride	10^6	4.2	Sodium perchlorate	6×10^7	5.4
Lead monoxide (15°C)	10^8	25.9	Sucrose (mean)	3×10^8	3.32
Lead nitrate	6×10^7	37.7	Sulfur (mean)	–	4.0
Lead oleate	4×10^8	3.27	Thallium chloride	10^6	46.9
Lead sulfate	10^6	14.3	p-Toluidine	4×10^8	3.0
Lead sulfide (15°)	10^6	17.9	Tourmaline ⊥ optic axis	10^4	7.10
Malachite (mean)	10^{12}	7.2			
Mercuric chloride	10^6	3.2	Tourmaline ‖ optic axis	10^4	6.3
Mercurous chloride	10^6	9.4			
Naphthalene	4×10^8	2.52	Urea	4×10^8	3.5
			Zircon ⊥, ‖	10^8	12

Table 10.31 Dielectric Constants of Glass (*From* [1]. *Used with permission.*)

Type	Dielectric constant at 100 MHz (20°C)	Volume resistivity (350°C M Ω · cm)	Loss factor[a]
Corning 0010	6.32	10	0.015
Corning 0080	6.75	0.13	0.058
Corning 0120	6.65	100	0.012
Pyrex 1710	6.00	2,500	0.025
Pyrex 3320	4.71	–	0.019
Pyrex 7040	4.65	80	0.013
Pyrex 7050	4.77	16	0.017
Pyrex 7052	5.07	25	0.019
Pyrex 7060	4.70	13	0.018
Pyrex 7070	4.00	1,300	0.0048
Vycor 7230	3.83	–	0.0061
Pyrex 7720	4.50	16	0.014
Pyrex 7740	5.00	4	0.040
Pyrex 7750	4.28	50	0.011
Pyrex 7760	4.50	50	0.0081
Vycor 7900	3.9	130	0.0023
Vycor 7910	3.8	1,600	0.00091
Vycor 7911	3.8	4,000	0.00072
Corning 8870	9.5	5,000	0.0085
G.E. Clear (silica glass)	3.81	4,000–30,000	0.00038
Quartz (fused)	3.75–4.1 (1 MHz)	–	0.0002 1 MHz

[a]Power factor × dielectric constant equals loss factor.

Table 10.32 Semiconducting Properties of Selected Materials (*From* [1]. *Used with permission.*)

Substance	Minimum energy gap, eV R.T.	0 K	$\frac{dE_g}{dT}$ ×10⁴, eV/°C	$\frac{dE_g}{dP}$ ×10⁶, eV·cm²/kg	Density of states electron effective mass m_{d_n} (m_o)	Electron mobility and temperature dependence μ_n, cm²/V·s	$-x$	Density of states hole effective mass m_{d_p}, (m_o)	Hole mobility and temperature dependence μ_p, cm²/V·s	$-x$
Si	1.107	1.153	−2.3	−2.0	1.1	1,900	2.6	0.56	500	2.3
Ge	0.67	0.744	−3.7	±7.3	0.55	3,800	1.66	0.3	1,820	2.33
αSn	0.08	0.094	−0.5		0.02	2,500	1.65	0.3	2,400	2.0
Te	0.33				0.68	1,100		0.19	560	
III–V Compounds										
AlAs	2.2	2.3				1,200			420	
AlSb	1.6	1.7	−3.5	−1.6	0.09	200	1.5	0.4	500	1.8
GaP	2.24	2.40	−5.4	−1.7	0.35	300	1.5	0.5	150	1.5
GaAs	1.35	1.53	−5.0	+9.4	0.068	9,000	1.0	0.5	500	2.1
GaSb	0.67	0.78	−3.5	+12	0.050	5,000	2.0	0.23	1,400	0.9
InP	1.27	1.41	−4.6	+4.6	0.067	5,000	2.0		200	2.4
InAs	0.36	0.43	−2.8	+8	0.022	33,000	1.2	0.41	460	2.3
InSb	0.165	0.23	−2.8	+15	0.014	78,000	1.6	0.4	750	2.1
II–VI Compounds										
ZnO	3.2		−9.5	+0.6	0.38	180	1.5			
ZnS	3.54		−5.3	+5.7		180			5(400°C)	
ZnSe	2.58	2.80	−7.2	+6		540			28	
ZnTe	2.26			+6		340			100	
CdO	2.5 ± 0.1		−6		0.1	120				
CdS	2.42		−5	+3.3	0.165	400		0.8		
CdSe	1.74	1.85	−4.6		0.13	650	1.0	0.6		
CdTe	1.44	1.56	−4.1	+8	0.14	1,200		0.35	50	
HgSe	0.30				0.030	20,000	2.0			
HgTe	0.15		−1		0.017	25,000		0.5	350	
Halite Structure Compounds										
PbS	0.37	0.28	+4		0.16	800		0.1	1,000	2.2
PbSe	0.26	0.16	+4		0.3	1,500		0.34	1,500	2.2
PbTe	0.25	0.19	+4	−7	0.21	1,600		0.14	750	2.2
Others										
ZnSb	0.50	0.56			0.15	10				1.5
CdSb	0.45	0.57	−5.4		0.15	300			2,000	1.5
Bi₂S₃	1.3					200			1,100	
Bi₂Se₃	0.27					600			675	
Bi₂Te₃	0.13		−0.95		0.58	1,200	1.68	1.07	510	1.95
Mg₂Si		0.77	−6.4		0.46	400	2.5		70	
Mg₂Ge		0.74	−9			280	2		110	
Mg₂Sn	0.21	0.33	−3.5		0.37	320			260	
Mg₃Sb₂		0.32				20			82	
Zn₃As₂	0.93					10	1.1		10	
Cd₃As₂	0.55				0.046	100,000	0.88			
GaSe	2.05		3.8						20	
GaTe	1.66	1.80	−3.6			14	−5			
InSe	1.8					9000				
TlSe	0.57		−3.9		0.3	30		0.6	20	1.5
CdSnAs₂	0.23				0.05	25,000	1.7			
Ga₂Te₃	1.1	1.55	−4.8							
α-In₂Te₃	1.1	1.2			0.7				50	1.1
β-In₂Te₃	1.0								5	
Hg₅In₂Te₈	0.5								11,000	
SnO₂									78	

Table 10.33 Resistivity of Semiconducting Minerals (*From* [1]. *Used with permission.*)

Mineral	ρ, $\Omega \cdot$ m	Mineral	ρ, $\Omega \cdot$ m
Diamond (C)	2.7	Gersdorffite, NiAsS	1 to 160 $\times 10^{-6}$
Sulfides		Glaucodote, (Co, Fe)AsS	5 to 100 $\times 10^{-6}$
Argentite, Ag$_2$S	1.5 to 2.0 $\times 10^{-3}$	Antimonide	
Bismuthinite, Bi$_2$S$_3$	3 to 570	Dyscrasite, Ag$_3$Sb	0.12 to 1.2 $\times 10^{-6}$
Bornite, Fe$_2$S$_3 \cdot n$Cu$_2$S	1.6 to 6000 $\times 10^{-6}$	Arsenides	
Chalcocite, Cu$_2$S	80 to 100 $\times 10^{-6}$	Allemonite, SbAs$_2$	70 to 60,000
Chalcopyrite, Fe$_2$S$_3 \cdot$ Cu$_2$S	150 to 9000 $\times 10^{-6}$	Lollingite, FeAs$_2$	2 to 270 $\times 10^{-6}$
Covellite, CuS	0.30 to 83 $\times 10^{-6}$	Nicollite, NiAs	0.1 to 2 $\times 10^{-6}$
Galena, PbS	6.8 $\times 10^{-6}$ to 9.0 $\times 10^{-2}$	Skutterudite, CoAs$_3$	1 to 400 $\times 10^{-6}$
Haverite, MnS$_2$	10 to 20	Smaltite, CoAs$_2$	1 to 12 $\times 10^{-6}$
Marcasite, FeS$_2$	1 to 150 $\times 10^{-3}$	Tellurides	
Metacinnabarite, 4HgS	2 $\times 10^{-6}$ to 1 $\times 10^{-3}$	Altaite, PbTe	20 to 200 $\times 10^{-6}$
Millerite, NiS	2 to 4 $\times 10^{-7}$	Calavarite, AuTe$_2$	6 to 12 $\times 10^{-6}$
Molybdenite, MoS$_2$	0.12 to 7.5	Coloradoite, HgTe	4 to 100 $\times 10^{-6}$
Pentlandite, (Fe, Ni)$_9$S$_8$	1 to 11 $\times 10^{-6}$	Hessite, Ag$_2$Te	4 to 100 $\times 10^{-6}$
Pyrrhotite, Fe$_7$S$_8$	2 to 160 $\times 10^{-6}$	Nagyagite, Pb$_6$Au(S, Te)$_{14}$	20 to 80 $\times 10^{-6}$
Pyrite, FeS$_2$	1.2 to 600 $\times 10^{-3}$	Sylvanite, AgAuTe$_4$	4 to 20 $\times 10^{-6}$
Sphalerite, ZnS	2.7 $\times 10^{-3}$ to 1.2 $\times 10^4$	Oxides	
Antimony-sulfur compounds		Braunite, Mn$_2$O$_3$	0.16 to 1.0
Berthierite, FeSb$_2$S$_4$	0.0083 to 2.0	Cassiterite, SnO$_2$	4.5 $\times 10^{-4}$ to 10,000
Boulangerite, Pb$_5$Sb$_4$S$_{11}$	2 $\times 10^3$ to 4 $\times 10^4$	Cuprite, Cu$_2$O	10 to 50
Cylindrite, Pb$_3$Sn$_4$Sb$_2$S$_{14}$	2.5 to 60	Hollandite, (Ba, Na, K)Mn$_8$O$_{16}$	2 to 100 $\times 10^{-3}$
Franckeite, Pb$_5$Sn$_3$Sb$_2$S$_{14}$	1.2 to 4	Ilmenite, FeTiO$_3$	0.001 to 4
Hauchecornite, Ni$_9$(Bi, Sb)$_2$S$_8$	1 to 83 $\times 10^{-6}$	Magnetite, Fe$_3$O$_4$	52 $\times 10^{-6}$
Jamesonite, Pb$_4$FeSb$_6$S$_{14}$	0.020 to 0.15	Manganite, MnO $\cdot$ OH	0.018 to 0.5
Tetrahedrite, Cu$_3$SbS$_3$	0.30 to 30,000	Melaconite, CuO	6000
Arsenic-sulfur compounds		Psilomelane, KMnO $\cdot$ MnO$_2 \cdot n$H$_2$O	0.04 to 6000
Arsenopyrite, FeAsS	20 to 300 $\times 10^{-6}$	Pyrolusite, MnO$_2$	0.007 to 30
Cobaltite, CoAsS	6.5 to 130 $\times 10^{-3}$	Rutile, TiO$_2$	29 to 910
Enargite, Cu$_3$AsS$_4$	0.2 to 40 $\times 10^{-3}$	Uraninite, UO	1.5 to 200

Source: Carmichael, R.S., ed. 1982. *Handbook of Physical Properties of Rocks*, Vol. I. CRC Press, Boca Raton, FL.

10.9 Magnetic Properties

Table 10.34 Saturation Constants and Curie Points of Ferromagnetic Elements (*From* [1]. *Used with permission.*)

Element	σ_s[a] (20°C)	M_s[b] (20°C)	σ_s (0 K)	n_B[c]	Curie point, °C
Fe	218.0	1,714	221.9	2.219	770
Co	161	1,422	162.5	1.715	1,131
Ni	54.39	484.1	57.50	0.604	358
Gd	0	0	253.5	7.12	16

[a] σ_s = saturation magnetic moment/gram.
[b] M_s = saturation magnetic moment/cm^3, in cgs units.
[c] n_B = magnetic moment per atom in Bohr magnetons.

Table 10.35 Characteristics of High-Permeability Materials (*From* [1]. *Used with permission.*)

Material	Form	Fe	Ni	Co	Mo	Other	Typical Heat Treatment, °C	Permeability at $B = 20$, G	Maximum Permeability	Saturation flux density B, G	Hysteresis loss[a], W_h, ergs/cm²	Coercive[a] force H_a, O	Resistivity[a], $\mu \cdot \Omega$cm	Density, g/cm³
Cold rolled steel	Sheet	98.5	–	–	–	–	950 Anneal	180	2,000	21,000	–	1.8	10	7.88
Iron	Sheet	99.91	–	–	–	–	950 Anneal	200	5,000	21,500	5,000	1.0	10	7.88
Purified iron	Sheet	99.95	–	–	–	–	1480 H_2 + 880	5,000	180,000	21,500	300	0.05	10	7.88
4% Silicon-iron	Sheet	96	–	–	–	4 Si	800 Anneal	500	7,000	19,700	3,500	0.5	60	7.65
Grain oriented[b]	Sheet	97	–	–	–	3 Si	800 Anneal	1,500	30,000	20,000	–	0.15	47	7.67
45 Permalloy	Sheet	54.7	45	–	–	0.3 Mn	1050 Anneal	2,500	25,000	16,000	1,200	0.3	45	8.17
45 permalloy[c]	Sheet	54.7	45	–	–	0.3 Mn	1200 H_2 Anneal	4,000	50,000	16,000	–	0.07	45	8.17
Hipernik	Sheet	50	50	–	–	–	1200 H_2 Anneal	4,500	70,000	16,000	220	0.05	50	8.25
Monimax	Sheet	–	–	–	–	–	1125 H_2 Anneal	2,000	35,000	15,000	–	0.1	80	8.27
Sinimax	Sheet	–	–	–	–	–	1125 H_2 Anneal	3,000	35,000	11,000	–	–	90	–
78 Permalloy	Sheet	21.2	78.5	–	–	0.3 Mn	1050 + 600 Q[d]	8,000	100,000	10,700	200	0.05	16	8.60
4–79 Permalloy	Sheet	16.7	79	–	4	0.3 Mn	1100 + Q	20,000	100,000	8,700	200	0.05	55	8.72
Mu metal	Sheet	18	75	–	–	2 Cr, 5 Cu	1175 H_2	20,000	100,000	6,500	–	0.05	62	8.58
Supermalloy	Sheet	15.7	79	–	5	0.3 Mn	1300 H_2+ Q	100,000	800,000	8,000	–	0.002	60	8.77
Permendur	Sheet	49.7	–	50	–	0.3 Mn	800 Anneal	800	5,000	24,500	12,000	2.0	7	8.3
2 V Permendur	Sheet	49	–	49	–	2 V	800 Anneal	800	4,500	24,000	6,000	2.0	26	8.2
Hiperco	Sheet	64	–	34	–	Cr	850 Anneal	650	10,000	24,200	–	1.0	25	8.0
2–81 Permalloy	Insulated powder	17	81	–	2	–	650 Anneal	125	130	8,000	–	<1.0	10^6	7.8
Carbonyl iron	Insulated powder	99.9	–	–	–	–	–	55	132	–	–	–	–	7.86
Ferroxcube III	Sintered powder	$MnFe_2O_4$ + $ZnFe_2O_4$					–	1,000	1,500	2,500	–	0.1	10^8	5.0

[a] At saturation.
[b] Properties in direction of rolling.
[c] Similar properties for Nicaloi, 4750 alloy, Carpenter 49, Armco 48.
[d] Q, quench or controlled cooling.

Table 10.36 Characteristics of Permanent Magnet Alloys (*From* [1]. *Used with permission.*)

Material	% composition (remainder Fe)	Heat treatment^a (temperature,°C)	Magnetizing force $H_{max.}$ O	Coercive force H_c, O	Residual induction B_r, G	Energy product $BH_{max.}$ $\times 10^{-6}$	Method of fabrication^b	Mechanical properties^c	Weight lb/In.^3
Carbon steel	1 Mn, 0.9 C	Q 800	300	50	10,000	0.20	HR, M, P	H, S	0.280
Tungsten steel	5 W, 0.3 Mn, 0.7 C	Q 850	300	70	10,300	0.32	HR, M, P	H, S	0.292
Chromium steel	3.5 Cr, 0.9 C, 0.3 Mn	Q 830	300	65	9,700	0.30	HR, M, P	H, S	0.280
17% Cobalt steel	17 Co, 0.75 C, 2.5 Cr, 8 W	–	1,000	150	9,500	0.65	HR, M, P	H, S	–
36% Cobalt steel	36 Co, 0.7 C, 4 Cr, 5 W	Q 950	1,000	240	9,500	0.97	HR, M, P	H, S	0.296
Remalloy or Comol	17 Mo, 12 Co	Q 1200, B 700	1,000	250	10,500	1.1	HR, M, P	H	0.295
Alnico I	12 Al, 20 Ni, 5 Co	A 1200, B 700	2,000	440	7,200	1.4	C, G	H, B	0.249
Alnico II	10 Al, 17 Ni, 2.5 Co, 6 Cu	A 1200, B 600	2,000	550	7,200	1.6	C, G	H, B	0.256
Alnico II (sintered)	10 Al, 17 Ni, 2.5 Co, 6 Cu	A 1300	2,000	520	6,900	1.4	Sn, G	H	0.249
Alnico IV	12 Al, 28 Ni, 5 Co	Q 1200, B 650	3,000	700	5,500	1.3	Sn, C, G	H	0.253
Alnico V	8 Al, 14 Ni, 24 Co, 3 Cu	AF 1300, B 600	2,000	550	12,500	4.5	C, G	H, B	0.264
Alnico VI	8 Al, 15 Ni, 24 Co, 3 Cu, 1 Ti	–	3,000	750	10,000	3.5	C, G	H, B	0.268
Alnico XII	6 Al, 18 Ni, 35 Co, 8 Ti	–	3,000	950	5,800	1.5	C, G	H, B	0.26
Vicalloy I	52 Co, 10 V	B 600	1,000	300	8,800	1.0	C, CR, M, P	D	0.295
Vicalloy II (wire)	52 Co, 14 V	CW + B 600	2,000	510	10,000	3.5	C, CR, M, P	D	0.292
Cunife (wire)	60 Cu, 20 Ni	CW + B 600	2,400	550	5,400	1.5	C, CR, M, P	D, M	0.311
Cunico	50 Cu, 21 Ni, 29 Co	–	3,200	660	3,400	0.80	C, CR, M, P	D, M	0.300
Vectolite	$30Fe_2O_3, 44Fe_3O_4, 26Co_2O_3$	–	3,000	1,000	1,600	0.60	Sn, G	W	0.113
Silmanal	86.8Ag, 8.8Mn, 4.4Al	–	20,000	6,000^d	550	0.075	C, CR, M, P	D, M	0.325
Platinum-cobalt	77 Pt, 23 Co	Q 1200, B 650	15,000	3,600	5,900	6.5	C, CR, M	D	–
Hyflux	Fine powder	–	2,000	390	6,600	0.97	–	–	0.176

^aQ, quenched in oil or water; A, air colled; B, baked; F, cooled in magnetic field; CW, cold worked.
^bHR, hot rolled or forged; CR, cold rolled or drawn; M, machined; G, must be ground; P, punched; C, cast; Sn, sintered.
^cH, hard; B, brittle; S, strong; D, ductile; M, malleable; W, weak.
^dValue given is intrinsic H_c.

Table 10.37 Properties of Antiferromagnetic Compounds (*From* [1]. *Used with permission.*)

Compound	Crystal Symmetry	$\theta_N{}^a$ K	$\theta_P{}^b$ K	$(P_A)_{eff}{}^c$ μ_B	$P_A{}^d$ μ_B
$CoCl_2$	Rhombohedral	25	−38.1	5.18	3.1 ± 0.6
CoF_2	Tetragonal	38	50	5.15	3.0
CoO	Tetragonal	291	330	5.1	3.8
Cr	Cubic	475			
Cr_2O_3	Rhombohedral	307	485	3.73	3.0
CrSb	Hexagonal	723	550	4.92	2.7
$CuBr_2$	Monoclinic	189	246	1.9	
$CuCl_2 \cdot 2H_2O$	Orthorhombic	4.3	4–5	1.9	
$CuCl_2$	Monoclinic	~70	109	2.08	
$FeCl_2$	Hexagonal	24	−48	5.38	4.4 ± 0.7
FeF_2	Tetragonal	79–90	117	5.56	4.64
FeO	Rhombohedral	198	507	7.06	3.32
$\alpha\text{-}Fe_2O_3$	Rhombohedral	953	2940	6.4	5.0
$\alpha\text{-}Mn$	Cubic	95			
$MnBr_2 \cdot 4H_2O$	Monoclinic	2.1	$\left\{ \begin{matrix} 2.5 \\ 1.3 \end{matrix} \right\}$	5.93	
$MnCl_2 \cdot 4H_2O$	Monoclinic	1.66	1.8	5.94	
MnF_2	Tetragonal	72–75	113.2	5.71	5
MnO	Rhombohedral	122	610	5.95	5.0
$\beta\text{-}MnS$	Cubic	160	982	5.82	5.0
MnSe	Cubic	~173	361	5.67	
MnTe	Hexagonal	310–323	690	6.07	5.0
$NiCl_2$	Hexagonal	50	−68	3.32	
NiF_2	Tetragonal	78.5–83	115.6	3.5	2.0
NiO	Rhombohedral	533–650	~2000	4.6	2.0
$TiCl_3$		100			
V_2O_3		170			

$^a\theta_N$ = Néel temperature, determined from susceptibility maxima or from the disappearance of magnetic scattering.
$^b\theta_P$ = a constant in the Curie-Weiss law written in the form $\chi_A = C_A/(T + \theta_P)$, which is valid for antiferromagnetic material for $T > \theta_N$.
$^c(P_A)_{eff}$ = effective moment per atom, derived from the atomic Curie constant $C_A = (P_A)^2_{eff}(N^2/3R)$ and expressed in units of the Bohr magneton, $\mu_B = 0.9273 \times 10^{-20}$ erg G^{-1}.
^dP_A = magnetic moment per atom, obtained from neutron diffraction measurements in the ordered state.

Table 10.38 Saturation Constants for Magnetic Substances (*From* [1]. *Used with permission.*)

Substance	Field Intensity	Induced Magnetization	Substance	Field Intensity	Induced Magnetization
		(For Saturation)			(For Saturation)
Cobalt	9,000	1,300	Nickel, hard	8,000	400
Iron, wrought	2,000	1,700	annealed	7,000	515
cast	4,000	1,200	Vicker's steel	15,000	1,600
Manganese steel	7,000	200			

Table 10.39 Magnetic Properties of Transformer Steels (*From* [1]. *Used with permission.*)

Ordinary Transformer Steel			High Silicon Transformer Steels		
B(Gauss)	H(Oersted)	Permeability $= B/H$	B	H	Permeability
2,000	0.60	3,340	2,000	0.50	4,000
4,000	0.87	4,600	4,000	0.70	5,720
6,000	1.10	5,450	6,000	0.90	6,670
8,000	1.48	5,400	8,000	1.28	6,250
10,000	2.28	4,380	10,000	1.99	5,020
12,000	3.85	3,120	12,000	3.60	3,340
14,000	10.9	1,280	14,000	9.80	1,430
16,000	43.0	372	16,000	47.4	338
18,000	149	121	18,000	165	109

10.10 Properties of Selected Materials

Table 10.40 Total Elongation at Failure of Selected Polymers (*From* [1]. *Used with permission.*)

Polymer	Elongation [a]
ABS	5–20
Acrylic	2–7
Epoxy	4.4
HDPE	700–1000
Nylon, type 6	30–100
Nylon 6/6	15–300
Phenolic	0.4–0.8
Polyacetal	25
Polycarbonate	110
Polyester	300
Polypropylene	100–600
PTFE	250–350

[a] % in 50 mm section.

Table 10.41 Melting Point of Selected Metals and Ceramics (*From* [1]. *Used with permission.*)

Metal	M.P. (°C)	Ceramic	M.P. (°C)
Ag	962	Al_2O_3	2049
Al	660	BN	2727
Au	1064	B_2O_3	450
Co	1495	BeO	2452
Cr	1857	NiO	1984
Cu	1083	PbO	886
Fe	1535	SiC	2697
Ni	1453	Si_3N_4	2442
Pb	328	SiO_2	1723
Pt	1772	WC	2627
Ti	1660	ZnO	1975
W	3410	ZrO_2	2850

Table 10.42 Tensile Strength of Selected Wrought Aluminum Alloys (*From* [1]. *Used with permission.*)

Alloy	Temper	TS (MPa)
1050	0	76
1050	H16	130
2024	0	185
2024	T361	495
3003	0	110
3003	H16	180
5050	0	145
5050	H34	195
6061	0	125
6061	T6, T651	310
7075	0	230
7075	T6, T651	570

Table 10.43 Density of Selected Materials, Mg/m^3 (*From* [1]. *Used with permission.*)

Metal		Ceramic		Glass		Polymer	
Ag	10.50	Al_2O_3	3.97–3.986	SiO_2	2.20	ABS	1.05–1.07
Al	2.7	BN (cub)	3.49	SiO_2 10 wt% Na_2O	2.291	Acrylic	1.17–1.19
Au	19.28	BeO	3.01–3.03	SiO_2 19.55 wt% Na_2O	2.383	Epoxy	1.80–2.00
Co	8.8	MgO	3.581	SiO_2 29.20 wt% Na_2O	2.459	HDPE	0.96
Cr	7.19	SiC(hex)	3.217	SiO_2 39.66 wt% Na_2O	2.521	Nylon, type 6	1.12–1.14
Cu	8.93	Si_3N_4 (α)	3.184	SiO_2 39.0 wt% CaO	2.746	Nylon 6/6	1.13–1.15
Fe	7.87	Si_3N_4 (β)	3.187			Phenolic	1.32–1.46
Ni	8.91	TiO_2 (rutile)	4.25			Polyacetal	1.425
Pb	11.34	UO_2	10.949–10.97			Polycarbonate	1.2
Pt	21.44	ZrO_2 (CaO)	5.5			Polyester	1.31
Ti	4.51	Al_2O_3 MgO	3.580			Polystyrene	1.04
W	19.25	$3Al_2O_3$ $2SiO_2$	2.6–3.26			PTFE	2.1–2.3

10.11 References

1. Whitaker, Jerry C. (ed.), *The Electronics Handbook*, CRC Press, Boca Raton, FL, 1996.

Index